Lecture Notes in Computer Science 7032

Commenced Publication in 1973
Founding and Former Series Editors:
Gerhard Goos, Juris Hartmanis, and Jan van Leeuwen

Editorial Board

David Hutchison
  Lancaster University, UK

Takeo Kanade
  Carnegie Mellon University, Pittsburgh, PA, USA

Josef Kittler
  University of Surrey, Guildford, UK

Jon M. Kleinberg
  Cornell University, Ithaca, NY, USA

Alfred Kobsa
  University of California, Irvine, CA, USA

Friedemann Mattern
  ETH Zurich, Switzerland

John C. Mitchell
  Stanford University, CA, USA

Moni Naor
  Weizmann Institute of Science, Rehovot, Israel

Oscar Nierstrasz
  University of Bern, Switzerland

C. Pandu Rangan
  Indian Institute of Technology, Madras, India

Bernhard Steffen
  TU Dortmund University, Germany

Madhu Sudan
  Microsoft Research, Cambridge, MA, USA

Demetri Terzopoulos
  University of California, Los Angeles, CA, USA

Doug Tygar
  University of California, Berkeley, CA, USA

Gerhard Weikum
  Max Planck Institute for Informatics, Saarbruecken, Germany

Lora Aroyo   Chris Welty
Harith Alani   Jamie Taylor
Abraham Bernstein   Lalana Kagal
Natasha Noy   Eva Blomqvist (Eds.)

The Semantic Web – ISWC 2011

10th International Semantic Web Conference
Bonn, Germany, October 23-27, 2011
Proceedings, Part II

 Springer

Volume Editors

Lora Aroyo
VU University Amsterdam, The Netherlands; l.m.aroyo@vu.nl

Chris Welty
IBM Research, Yorktown Heights, NY, USA; cawelty@gmail.com

Harith Alani
The Open University, Milton Keynes, UK; h.alani@open.ac.uk

Jamie Taylor
Google, Mountain View, CA, USA; jamietaylor@google.com

Abraham Bernstein
University of Zurich, Switzerland; Bernstein@ifi.uzh.ch

Lalana Kagal
Massachusetts Institute of Technology, Cambridge, MA, USA; lkagal@csail.mit.edu

Natasha Noy
Stanford University, CA, USA; noy@stanford.edu

Eva Blomqvist
Linköping University, Sweden; eva.blomqvist@liu.se

ISSN 0302-9743   e-ISSN 1611-3349
ISBN 978-3-642-25092-7   e-ISBN 978-3-642-25093-4
DOI 10.1007/978-3-642-25093-4
Springer Heidelberg Dordrecht London New York

Library of Congress Control Number: 2011939851

CR Subject Classification (1998): C.2, H.4, H.3, H.5, J.1, K.4

LNCS Sublibrary: SL 3 – Information Systems and Application, incl. Internet/Web
and HCI

© Springer-Verlag Berlin Heidelberg 2011
This work is subject to copyright. All rights are reserved, whether the whole or part of the material is
concerned, specifically the rights of translation, reprinting, re-use of illustrations, recitation, broadcasting,
reproduction on microfilms or in any other way, and storage in data banks. Duplication of this publication
or parts thereof is permitted only under the provisions of the German Copyright Law of September 9, 1965,
in its current version, and permission for use must always be obtained from Springer. Violations are liable
to prosecution under the German Copyright Law.
The use of general descriptive names, registered names, trademarks, etc. in this publication does not imply,
even in the absence of a specific statement, that such names are exempt from the relevant protective laws
and regulations and therefore free for general use.

Typesetting: Camera-ready by author, data conversion by Scientific Publishing Services, Chennai, India

Printed on acid-free paper

Springer is part of Springer Science+Business Media (www.springer.com)

Preface

Ten years ago, several researchers decided to organize a workshop to bring together an emerging community of scientists who were working on adding machine-readable semantics to the Web, the Semantic Web. The organizers were originally planning for a few dozen researchers to show up. When 200 of them came to Stanford in August 2001, the Semantic Web Workshop became the Semantic Web Working Symposium, and the International Semantic Web Conference (ISWC) was born. Much has changed in the ten years since that meeting. The Semantic Web has become a well-recognized research field in its own right, and ISWC is a premier international research conference today. It brings together researchers, practitioners, and users in artificial intelligence, databases, social networks, distributed computing, Web engineering, information systems, human–computer interaction, natural-language processing, and others. Companies from Facebook to Google to the *New York Times* rely on Semantic Web technologies to link and organize their data; governments in the United States, United Kingdom, and other countries open up their data by making it accessible to Semantic Web tools; scientists in many domains, from biology, to medicine, to oceanography and environmental sciences, view machine-processable semantics as key to sharing their knowledge in today's data-intensive scientific enterprise; semantic technology trade shows attract more than a thousand attendees. The focus of Semantic Web research has moved from issues of representing data on the Web and the growing pains of figuring out a common format to share it, to such challenges as handling billions of statements in a scalable way to making all this data accessible and usable to regular citizens.

This volume contains the main proceedings of the 10th International Semantic Web Conference (ISWC 2011), which was held in Bonn, Germany, in October 2011. We received tremendous response to our calls for papers from a truly international community of researchers and practitioners. Indeed, every track of the conference received a record number of submissions this year. The careful nature of the review process, and the breadth and scope of the papers finally selected for inclusion in this volume, speak to the quality of the conference and to the contributions made by researchers whose work is presented in these proceedings.

The Research Track of the conference attracted 264 submissions. Each paper received at least three, and sometimes as many as five, reviews from members of the Program Committee. After the first round of reviews, authors had the opportunity to submit a rebuttal, leading to further discussions among the reviewers, a meta-review and a recommendation from a member of the Senior Program Committee. Every paper that had at least one recommendation for acceptance was discussed in a virtual meeting of the Senior Program Committee.

As the Semantic Web develops, we find a changing variety of subjects that emerge. This year the keywords of accepted papers were distributed as follows (frequency in parentheses): ontologies and semantics (15), database, IR, and AI technologies for the Semantic Web (14), management of Semantic Web data (11), reasoning over Semantic Web data (11), search, query, integration, and analysis on the Semantic Web (10), robust and scalable knowledge management and reasoning on the Web (10), interacting with Semantic Web data (9), ontology modularity, mapping, merging, and alignment (8), languages, tools, and methodologies for representing and managing Semantic Web data (8), ontology methodology, evaluation, reuse, extraction, and evolution (7), evaluation of Semantic Web technologies or data (7), specific ontologies and ontology patterns for the Semantic Web (6), new formalisms for the Semantic Web (4), user interfaces to the Semantic Web (3), cleaning, assurance, and provenance of Semantic Web data, services, and processes (3), social Semantic Web (3), evaluation of Semantic Web technology (3), Semantic Web population from the human Web (3).

Overall, the ISWC Program Committee members adopted strict standards for what constitutes high-quality Semantic Web research and what papers must deliver in terms of theory, practice, and evaluation in order to be accepted to the Research Track. Correspondingly, the Program Committee accepted only 50 papers, 19% of the submissions.

The Semantic Web In-Use Track received 75 submissions. At least three members of the In-Use Program Committee provided reviews for each paper. Seventeen papers were accepted – a 23% acceptance rate. The large number of submissions this year demonstrated the increasingly diverse breadth of applications of Semantic Web technologies in practice. Papers demonstrated how semantic technologies could be used to drive a variety of simulation and test systems, manage distributed content and operate within embedded devices. Several papers tapped the growing amount of semantically enriched environmental data available on the Web allowing communities to visualize, organize, and monitor collections for specific purposes.

The Doctoral Consortium has become a key event at the conference over the years. PhD students get an opportunity to present their thesis proposals and to get detailed feedback on their research topics and plans from the leading academic and industrial scientists in the field. Out of 31 submissions to the Doctoral Consortium, 6 were accepted as long papers for presentation at the conference, and 9 were accepted for presentation at the special Consortium-only poster session. Each student was assigned a mentor who led the discussion following the presentation of their proposal, and provided extensive feedback and comments.

A unique aspect of the ISWC conference is the Semantic Web Challenge. In this competition, the ninth to be held at the conference, practitioners and scientists showcase useful and leading-edge applications of Semantic Web technology. Diana Maynard and Chris Bizer organized the Semantic Web Challenge this year.

The keynote talks given by leading scientists in the field further enriched the ISWC program. Alex (Sandy) Pentland, the director of the Human Dynamics Laboratory and the Media Lab Entrepreneurship Program at the Massachusetts

Institute of Technology, discussed the New Deal on Data—a new data ecosystem that can allow personal data to become an accessible asset for the new generation of systems in health, finance, logistics, and transportation. Gerhard Weikum, a Research Director at the Max Planck Institute for Informatics, discussed the issues and approaches to extending and enriching linked data, in order to improve its scope, quality, interoperability, cross-linking, and usefulness. Frank van Harmelen, a professor at the VU University Amsterdam, and a participant and leader in Semantic Web research, provided his analysis of the past ten years, discussing whether any universal patterns have emerged in the way we built the Semantic Web. Nigel Shadbolt, Deputy Head of the School of Electronics and Computer Science at the University of Southampton, gave a lively dinner talk.

As in previous ISWC editions, the conference included an extensive Tutorial and Workshop program. Tania Tudorache and Heiner Stuckenschmidt, the Chairs of this track, created a stellar and diverse collection of 7 tutorials and 16 workshops, where the only problem that the participants faced was which of the many exciting workshops to attend.

We would like to thank Marta Sabou and Guilin Qi for organizing a lively Poster and Demo Session. This year, the Posters and Demos were introduced in a Minute Madness Session, where every presenter got 60 seconds to provide a teaser for their poster or demo. Marco Neumann coordinated an exciting Industry Track with presentations both from younger companies focusing on semantic technologies and software giants, such as Yahoo! and Microsoft.

As we look forward to the next ten years of Semantic Web research, we organized an Outrageous Ideas Session, with a special award sponsored by the Computing Community Consortium. At this track, we invited scientists to submit short papers describing unconventional and innovative ideas that identify new research opportunities in this field. A Program Committee of established Semantic Web researchers judged the submissions on the extent to which they expand the possibilities and horizons of the field. After presentation of short-listed papers at the conference both the PC members and the audience voted for the prize winners.

We are indebted to Eva Blomqvist, our Proceedings Chair, who provided invaluable support in compiling the volume that you now hold in your hands (or see on your screen) and exhibited super-human patience in allowing the other Chairs to stretch deadlines to the absolute limits. Many thanks to Jen Golbeck, the Fellowship Chair, for securing and managing the distribution of student travel grants and thus helping students who might not have otherwise attended the conference to come to Bonn. Mark Greaves and Elena Simperl were tireless in their work as Sponsorship Chairs, knocking on every conceivable virtual 'door' and ensuring an unprecedented level of sponsorship this year. We are especially grateful to all the sponsors for their generosity.

As has been the case in the past, ISWC 2011 also contributed to the linked data cloud by providing semantically annotated data about many aspects of the conference. This contribution would not have been possible without the efforts of Lin Clark, our Metadata Chair.

Juan Sequeda, our Publicity Chair, was tirelessly twittering, facebooking, and sending old-fashioned announcements on the mailing lists, creating far more lively 'buzz' than ISWC ever had.

Our very special thanks go to the Local Organization Team, led by Steffen Staab and York Sure-Vetter. They did a fantastic job of handling local arrangements, thinking of every potential complication way before it arose, often doing things when members of the Organizing Committee were only beginning to think about asking for them. Special thanks go to Ruth Ehrenstein for her enormous resourcefulness, foresight, and anticipation of the conference needs and requirements. We extend our gratitude to Silke Werger, Holger Heuser, and Silvia Kerner.

Finally, we would like to thank all members of the ISWC Organizing Committee not only for handling their tracks superbly, but also for their wider contribution to the collaborative decision-making process in organizing the conference.

October 2011

Lora Aroyo
Chris Welty
Program Committee Co-chairs
Research Track

Harith Alani
Jamie Taylor
Program Committee Co-chairs
Semantic Web In-Use Track

Abraham Bernstein
Lalana Kagal
Doctoral Consortium Chairs

Natasha Noy
Conference Chair

Conference Organization

Organizing Committee

Conference Chair

Natasha Noy   Stanford University, USA

Program Chairs–Research Track

Lora Aroyo   VU University Amsterdam, The Netherlands
Chris Welty   IBM Watson Research Center, USA

Semantic Web In-Use Chairs

Harith Alani   KMI, Open University, UK
Jamie Taylor   Google, USA

Doctoral Consortium Chairs

Abraham Bernstein   University of Zurich, Switzerland
Lalana Kagal   Massachusetts Institute of Technology, USA

Industry Track Chair

Marco Neumann   KONA, USA

Posters and Demos Chairs

Guilin Qi   Southeast University, China
Marta Sabou   MODUL University, Austria

Workshops and Tutorials Chairs

Heiner Stuckenschmidt   University of Mannheim, Germany
Tania Tudorache   Stanford University, USA

Semantic Web Challenge Chairs

Christian Bizer   Free University Berlin, Germany
Diana Maynard   University of Sheffield, UK

Metadata Chair

Lin Clark   DERI Galway, Ireland

Local Organization Chairs

Steffen Staab	University of Koblenz-Landau, Germany
York Sure-Vetter	GESIS and University of Koblenz-Landau, Germany

Local Organization

Ruth Ehrenstein	University of Koblenz-Landau, Germany

Sponsorship Chairs

Mark Greaves	Vulcan, USA
Elena Simperl	Karlsruhe Institute of Technology, Germany

Publicity Chair

Juan Sequeda	University of Texas at Austin, USA

Fellowship Chair

Jen Golbeck	University of Maryland, USA

Proceedings Chair

Eva Blomqvist	Linköping University and ISTC-CNR, Sweden/Italy

Webmaster

Holger Heuser	GESIS, Germany

Senior Program Committee – Research

Mathieu d'Aquin	Open University, UK
Philippe Cudré-Mauroux	University of Fribourg, Switzerland
Jérôme Euzenat	INRIA and LIG, France
Aldo Gangemi	STLab, ISTC-CNR, Italy
Jeff Heflin	Lehigh University, USA
Ian Horrocks	University of Oxford, UK
Geert-Jan Houben	Delft University of Technology, The Netherlands
Aditya Kalyanpur	IBM Research, USA
David Karger	MIT, USA
Manolis Koubarakis	National and Kapodistrian University of Athens, Greece
Diana Maynard	University of Sheffield, UK
Peter Mika	Yahoo! Research, Spain
Peter F. Patel-Schneider	Bell Labs, USA
Axel Polleres	Siemens AG/DERI Galway, Austria/Ireland

Jie Tang   Tsinghua University, China
Paolo Traverso   FBK, Italy
Lei Zhang   IBM China Research, China

Program Committee – Research

Fabian Abel   Xiaoyong Du
Sudhir Agarwal   Dieter Fensel
Faisal Alkhateeb   Achille Fokoue
Yuan An   Fabien Gandon
Melliyal Annamalai   Zhiqiang Gao
Grigoris Antoniou   Fausto Giunchiglia
Kemafor Anyanwu   Birte Glimm
Knarig Arabshian   Bernardo Grau
Lora Aroyo   Alasdair Gray
Manuel Atencia   Paul Groth
Sören Auer   Tudor Groza
Christopher Baker   Michael Grüninger
Jie Bao   Gerd Gröner
Michael Benedikt   Nicola Guarino
Sonia Bergamaschi   Volker Haarslev
Eva Blomqvist   Peter Haase
Piero Bonatti   Willem van Hage
Kalina Bontcheva   Harry Halpin
Aidan Boran   Andreas Harth
John Breslin   Michael Hausenblas
Paul Buitelaar   Sandro Hawke
Diego Calvanese   Tom Heath
Mari Carmen   Nathalie Hernandez
Enhong Chen   Stijn Heymans
Key-Sun Choi   Michiel Hildebrand
Benoit Christophe   Kaoru Hiramatsu
Lin Clark   Pascal Hitzler
Oscar Corcho   Aidan Hogan
Isabel Cruz   Laura Hollink
Richard Cyganiak   Matthew Horridge
Claudia D'Amato   Wei Hu
Theodore Dalamagas   Jane Hunter
Mike Dean   David Huynh
Stefan Decker   Eero Hyvönen
Renaud Delbru   Giovambattista Ianni
Ian Dickinson   Lalana Kagal
Stefan Dietze   Yevgeny Kazakov
Li Ding   Anastasios Kementsietsidis
John Domingue   Teresa Kim

Yasuhiko Kitamura
Vladimir Kolovski
Kouji Kozaki
Thomas Krennwallner
Markus Krötzsch
Oliver Kutz
Ora Lassila
Georg Lausen
Juanzi Li
Shengping Liu
Carsten Lutz
Christopher Matheus
Jing Mei
Christian Meilicke
Alessandra Mileo
Knud Moeller
Boris Motik
Yuan Ni
Daniel Oberle
Jacco van Ossenbruggen
Sascha Ossowski
Jeff Pan
Bijan Parsia
Alexandre Passant
Chintan Patel
Terry Payne
Jorge Perez
Giuseppe Pirrò
Valentina Presutti
Guilin Qi
Yuzhong Qu
Anand Ranganathan
Riccardo Rosati
Marie-Christine Rousset
Sebastian Rudolph
Tuukka Ruotsalo
Alan Ruttenberg
Marta Sabou
Uli Sattler
Bernhard Schandl
François Scharffe
Ansgar Scherp
Daniel Schwabe
Luciano Serafini

Yidong Shen
Amit Sheth
Pavel Shvaiko
Michael Sintek
Sergej Sizov
Spiros Skiadopoulos
Kavitha Srinivas
Giorgos Stamou
Johann Stan
George Stoilos
Umberto Straccia
Markus Strohmaier
Heiner Stuckenschmidt
Rudi Studer
Xingzhi Sun
Valentina Tamma
Sergio Tessaris
Philippe Thiran
Christopher Thomas
Cassia Trojahn dos Santos
Raphaël Troncy
Christos Tryfonopoulos
Tania Tudorache
Anni-Yasmin Turhan
Octavian Udrea
Victoria Uren
Stavros Vassos
Maria Vidal
Ubbo Visser
Denny Vrandecic
Holger Wache
Haofen Wang
Fang Wei
Chris Welty
Max L. Wilson
Katy Wolstencroft
Zhe Wu
Guotong Xie
Bin Xu
Yong Yu
Ondrej Zamazal
Ming Zhang
Antoine Zimmermann

Additional Reviewers – Research

Saminda Abeyruwan
Pramod Ananthram
Marcelo Arenas
Ken Barker
Thomas Bauereiß
Domenico Beneventano
Meghyn Bienvenu
Nikos Bikakis
Holger Billhardt
Victor de Boer
Georgeta Bordea
Stefano Borgo
Loris Bozzato
Volha Bryl
Carlos Buil-Aranda
Federico Caimi
Delroy Cameron
Davide Ceolin
Melisachew Wudage Chekol
Yueguo Chen
Gong Cheng
Alexandros Chortaras
Mihai Codescu
Anna Corazza
Minh Dao-Tran
Jérôme David
Brian Davis
Jianfeng Du
Liang Du
Songyun Duan
Michel Dumontier
Jinan El-Hachem
Cristina Feier
Anna Fensel
Alberto Fernandez Gil
Roberta Ferrario
Pablo Fillottrani
Valeria Fionda
Haizhou Fu
Irini Fundulaki
Sidan Gao
Giorgos Giannopoulos
Kalpa Gunaratna

Claudio Gutierrez
Ollivier Haemmerlé
Karl Hammar
Norman Heino
Cory Henson
Ramon Hermoso
Daniel Herzig
Matthew Hindle
Joana Hois
Julia Hoxha
Matteo Interlandi
Prateek Jain
Ernesto Jiménez-Ruiz
Mei Jing
Fabrice Jouanot
Martin Junghans
Ali Khalili
Sheila Kinsella
Szymon Klarman
Matthias Knorr
Srdjan Komazec
Jacek Kopecky
Adila Alfa Krisnadhi
Sarasi Lalithsena
Christoph Lange
Danh Le Phuoc
Ning Li
John Liagouris
Dong Liu
Nuno Lopes
Uta Lösch
Theofilos Mailis
Alessandra Martello
Michael Martin
Andrew Mccallum
William J. Murdock
Vijayaraghava Mutharaju
Nadeschda Nikitina
Andriy Nikolov
Philipp Obermeier
Marius Octavian Olaru
Matteo Palmonari
Catia Pesquita

Sponsors

Platinum Sponsors

AI Journal
Elsevier
fluid Operations
OASIS
THESEUS

Gold Sponsors

Microsoft Research
ontoprise
Ontotext
PoolParty
SoftPlant
Yahoo! Research

Silver Sponsors

Computing
  Community
  Consortium
IBM Research
IEEE Intelligent
  Systems
IGI Global
IOS Press
LATC
Monnet
Planet Data
RENDER
Springer

Table of Contents – Part II

Semantic Web In-Use Track

Doctoral Consortium

Invited Talks—Abstracts

Table of Contents – Part I

Research Track

KOIOS: Utilizing Semantic Search for Easy-Access and Visualization of Structured Environmental Data[*]

Veli Bicer[1], Thanh Tran[2], Andreas Abecker[3], and Radoslav Nedkov[3]

[1] FZI Forschungszentrum Informatik, Haid-und-Neu-Str. 10-14,
D-76131 Karlsruhe, Germany
bicer@fzi.de

[2] Institute AIFB, Geb. 11.40 KIT-Campus Sd, D-76128 Karlsruhe, Germany
duc.tran@kit.edu

[3] disy Informationssysteme GmbH, Erbprinzenstr. 4-12, D-76133 Karlsruhe, Germany
firstname.lastname@disy.net

Abstract. With the increasing interest in environmental issues, the amount of publicly available environmental data on the Web is continuously growing. Despite its importance, the uptake of environmental information by the ordinary Web users is still very limited due to intransparent access to complex and distributed databases. As a remedy to this problem, in this work, we propose the use of semantic search technologies recently developed as an intuitive way to easily access structured data and lower the barriers to obtain information satisfying user information needs. Our proposed system, namely KOIOS, enables a simple, keyword-based search on structured environmental data and built on top of a commercial Environmental Information System (EIS). A prototype system successfully shows that applying semantic search techniques this way provides intuitive means for search and access to complex environmental information.

1 Introduction

As environmental issues become a hot topic for the general public, we perceive that the amount of publicly available environmental data is also continuously growing on the Web. Over the last ten years, public access to environmental data is highly encouraged by the governments since they have recognized that environmental information could have a profound impact on our ability to protect the environment [7]. Starting from the Directive 2003/4/EC, for instance, the European Union grants public access to environmental data. *PortalU*[1] in

---

[*] This research was funded by means of the German Federal Ministry of Economy and Technology under the promotional reference 01MQ07012. The authors take the responsibility for the contents.

[1] http://www.portalu.de/

L. Aroyo et al. (Eds.): ISWC 2011, Part II, LNCS 7032, pp. 1–16, 2011.
© Springer-Verlag Berlin Heidelberg 2011

Germany, *Envirofacts*[2] in the USA or *EDP*[3] in the UK are just few examples
that provide access to large volumes of environmental data as a result of recent activities. Besides, environmental data are also made accessible as a part of
the Linking Open Data (LOD) project in a structured format (RDF) with the
idea of linking environmental data in an international context of cooperating
governmental authorities [15]. Thus, previously local databases of environmental data have become part of the LOD cloud of datasets, enabling the active
dissemination of environmental information to the masses.

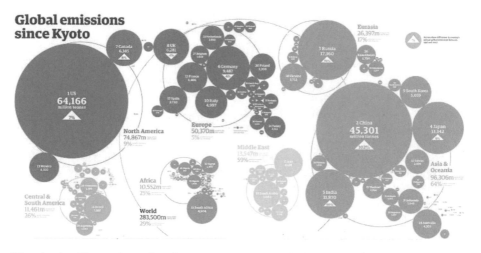

Fig. 1. An illustration of carbon emissions around the World (source: Copenhagen
Climate Council)

Despite the increasing importance and availability, the commercial and soci-
etal impact of open environmental data is still very limited to the end-users. This
has a number of reasons, but one of them is certainly the largely intransparent
access to complex databases. This obviously holds true for interested citizen and
companies, but even for employees of public authorities in a different domain,
the heterogeneity and distribution of environmental data is often overwhelming. Hence, user-friendly and powerful search interfaces are a must-have in this
area. For example, assume for a moment that you are searching carbon (CO)
emission values that are highly critical figures considering their impact on the
climate change and global warming. Each year several organizations review the
underlying consumption data for petroleum, natural gas and coal worldwide and
estimate our CO emission values. By using a classical Web search and keyword
queries, it is quite easy to obtain the country-specific figures as shown in Figure
1 as these are just parts of the documents in which the keywords occur. However, a more detailed search on the CO emission values such as "CO emission
values around Karlsruhe area in Germany" or obtaining more analytical results

---

[2] http://www.epa.gov/enviro/index.html
[3] http://www.edp.nerc.ac.uk

based on a particular year, or emission type (e.g. by industry, by transportation) requires more advanced search on structured data. This search paradigm however, assumes users to be an expert of the underlying data and domain, or design of proprietary interfaces to access data. The main challenge here is to provide ordinary users an easy-to-formulate queries (i.e. keyword query) and provide complex structured results in return to satisfy their information needs. According to the survey in [6], the percentage of people who used the Internet to find environmental information (49%) is significantly lower than those with frequent access to it (69%).

Recently, semantic search approaches to enable keyword search on structured data has gained a lot of interests as keywords have proven to be easy for the user to specify, and intuitive for accessing information. By utilizing lightweight semantics of the underlying RDF data, keyword search can help to circumvent the complexity of structured query languages, and hide the underlying data representation. Without knowledge of the query syntax and data schema, even the non-technical end-users can obtain complex structured results, including complex answers generated from RDF resources [17,19]. To this end, semantic search in that sense can provide the means to simplify the search on environmental data, allowing the users to access rich information with less effort, and lower the access barriers resulting from the complex interfaces of current systems.

In this work, we present a novel semantic search system, namely KOIOS, that provides semantic search capabilities on structured data with the aim of easy-access to rich environmental information. Our contributions mainly include: 1) enabling keyword search as an intuitive mechanism to search environmental data, 2) interpretation of user's possible information needs via the query translation from keyword query to structured query by utilizing the underlying schema structure and data, 3) offering dynamic and flexible faceted search to allow the user to specify his/her further preferences, and 4) integration to a commercial, Web-based Environmental Information System (EIS) for user-friendly presentation of information using visualization components such as maps, charts, tables, etc.

Structure. In Section 2, we introduce an overview of KOIOS system. Section 3 presents the KOIOS semantic search process detailing the steps starting from the specification of the keyword query to the final display of the search results. In Section 4, a prototype implementation of the system is presented. After a discussion of related work in Section 5, we conclude in Section 6.

2   KOIOS Overview

The KOIOS system provides semantic search capabilities over structured data that is either present in relational databases or RDF repositories. Figure 2 depicts an overview of the KOIOS system. KOIOS operations can be divided into two main stages: *preprocessing* and *search*. Preprocessing is an offline stage not visible to the users that mainly creates three special search indexes out of the structured data of a given database or RDF store. The core index directly created from the data is called *data index*, which is a graph-based representation

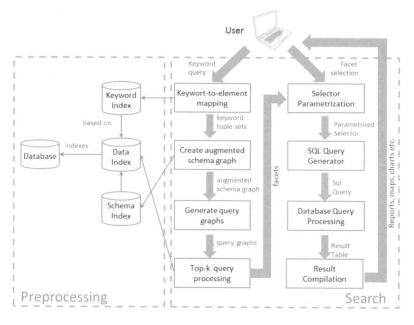

Fig. 2. Overview of KOIOS System

of the data implemented as an inverted index, and optimized for efficient access. Based on the data index, two other indexes are also created: the *keyword index* is mainly designed for IR-style access that captures the unstructured part of the data, and the *schema index* is extracted from the data, representing classes and relationships among them.

The second stage, search, is the actual part in which the system interacts with the user. User specifies his/her information with a short keyword query and considered as a set of keywords, $Q = \{q_1, q_2, ..., q_n\}$. Based on this query, the system first discovers possible keyword elements using the keyword index to find particular tuples (entities) in the data in which one of the keywords occur. A number of keyword tuple sets are created for each keyword in the query. These sets are then combined with the schema information resulting in an augmented schema graph, which represents the query space. By exploring this graph, a number of structured query graphs are constructed, each of which can be executed on the data index to find relevant results. This part of the search stage mainly interprets the user's possible information needs in terms of structured query graphs, and computes their corresponding result sets.

Based on the outputs of this stage, KOIOS generates a number of facets to facilitate further interactions and refinements of query and results. It uses a faceted search interface to present the possible categories (facets) and values generated from the underlying results. This helps the user to refine his/her query. Further, additional user preferences can be incorporated in order to obtain more precise results. In particular, KOIOS maps user choices and preferences to an internal representation, called *selectors*, which are coarse-granular query templates that are used to generate more precise queries to be executed over the database. This selector mechanism is also employed to select appropriate presentation components. The results are presented to the user via an integrated EIS interface.

In overall, KOIOS minimizes the inherent complexity of searching structured data by guiding the user through the search process via analyzing the underlying data and schema structure. This significantly minimizes the effort and cognitive complexity in the search process.

3   KOIOS Semantic Search Process

In this section, we present technical details of the semantic search process over environmental data. For the sake of presentation, we decompose the overall process into three steps: 1) Indexing, 2) keyword query interpretation and structured query generation, and 3) faceted search and selectors.

3.1   Indexing

In KOIOS, we apply a preprocessing step on the data to create index structures that help to perform the search functionalities more efficiently. Generally speaking, the underlying data can be conceived as a directed labeled data graph $G = (V, E)$, where V is a disjoint union ($V = V_R \uplus V_A$) of resource nodes (V_R), and attribute nodes (V_A), and $E = E_F \uplus E_A$ represents a disjoint union of relation edges also called foreign key edges (E_F) that connect resource nodes, and attribute edges (E_A) that link between a resource node and an attribute node. This model closely resembles the graph-structured RDF data model (omitting special features such as RDF blank nodes). The intuitive mapping of this model to relational data is as follows: a database tuple captures a resource, its attributes, and references to related resources in the form of foreign keys; the column names correspond to edge labels.

In order to perform the search steps in an efficient way, we preprocess the data graph to obtain the data index, which is actually a number of inverted indexes that store data in the form of *subject* → *predicate* → *object* triples. In particular, each inverted index returns a specific element of the triple given the other pair of elements. Based on this index, the keyword index that is used for keyword-to-element mapping is created. Conceptually, the keyword index is a keyword-resource map and used for the evaluation of a multi-valued function $f : V \to 2^{V_R}$, which for each keyword $k_i \in V$ in the vocabulary, returns the set of corresponding graph resources V_R (i.e. keyword elements). In addition, a lexical analysis (stemming, removal of stopwords) as supported by standard IR engines is performed on the attributes of resources in order to obtain terms. Processing attributes consisting of more than one word might result in many terms. Then, a list of references to the corresponding graph elements is created for every term.

For exploration, a schema index is constructed, which is basically a summary of the original data graph containing structural (schema) elements only. In essence, we attempt to obtain such a schema from the data graph instead of assuming a pre-given schema. The computation of the schema index follows straightforwardly from and is accomplished by a set of aggregation rules presented in previous work [18], which compute the equivalence classes of all resources belonging to one class and project all edges to corresponding edges at the schema level with the result that for every path in the data graph, there is at least one path in the schema index (while this is not the other way round).

At the time of query processing, this schema index is augmented with keyword elements obtained from the keyword-to-element mapping. Since we are interested in the top-k results, the index elements are also augmented with scores. While scores associated with structure elements can be computed off-line, scores of keyword elements are specific to the query and thus can only be processed at query computation time.

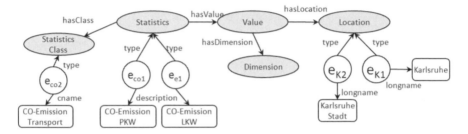

Fig. 3. Augmented schema graph showing schema elements (in gray), associated keyword elements (circles) and their corresponding attributes (rectangles)

3.2 Keyword Query Interpretation and Structured Query Generation

The goal of this step is to interpret the keywords entered by the user using the index structures created before, and to generate a number of structured queries. These queries are basically conjunctions of triple patterns, forming graph patterns corresponding to the Basic Graph Pattern feature of SPARQL. Typically, due to the ambiguity inherent in keyword queries, such an interpretation is not unique. Therefore, we rely on a top-k procedure to generate candidate interpretations and obtain the possible results that best match the user information need. We build on our previous work [17] on translating keyword queries into structured queries based on a graph-exploration technique. For this purpose, we consider available knowledge bases and data as data graphs as defined previously. (Intuitively, this graph).

The computation of structured queries as interpretations of the user keywords involves three tasks: 1) detecting the keyword elements as resources in the graph containing at least one of the query keywords, 2) creation of an augmented schema graph, 3) graph exploration and structured query generation and 4) top-k query processing to compute the best queries. Specific concepts and algorithms for these tasks have been introduced in [17].

Keyword-to-element mapping: We rely on the keyword index to map keywords to elements of the data graph, which in our approach might be classes, foreign key relationships, attributes, and attribute values. IR concepts are adopted to support an imprecise matching that incorporates syntactic and semantic similarities. As a result, the user does not need to know the exact labels of the data elements while doing keyword search. Each element finally returned is associated with a score measuring the degree of matching, which is later used for ranking possible interpretations. For each query keyword $q_i \in Q$, we define a set of candidate keyword elements $V_i = \{v_1, .., v_n\}$ that are potential elements in the data graph that the user is looking for.

Augmented schema graph: Given the sets of keyword elements for the query keywords, the query search space contains all the elements that are necessary for the computation of possible interpretations. This mainly includes all keyword elements and the corresponding classes, foreign key edges, and attribute edges of the data graph. It has been shown that keyword search is most efficient when the exploration for possible interpretations is performed on an augmented schema graph, instead of using the entire data graph (c.f. [17]). The schema graph can be trivially obtained from the class and property definitions in the data, or might be pre-given as an ontology. From experiences with Web data, we know that pre-given schemas are typically incomplete and often do not reflect the underlying data as some schema elements may actually not be instantiated in the data. Therefore, we additionally apply techniques for computing schema graphs automatically. In particular, a schema graph is derived from the data using the aggregation rules as described in [17] (during preprocessing).

Figure 3 illustrates the query space constructed for our example keyword query "karlsruhe co emission". It consists of (the fragment of) a schema graph (nodes in gray), keyword elements found in the previous step for different keywords, and the corresponding attributes of those elements. Note that the augmented schema graph consists of all keyword elements and all the possible paths between them that can be found in the schema.

Graph exploration and query translation: Given the augmented schema graph, the remaining task is to search for the minimal query graphs in this space. Informally, a query graph is a matching subgraph of the augmented schema graph, such that for every keyword of the user query, it contains at least one representative keyword-matching element, and (2) the graph is connected, i.e. there exists a path from every graph element to every other graph element. A matching query graph is minimal if there exists no other query graph with a lower score. This procedure starts from the keyword elements and iteratively explores the query space for all distinct paths beginning from these elements. During this procedure, the path with the highest score so far is selected for further exploration. At some point, an element might be discovered to be a connecting element, i.e. there is a path from that element to at least one keyword element, for every keyword in the user query. These paths are merged to form a query graph. The explored graphs are added to the candidate list. The process continues until the upper bound score for the query graphs yet to be explored is lower than the score of the k-ranked query graph in the candidate list. An example query graph that can be found through exploration along these paths is shown in Figure 4.

Top-k Query Processing: Query translation results in a set of query graphs (not only one), each of which can be a potential representation of the user's information need. In fact, at this point we are not interested in all the final results to be retrieved from the database using the computed queries, but only in the top-k results, based on which facets will be constructed in the next step. Thus, ranking the top results (e.g. answers to SPARQL query) w.r.t. their relevance to the initial keyword query Q is more important. The goal is to summarize these results as facets and to use the resulting facets for enabling additional refinement. For this, we query our data index with a retrieval algorithm as sketched in our previous work [17] to get top results for each query graph. In the next step, we discuss how facet values are generated from this initial query run on the data.

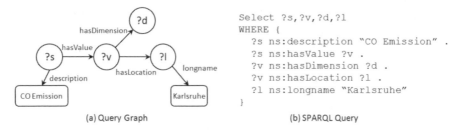

(a) Query Graph   (b) SPARQL Query

Fig. 4. a) An example query graph and b) the corresponding SPARQL query

3.3   Faceted Search and Selectors

The query graphs and their corresponding result sets are possible interpretations of the query. However, instead of assuming that a query graph is a direct representation of the user's information need, we utilize it in KOIOS to bridge the gap between the keyword query and the actual information need. For this purpose, we support a second round of user interaction as shown in Figure 2. Basically, we aim to increase the precision of the final search results, while providing an intuitive and easy-to-use way for the user to specify further preferences. We propose to use 1) facets generated from a number of query graphs and their result sets, as well as 2) selectors that are parameterized, pre-defined query templates, run against the database in the back-end, and map to final visualization elements to be displayed to the user.

Facets: In IR, faceted search mostly refers to techniques for accessing a collection of information represented using facets, allowing users to explore by filtering available information [9]. In this setting, facets correspond to the attributes (called facet categories) as well as possible attribute values (facet values) that are common to a set of resources (top-k results computed previously). Traditionally, they are mostly derived by an analysis of the text or from pre-defined fields in a database table. A shortcoming of this sort of faceted search is its basic data model, where facets are associated with sets of values from independent facet hierarchies. This model is too restrictive for some real-world data. A more appropriate faceted search solution for our environmental data scenario should provide richer insights into the data and the ability to perform flexible and dynamic aggregation over faceted data [1].

In our approach, we use the generated query graphs to construct dynamic facets for a particular search session to realize such a functionality. Given a query graph, we consider every variable binding (e.g. $?s$, $?v$, $?l$, $?d$ in Figure 4-a) as a possible candidate for a facet category. We utilize the corresponding types of these variables to retrieve their particular descriptive attributes. For example, the *Statistics* class of the variable $?s$ is used to retrieve the descriptive attribute "description". Then a facet category is created with the name *Statistics.description* for this attribute. As we obtain the top-k entities of this query graph in the last step of the previous section, the facet values are generated by clustering values that are specific to the *description* attribute of those entities. For this particular attribute, the possible facet values are shown by our prototype system in Figure 6. Note that an attribute does not need to be a part of the query graph in order to be utilized as a facet category. Also, for example, for the *Value.Year* attribute of the *?v* variable, we further generate facet values representing possible years (e.g. 2000, 2005 etc.), which help the user to refine the search based on a restricted set of relevant options.

Selectors: Using the facet categories and values chosen by the user, the system can identify a number of stored, semantically indexed *selectors*, which are parameterized, pre-defined query templates finally used for accessing the data sources in the back-end. In our architecture, a selector contains a variety of information and have the core functionality to map the results to visualization elements such as tables, charts, or maps. Basically, a selector has three types of parameters:

1. *Data parameters.* These parameters represent particular attributes to specify the scope of the selector for a particular information need. Some of the parameters (if not all) can be initialized to a particular value via the facets. That is, facets categories and values are mapped to data parameters of the selector that correspond to the same attributes. For the other data parameters that are not initialized this way, a SQL query is constructed by the system to retrieve the corresponding data directly from the database.
2. *Query parameters.* These are parameters pre-defined for each selector, indicating GROUP-BY and SORT statements, which are finally used for generating the SQL query.
3. *Visualization parameters.* In our system, the query results are visualized using different means, each one of them is derived from the visualization parameters of the selector. In particular, these parameters capture the visualization or presentation type (data value, data series, data table, map-based visualization, specific diagram type, etc.) for selectors' results.

(a) Facet-Selector Mapping   (b) SQL Query

Fig. 5. (a) Mapping the facet values to the corresponding selector parameters, (b) the generated SQL query

For example, Figure 5-a illustrates some possible facet values for our running example and their mappings to one of the selectors. As indicated, a selector may include more parameters than the facets. Thus, a selector can be flexibly initialized in different ways. Based on different facet categories and values, the system can generate a variety of initialized selectors, each corresponds to a different information need of the user. This way, with a relatively small number of selectors, the system can respond to a large number of queries. In addition, we also check the conformity of a possible facet selection of the user to a number of selectors available in the system to dynamically eliminate the selectors that can not answer a particular query. A selector is considered as non-conforming if it does not include any one of the facet categories (attributes) in its data parameters, and therefore is removed. In our prototype system, a number of selectors and their corresponding visualization components are automatically displayed to the user based on his/her selection of facets as shown in Figure 6.

Based on a particular initialization of selector, the system generates a SQL query, accesses the database and retrieves the results. An example SQL query is shown in Figure 5-b. Note that all the initialized values are captured in the WHERE clause whereas the non-initialized parameters are included in the SELECT statement of the query. The retrieved data are then displayed to the user as a final result based on his/her selection of visualization type (e.g. tables, maps, charts etc.).

4   Implementation

A prototypical KOIOS system has been implemented in Java within the scope of the German Internet project *THESEUS*[4]. For indexing the data, we use the open-source IR engine Lucene Framework[5] from Apache Software Foundation[6]. In addition, the system is integrated with a commercial EIS application Cadenza[7] for the management of selectors and for visualizing the results. In the current prototype, we are using the environmental data of the German state Baden-Württemberg, collected during the years from 1990 to 2006. The demo of the system is reachable at http://krake05.fzi.de:8888/koios.

---

[4] http://theseus-programm.de/
[5] http://lucene.apache.org/
[6] http://www.apache.org/
[7] http://www.disy.net/produkte/cadenza.html

Fig. 6. KOIOS search interface and facets generated for the query "karlsruhe co emission"

In Figure 6, we can see the start page (in German) of the KOIOS semantic search engine. The user can type a keyword query using a text field similar to the one provided by classical Web search engines that most users are familiar with. In the initial execution of the query, the system performs the aforementioned steps of query translation and displays users the facets on the left hand side of the page together with the possible visualization options on the right. Common types of visualization options are grouped together as tables, charts, or maps. Besides, the system also displays the base selectors suitable to the query and selected facets in order to give the user the flexibility to specify his/her information need in a more fine-grained way using the functionalities of the underlying EIS Cadenza system.

After the selection of facets and visualization type, the system displays the final results to the user using the Web version of the Cadenza system. Figure 7 shows the search result displayed as a chart for our running example query "karlsruhe co emission". As shown in the figure, the parameters selected include a number of possible emission types (e.g. industry, transportation PKW, LKW etc.), the year 2005, and Karlsruhe and Karlsruhe Stadt (i.e. Karlsruhe city center) for location. The system offers a number of visualization capabilities plus the option to store the chart into a file for later use. In addition, the resulting data can easily be visualized by another type of presentation module without the need for re-issuing the query thanks to the selector mechanism that can generate a variety of options from a single source query.

Another visualization option that displays the aggregated data as a map is shown in Figure 8. As in the case of chart display, the parameterization of this

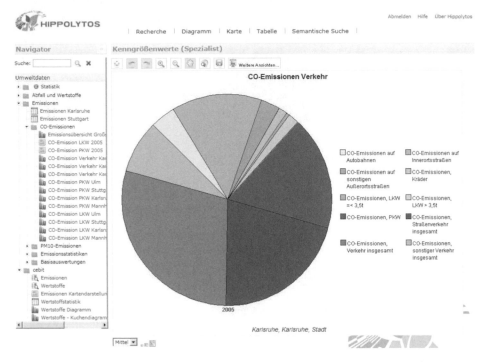

Fig. 7. Chart showing the CO emission values around Karlsruhe in 2005

type of visualization is also done via facets in a dynamic fashion. Starting from a keyword query "karlsruhe co emission", the user can obtain a variety of results when employing KOIOS's semantic search capabilities. In this type of visualization, the configuration options such as coloring scheme on the right hand side of the map also provides additional capabilities to the user to perform further refinements.

5   Related Work

Our research relates to work from three major areas, namely (1) environmental data, (2) semantic search on structured data, and (3) faceted search.

Environmental data on the Web: As environmental issues secured their position on a global level, environmental information has started to be considered as a public asset. As a consequence, governments and other administrative units have become more active in promoting access to environmental data as a mean to improve public participation in environmental decision making and awareness of environmental issues [7]. According to the survey conducted in [6], the major environmental information needs relate mainly to the people's everyday activities, which can be seen as a mix between livelihood issues, quality of life, and health issues. This mainly includes easy-access to information about public transport,

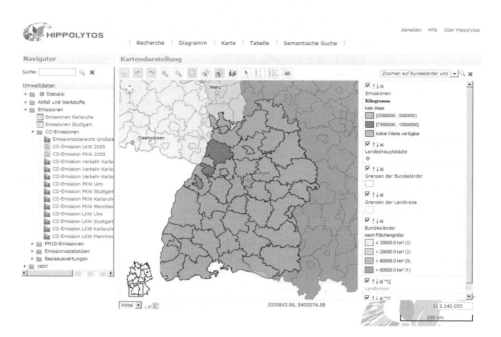

Fig. 8. Map showing the CO emission values around Karlsruhe in 2005

air quality, water quality, traffic, noise, and toxicity. Traditionally, environmental data are managed with the help of Environmental Information Systems and Environmental Decision Support Systems (EDSS) [3,4]. Recently, there is an abundance of environmental information on the Internet, from raw data which are broadcasted directly from monitoring stations, to politically charged information made available by specific interest groups [11,20,16,14,13]. Although the use of EIS and EDSS has a long history, easy-access to environmental data by the mass users has become a more prevalent problem, as public interest in these data increases. Previous works mostly provide proprietary interfaces to search and visualize data that are geared towards expert usage, and less tailored to the needs of the non-technical end users.

Semantic search on structured data: Finding and ranking relevant resources is the core problem in the Information Retrieval community, for which different approaches have been investigated. Clearly, the main difference of keyword search on structured data and the traditional keyword search on documents is that instead of one single document, structured data may encompass several resources (e.g. database tuples, documents, RDF resource descriptions) that are connected over a possibly very long path of relationships. [19] provides a review of different semantic search tools and focuses on different modes of user interaction. Compared with other modes of interaction (form-based, view-based, or natural language), the advantages of keyword-based querying lie in its simplicity and the familiarity most users already have with it. The problem of keyword queries on structured data has been studied from two different directions: 1) computing answers directly through exploration of substructures on the data graph [10,8,12] and 2) computing queries through exploration of a query space [17]. It has been shown in [17] that keyword translation operates on a much smaller query space, and is thus efficient. Besides, the structured queries presented to the user help in understanding the underlying data (answer) and allow for more precise query refinement. We follow the second line of work to keyword search and adapt it to the problem of searching environmental data. In addition, in order to rank the final retrieved results from the databases, we recently proposed a relevance model based ranking support [2].

Faceted Search: Faceted search is increasingly used in search applications, and many Websites already feature some sort of faceted search to improve the precision of search results. A crucial aspect of faceted search is the design of a user interface, which offers these capabilities in an intuitive way. This has been studied by [9,21,1] and applied in systems like Flamenco[8], Exhibit[9] or Parallax[10]. In a Semantic MediaWiki context, this paradigm has been applied by "Ask the Wiki" for browsing Wiki pages along semantic facets. [5]. Another cornerstone of faceted search is the question what is actually used as facets and if they are hierarchical or multidimensional, which obviously depends on the data corpus and its structure. Flamenco and Exhibit require a predefined set of properties for every data item, and then allows browsing along the values of these properties. We actually use the schema and query graphs for facet construction, and thus dynamically determine which facets should be present in a generic fashion. The diversity of query graphs results in a multidimensional facet generation, that user can refine by considering different aspects (e.g. time, location, category etc.). We also precompute the possible values of facets, which serve as feedback to the user and offer guidance for the underlying data and possible selectors to be chosen.

6   Conclusion and Future Work

Summary. We have sketched the functionalities and discussed the realization of KOIOS, a semantic search engine over structured environmental data. The goal is an "intuitive and simple way for information access" for emerging environmental data on the Web. KOIOS provides a Google-like, simple keyword-based query interface, which automatically finds and instantiates available selectors and thus automatically configures appropriate structured queries to be processed against the back-end data sources. It does not require pre-specified knowledge in the form of a schema or ontology, but instead, automatically computes schema graphs from the underlying data in order to find possible query interpretations. The results are provided to the user via a faceted-search interface, which facilitates further refinement in order to obtain more precise search results.

In fact, our approach differs from the use of a full semantic infrastructure that comes with expressive ontologies, SPARQL query processing and reasoning

---

[8] http://flamenco.berkeley.edu/
[9] http://www.simile-widgets.org/exhibit/
[10] http://www.freebase.com/labs/parallax/

capabilities. In this regard, our work mainly incorporates an IR-based approach to infer the users' information needs, and to retrieve, rank and visualize the relevant data for those needs. We consider that in comparison, our approach offers a number of benefits: First, it removes the need to specify an expressive ontology capturing all the domain knowledge which is not easily attainable for various scenarios in the environmental domain. In addition, most of the Semantic Web approaches still rely on the definition of structured queries (e.g. SPARQL) that is not so easy-to-formulate for ordinary users as we discussed above. Finally, ranking support is an important element of IR-based approaches that we incorporate as a further extension to our approach [2].

Status and Future Work. The prototype system is built on top of a commercial EIS called Cadenza and demonstrated over real-world environmental data of the German state Baden-Württemberg. It shows for realistic data volumes and schema sizes that it is possible to deliver reasonable results with acceptable performance. The evaluation of the result quality is yet to be conducted through further experiments. The ranking heuristics are most crucial, and particularly requires attention and detailed investigation in future work. Further, while the prototype can be seen as a study of technical feasibility, this work yet lacks evaluation from a usability point of view.

References

1. Ben-Yitzhak, O., Golbandi, N., Har'El, N., Lempel, R., Neumann, A., Ofek-Koifman, S., Sheinwald, D., Shekita, E., Sznajder, B., Yogev, S.: Beyond basic faceted search. In: Proceedings of the International Conference on Web Search and Web Data Mining, pp. 33–44. ACM (2008)
2. Bicer, V., Tran, T., Nedkov, R.: Ranking support for keyword search on structured data using relevance models. In: Proceedings of the 20th ACM Conference on Information and Knowledge Management (CIKM 2011). ACM (2011)
3. Denzer, R.: Generic integration of environmental decision support systems-state-of-the-art. Environmental Modelling & Software 20(10), 1217–1223 (2005)
4. El-Gayar, O., Fritz, B.: Environmental management information systems (emis) for sustainable development: a conceptual overview. Communications of the Association for Information Systems 17(1), 34 (2006)
5. Haase, P., Herzig, D., Musen, M., Tran, T.: Semantic Wiki Search. In: Aroyo, L., Traverso, P., Ciravegna, F., Cimiano, P., Heath, T., Hyvönen, E., Mizoguchi, R., Oren, E., Sabou, M., Simperl, E. (eds.) ESWC 2009. LNCS, vol. 5554, pp. 445–460. Springer, Heidelberg (2009)
6. Haklay, M.: Public environmental information: understanding requirements and patterns of likely public use. Area 34(1), 17–28 (2002)
7. Haklay, M.: Public access to environmental information: past, present and future. Computers, Environment and Urban Systems 27(2), 163–180 (2003)
8. He, H., Wang, H., Yang, J., Yu, P.S.: Blinks: ranked keyword searches on graphs. In: SIGMOD Conference, pp. 305–316 (2007)
9. Hearst, M.: Design recommendations for hierarchical faceted search interfaces. In: ACM SIGIR Workshop on Faceted Search, Citeseer, pp. 1–5 (2006)

10. Kacholia, V., Pandit, S., Chakrabarti, S., Sudarshan, S., Desai, R., Karambelkar, H.: Bidirectional expansion for keyword search on graph databases. In: VLDB, pp. 505–516 (2005)
11. Kruse, F., Uhrich, S., Klenke, M., Lehmann, H., Giffei, C., Töpker, S.: Portalu®, a tool to support the implementation of the shared environmental information system (seis) in germany. In: European Conference of the Czech Presidency of the Council of the EU TOWARDS eENVIRONMENT-Opportunities of SEIS and SISE: Integrating Environmental Knowledge in Europe, Prague (2009)
12. Li, G., Ooi, B.C., Feng, J., Wang, J., Zhou, L.: Ease: an effective 3-in-1 keyword search method for unstructured, semi-structured and structured data. In: SIGMOD Conference, pp. 903–914 (2008)
13. Mayer-Foll, R., Keitel, A., Geiger, W.: Uis baden-wuterttemberg. projekt aja. anwendung java-basierter und anderer leistungsfahiger losungen in den bere-ichen umwelt, verkehr und verwaltung. phase v 2004. Wissenschaftliche Berichte, FZKA-7077 (2004)
14. Pillmann, W., Geiger, W., Voigt, K.: Survey of environmental informatics in europe. Environmental Modelling & Software 21(11), 1519–1527 (2006)
15. Ruther, M., Bandholtz, T., Logean, A.: Linked environment data for the life sciences. Arxiv preprint arXiv:1012.1620 (2010)
16. Shrode, F.: Environmental resources on the world wide web. Electronic Green Journal 1(8) (1998)
17. Tran, T., Wang, H., Rudolph, S., Cimiano, P.: Top-k exploration of query candidates for efficient keyword search on graph-shaped (rdf) data. In: IEEE 25th International Conference on Data Engineering, ICDE 2009, pp. 405–416. IEEE (2009)
18. Tran, T., Zhang, L., Studer, R.: Summary Models for Routing Keywords to Linked Data Sources. In: Patel-Schneider, P.F., Pan, Y., Hitzler, P., Mika, P., Zhang, L., Pan, J.Z., Horrocks, I., Glimm, B. (eds.) ISWC 2010, Part I. LNCS, vol. 6496, pp. 781–797. Springer, Heidelberg (2010)
19. Uren, V., Lei, Y., Lopez, V., Liu, H., Motta, E., Giordanino, M.: The usability of semantic search tools: a review. The Knowledge Engineering Review 22(04), 361–377 (2007)
20. Vogele, T., Klenke, M., Kruse, F., Lehmann, H., Riegel, T.: Easy access to environmental information with portalu. In: Proceedings of EnviroInfo 2006, Graz, Austria (2006)
21. Yee, K., Swearingen, K., Li, K., Hearst, M.: Faceted metadata for image search and browsing. In: Proceedings of the SIGCHI Conference on Human factors in Computing Systems, pp. 401–408. ACM (2003)

Wiki-Based Conceptual Modeling:
An Experience with the Public Administration

Cristiano Casagni[1], Chiara Di Francescomarino[2], Mauro Dragoni[2], Licia Fiorentini[3],
Luca Franci[3], Matteo Gerosa[2], Chiara Ghidini[2], Federica Rizzoli[3], Marco Rospocher[2],
Anna Rovella[5], Luciano Serafini[2], Stefania Sparaco[4], and Alessandro Tabarroni[3]

[1] Polo Archivistico Regionale, Istituto per i beni artistici, culturali e naturali della Regione
Emilia-Romagna, Viale Aldo Moro 64 - Bologna, Italy
[2] FBK-irst, Via Sommarive 18 Povo, I-38123,Trento, Italy
[3] SCS Consulting, Via Marco Emilio Lepido182/3, I-40132, Bologna, Italy
[4] Servizio sviluppo amministrazione digitale e sistemi informativi geografici, Regione
Emilia-Romagna, Viale Silvani 4/3, I-40122, Bologna, Italy
[5] Dipartimento Sistemi di Produzione, Consiglio Nazionale delle Ricerche, Italy

Abstract. The dematerialization of documents produced within the Public Administration (PA) represents a key contribution that Information and Communication Technology can provide towards the modernization of services within the PA. The availability of proper and precise models of the administrative procedures, and of the specific "entities" related to these procedures, such as the *documents* involved in the procedures or the *organizational roles* performing the activities, is an important step towards both (1) the replacement of paper-based procedures with electronic-based ones, and (2) the definition of guidelines and functions needed to safely store, catalogue, manage and retrieve in an appropriate archival system the electronic documents produced within the PA. In this paper we report the experience of *customizing* a semantic wiki based tool (MoKi) for the modeling of administrative procedures (processes) and their related "entities" (ontologies). The tool has been *used* and *evaluated* by several domain experts from different Italian regions in the context of a national project. This experience, and the reported evaluation, highlight the potential and criticality of using semantic wiki-based tools for the modeling of complex domains composed of processes and ontologies in a real setting.

1   Introduction

In the last few years, the Public Administrations (PA) of several countries around the world have invested effort and resources into modernizing their services, in order to improve labor productivity as well as PA efficiency and transparency. The recent contributions and developments in ICT (Information and Communication Technology) can boost this modernization process, as shown by the dematerialization of documents produced within a PA. The availability of proper and precise models of the administrative procedures of the PA and of specific "entities" related to these procedures, such as the *documents* involved in the procedures or the *organizational roles* performing the activities, is a a key factor towards both (1) the re-design of the administrative procedures in

L. Aroyo et al. (Eds.): ISWC 2011, Part II, LNCS 7032, pp. 17–32, 2011.
© Springer-Verlag Berlin Heidelberg 2011

order to replace paper-based documents with electronic-based ones, and (2) the definition of guidelines and functions needed to safely store, catalogue, manage and retrieve in an appropriate archival system the electronic documents produced within the PA. The definition of these models requires the collaborative interplay of several actors with different competencies:

- *specific knowledge of the administrative procedures and their related documents in different domains.* Examples are the administrative procedures for business-to-government purchase and sale of goods and services, the ones for the management of personnel, those for the services that the PA offers to individual citizens, and so on. This knowledge is provided by domain experts working in the PA.
- *specific knowledge in archival science.* This knowledge is needed to identify what aspects of the administrative procedures have to be modeled in order to design the functionalities of an appropriate document management system. This knowledge is provided by experts in archival science.
- *specific knowledge in conceptual modeling (including process modeling).* This knowledge is necessary to help the construction of proper, precise and unambiguous models that can facilitate the analysis of the procedures, and of the "entities" related to these procedures. This knowledge is provided by knowledge engineers.

In this paper we report the experience of making these three groups of actors share their competences and collaborate in the modeling activities using the wiki-based MoKi tool in the context of the ProDe Italian national project. The reasons behind the choice of MoKi were its ability to involve domain experts in the modeling process as well as its ability to model both procedural aspects (the administrative procedures) and ontological aspects (the "entities" related to the procedures) in an integrated manner. While the general version of MoKi enables people to model generic processes and ontologies, a customization of the tool (ProDeMoKi) was developed for ProDe, in order to guide the domain experts working in the PA in modeling precisely the elements of the domain at hand. Thus, the entire modeling process we report in this paper consists on:

- an identification of the main entities to be modeled and their relations (conceptual schema). This activity was driven by experts in archival science, with the help of knowledge engineers;
- a customization of MoKi for building models coherent with the conceptual schema proposed, which led to the development of ProDeMoKi. This activity was driven by knowledge engineers, with feedback from experts in archival science;
- the final modeling activity, which was performed by domain experts from the PA with some supervision from knowledge engineers and experts in archival science.

The contribution of the paper is therefore twofold: (1) it provides an empirical evidence of how to customize a generic wiki-based modeling tool for a specific complex scenario (Section 4) on the basis of its conceptual description (Section 2.1); and, (2) it provides an evaluation of the main features of the tool (Section 5) and an analysis of the lessons learned. This experience, and the reported evaluation, highlight the potential and criticality of using semantic wiki-based tools for the modeling of complex domains composed of processes and ontologies in a real setting, and can provide the basis for future customizations of wiki-based modeling tools to specific complex domains.

2 The ProDe Project

ProDe[1] is an Italian inter-regional project with the aim of defining a national reference model for the management of electronic documentation (dematerialized document) in the Public Administration. This reference model follows an archival science perspective, and can be used for the identification of guidelines and functions needed to safely store, classify, manage, and retrieve, electronic documents produced within the PA in an archival system. The project has a duration of 30 months, from May 2010 to October 2012, and is composed of 11 tasks assigned to 11 teams (*task-teams*) coming from 10 regions (Piemonte, Lombardia, Liguria, Emilia-Romagna, Marche, Abruzzo, Campania, Puglia, Sicilia and Trentino). The 11 tasks are divided in 4 **central tasks**, that represent the core part of the project and are in charge of guiding the activities on specific topics such as document management and digital preservation; and 7 **peripheral tasks**, that provide the specific expertise on specific sectors of the PA (e.g., the administrative procedures for business-to-government purchase and sale of goods and services, the ones for the management of personnel). Thus, the central tasks provided the main expertise in archival science, while the peripheral tasks provided domain expertise in different fields of the PA.

2.1 Modeling Flows of Documents: The Conceptual Schema

In this section we illustrate the conceptual schema that was used in the customization of the ProDeMoKi and was proposed to the domain experts to guide the modeling of their administrative procedures. This conceptual schema, whose simplified version is graphically depicted in Figure 1 using an Entity-Relationship notation, was developed by the experts in archival, computer, and organizational sciences working in the central tasks of the ProDe project.

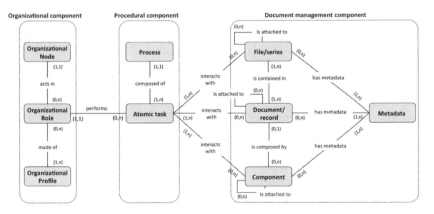

Fig. 1. The conceptual schema

As shown in Figure 1, the entities of the model can be clustered in three different components, each of which plays an important role in the production and management of documents in administrative procedures:

[1] http://www.progettoprode.it/Home.aspx

1. **Document management** component. This component describes the archiving-related aspects of the domain and constitutes the central part of the model.
2. **Procedural** component. This component describes the activities which produce (manage, consume) documents.
3. **Organizational structure** component. This component describes the structure of the offices involved in the dematerialized procedures and the different profiles which interact with documents, possibly with different permissions.

The document management component. The entities *document* and *record*[2], hereafter named document/record, or, for the sake of simplicity only document, constitute the central entities of the conceptual model. In ProDe, the description of the document/record entity is mainly devoted to the life-cycle of the document, which consists of the 5 following actions: **Create** (how the document is created); **Capture** (how the document is acquired in the document management system); **Manage** (how the document is managed inside the document management system); **Store/preserve** (how the document is stored in the document management system and preserved in the long run); and **Deliver** (how the document is distributed and made available by the system). These actions are used to identify the services and functionalities needed in the document management system to handle documents and records in a correct and appropriate manner. Furthermore, the description of document/record is characterized by a set of attributes such as the name of the document/record, the type of document, the origin and destination of the entity, and so on. The entity *component* identifies the set of bits which (possibly together with other components) composes a document/record. Moreover, each document/record is classified according to a filing plan and inserted in an appropriate *file/series*. According to the type of file/series, different criteria of management, storage, preservation and access can be granted to the document/record. Finally, *Metadata* are used to provide information about the context, content, structure, and management of the different entities described so far. To support the construction of an homogeneous model and the compatibility with the main standards for metadata such as Moreq2[3] and EAD[4], a common dictionary of metadata was also provided by the central teams of the project, and inserted in ProDeMoKi.

The procedural component. Usually a business process is composed of a set of related activities, which happen inside an organization and transform resources in products or services for an (internal or external) customer. Within the ProDe project the entity *process* has been used also to include all those (complex) activities carried out by document management systems. *Atomic task* is instead used to describe an atomic action within a process. The relations with other entities of the document management component emphasize the fact that activities can perform actions over these entities (such as the creation of a document or the modification of a file). The relations with the

---

[2] By *record* we refer to an archival document, in a final and correct state, registered into a document management system and not modifiable or deletable. The only operation possible on a record is the modification of its metadata. By *document* we instead refer to an artifact which still requires modifications and is amenable to cancellation.

[3] http://www.dlmforum.eu/

[4] http://www.loc.gov/ead/

organizational role entity in the organizational structure component emphasize the fact that actions are performed by specific roles within the organization.

The organizational structure component. The main entity of the organizational structure is the *organizational role*. A role refers to an *organizational node*, which defines the atomic component of an organization (e.g., finance office), and is associated to an *organizational profile*, which instead defines the different profiles of permissions available within the organization.

3   The MoKi Architecture and Tool

MoKi[5] is a collaborative MediaWiki-based [8] tool for modeling ontological and procedural knowledge. The main idea behind MoKi is to associate a wiki page, containing both unstructured and structured information, to each entity of the ontology and process model. From a high level perspective, the main features of MoKi[6] are:

- the capability to model different types of conceptual models in an integrated manner. In particular the current version of MoKi is tailored to the integrated modeling of ontological and procedural knowledge;
- the capability to support on-line collaboration between members of the modeling team, including collaboration between domain experts and knowledge engineers.

These features have been proved extremely important in the context of the ProDe project. In fact, as we can see from the ER model depicted in Figure 1, the scenario addressed in the ProDe project required the modeling of administrative procedures, usually better described using a business process modeling notation, enriched with knowledge which typically resides in an ontology, such as the classification of document types, organizational roles, and so on. Moreover, the modeling team was composed by an heterogeneous group of domain experts and knowledge engineers situated in different Italian geographical regions. In the following we illustrate how these features are realized in the generic MoKi architecture. In Section 4 we illustrate the ad-hoc customization we performed for the ProDe project and how these features were realized for the specific MoKi used in ProDe.

Modeling integrated ontological and procedural knowledge. The capability of modeling integrated ontological and procedural knowledge is based on two different characteristics of MoKi. First of all, MoKi associates a wiki page to each *concept*, *property*, and *individual* in the ontology, and to each (complex or atomic) *process* in the process model. Special pages enable to visualize (edit) the ontology and process models organized according to the generalization and the aggregation/decomposition dimensions respectively. The ontological entities are described in Web Ontology Language (OWL [11]), while the process entities are described in Business Process Modeling Notation (BPMN [9]). Second, MoKi has extended the functionalities of the BPMN Oryx editor [3], used in MoKi to represent BPMN diagrams, to annotate tasks with concepts

---

[5] See http://moki.fbk.eu

[6] A comprehensive description of MoKi can be found in [6].

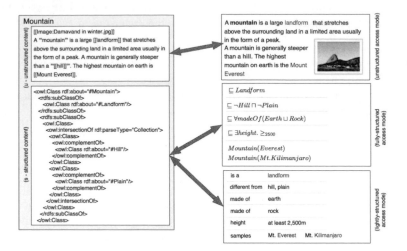

Fig. 2. Multi-mode access to a wiki page

described in the ontology, or to incorporate data objects formalized in the ontology. The integrated procedural and ontological knowledge is then exported in a comprehensive OWL model following the approach described in [4].

Supporting collaboration between domain experts and knowledge engineers. MoKi is an on-line tool based on MediaWiki, thus inheriting all the collaborative features provided by it. In addition MoKi facilitates the collaboration between domain experts and knowledge engineers by providing different access modes to the elements described on the model, as illustrated in Figure 2 for the ontology concept "Mountain".

MoKi allows to store both *unstructured* and *structured* descriptions of the elements of the models, as shown on the left hand side of Figure 2. The unstructured part contains a rich and often exhaustive description of knowledge better suited to humans, usually provided with linguistic and pictorial instruments. Instead, the structured part is the one which is used to provide the portion of knowledge which will be directly encoded in the modeling language used to describe the specific element (OWL in the case of the concept "Mountain"). The advantage of storing the unstructured and structured descriptions in MoKi is twofold. First, informal descriptions are usually used to provide the initial description upon which the formal model is built, and to document the elements of the model (e.g., for future access and revisions). Storing the unstructured and structured descriptions in the same tool can facilitate the interplay between these parts. Second, domain experts, who usually create, describe, and review knowledge at a rather informal/human intelligible level, may find the unstructured part their preferred portion of page where to describe knowledge, while knowledge engineers should be mainly focused on the descriptions contained in the structured part. Nevertheless, by using the same tool and accessing the same pages, all of them can be notified of what the others are focused at. Moreover, the discussion facilities of wikis, together with special fields for comments, can be used by both roles to discuss on specific parts of the model.

The organization of a page in an unstructured and a structured part is a first important collaborative feature, but may not be enough in the case of complex conceptual modeling languages, such as OWL or BPMN. In this case the structured part of the page will contain very precise, and often logic based, descriptions of the knowledge, preventing domain experts from accessing the domain knowledge encoded in the conceptual model. To overcome this problem, MoKi associates different *access modes* to each part of the page, as depicted in the right hand side of Figure 2. The current general version of MoKi is based on three different access modes:

- an *unstructured access mode* to view/edit the unstructured content;
- a *fully-structured access mode* to view/edit the complete structured content; and
- a *lightly-structured access mode* to view/edit (part of) the structured content via simple templates.

As shown in Figure 2, the access mode to the unstructured part can be provided by means of the regular view/edit facilities of wikis, while the access to the structured content can be provided by means of two different modes: one based on a translation of the OWL content in, e.g., DL axioms or in the Manchester OWL syntax, and another based on a structured, but semi-formal rendering of the OWL content in a pre-defined template. This way, the knowledge engineers can formally describe the ontology concept "Mountain" in OWL by using a highly formal access mode, while the domain experts can access a simplified version of the same content using a different, simpler, mode. A similar structure is provided also for the description of BPMN elements. By providing distinct modalities to access the structured content of a wiki page domain experts can not only have access to the knowledge inserted by knowledge engineers, but can also comment or directly modify part of it. Therefore, the design of appropriate access modes, according to the conceptual modeling language used and the degree of complexity handled by the domain experts, is a key aspect of the implementation of a wiki-based tool for conceptual modeling.

4   The MoKi Customization for the ProDe Project

ProDeMoKi[7] is the customization of MoKi that has been developed for the ProDe project. The general version of the tool has been adapted to support both the representation of the conceptual models needed in the project, and the skills of the modeling actors, who had a good expertise in the design of business processes, but had no experience in ontology design. Therefore, a first personalization of MoKi consisted in using only the unstructured and the lightly-structured access modes for the definition of ontological entities (the ones in the Document management component, and Organizational structure component in Figure 1), and only the unstructured and the fully-structured access modes for the modeling of the procedural component.

The second, important, personalization involved the templates used in the lightly-structured access mode of the ontological entities. In particular we have created an ad-hoc template for each entity shown in 1. In this paper we focus on the template for the document entity, whose main parts are shown in Figure 3, as it provides a representative and exhaustive example of the customizations implemented in ProDeMoKi.

---

[7] Available at https://dkmtools.fbk.eu/moki/prode/tryitout/index.php/Main_Page

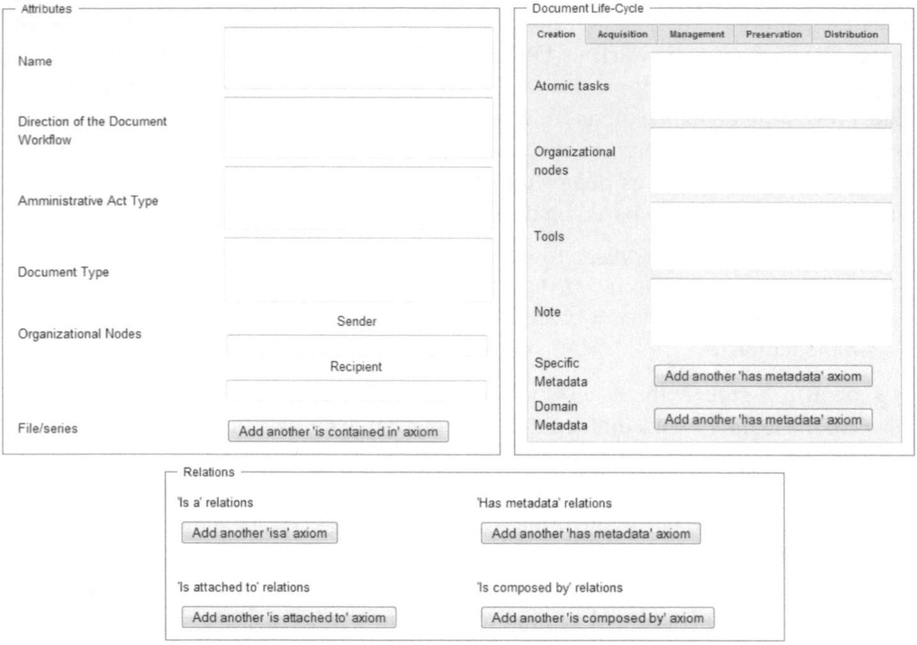

Fig. 3. The template used to insert document information

The *Attributes* box in Figure 3 allows the insertion of general information through the first four text areas. Besides this general information, the user is able to insert the relations between a document and the organizational nodes who perform the actions which involve a document. The bottom part describes the relations between a document and the file/series that contain it. The *Document life-cycle* box in Figure 3 is focused on the steps which involve a document during a process, from its creation to its preservation in the document management system; for example a document may be classified in file/series during an "Acquisition" task and it may be electronically signed during a "Management" task. The user is able to describe these actions in the text boxes contained in the template, and to define the metadata that are required in each task via the "has-metadata" relation. Finally, the *Relations* box in Figure 3 is used to specify additional relations between the document and other entities in the ER model, namely, the "is-a" relation with other documents; the "is-attached-to"relation which expresses the fact that a document may be an annex of another document; the "has-metadata" relation which is used to express metadata that hold for the document in general, independently from the specific phase in the lifecycle; and the "is-composed-by" which expresses the fact that a document is composed of a certain set of bits in a certain electronic format (e.g., a pdf file).

By combining the general features of MoKi and the customized ones discussed in this section, ProDeMoKi enables the following macro-functionalities: *model overview*, *model navigation*, and *entity modeling* (i.e., creation, revision, deletion and renaming). In detail, the global view of the model is provided both in the form of an unstructured

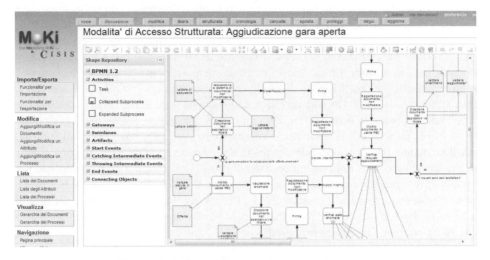

Fig. 4. The BPMN diagram enriched with documents

list and of a hierarchical taxonomy of documents and processes. Similarly, the model can be navigated starting from an unstructured list, a hierarchical view, and also directly from the graphical representation of BPMN diagrams, enriched with data objects representing documents taken from the ontology, as shown in Figure 4[8]. Such a graphical representation can also be exploited for modeling processes and documents, together with the side-bar "add" commands (for both documents and processes) and the "add"commands available in the hierarchical view (only for documents).

The customization of MoKi required about 1 person-week for the definition of the conceptual model and 3 person-weeks for the implementation and testing of ProDeMoKi.

5   Evaluation

With the aim of evaluating the usage of ProDeMoKi for supporting domain experts in the collaborative modeling of specific knowledge, we investigated the following research questions:

- (*RQ1*) Is ProDeMoKi easy to use for domain experts?
- (*RQ2*) Is ProDeMoKi useful for collaboratively modeling domain knowledge?
- (*RQ3*) Are all the provided views useful or is there a "best" view among the different interface views provided by ProDeMoKi for: (a) getting the model overview? (b) navigating the model? (c) creating new entities?

In order to answer these questions we performed two types of analysis: a quantitative and a qualitative one. In the former, data about tasks performed by ProDeMoKi users in a use case have been analyzed, while in the latter, ProDeMoKi users have been asked to answer a questionnaire in order to capture their perceptions about the ease of use and usefulness of the tool according to their experience.

---

[8] This feature has been added in ProDeMoKi in January 2011. Documents are represented by means of green data objects.

The Use Case. The considered use case consists in the actual usage of ProDeMoKi in the first phase of the ProDe project, where participating regions have been asked to produce two types of models: the "Modello di riferimento" and "Modello di gestione". While the "Modello di riferimento" is an abstract model of administrative procedures and documents, i.e., a kind of meta-model of concrete models, the "Modello di gestione" refines processes of the "Modello di riferimento" into concrete and more specific tasks, thus creating a link with the software applications to be used in the process realization. The ProDeMoKi users involved are PA employees distributed across the 7 peripheral tasks of the project. All of them are domain experts, i.e., they have knowledge of the PA domain and had previous experiences in the analysis of administrative procedures and documents as well as in documentation drafting. However, not all the users had the opportunity to model processes before and none to use ontologies. Before the beginning of the modeling activities (February 2011), the PA employees have been trained with a learning session, in which all the features of ProDeMoKi have been illustrated, and hands-on exercises have been proposed.

The Questionnaire. The PA employees have been asked to fill an on-line questionnaire, requiring about 15 minutes and including 31 questions[9]. The questionnaire mainly aimed at investigating the users' background, their perception about three macro-functionalities of the tool (model overview, navigation and entity modeling), and their overall subjective evaluation about ProDeMoKi. Some of the questions were provided in the form of open questions, while most of them were closed questions. The latter mainly concern the user evaluation of the tool on a scale from 1 to 5 (e.g., 1 = *difficult to use*, ... , 5 = *intuitive to use*) or the direct comparison between two alternatives (e.g., 1 = *I prefer A*, 2 = *I prefer B*, and 3 = *I equally evaluate A and B*).

5.1   Quantitative Evaluation Results

We analyzed the data on the usage of ProDeMoKi (October 2010 - June 2011), that have been obtained by combining the information stored in the ProDeMoKi database, and the access logs of the web server hosting the tool. Table 1 shows the overall number of page creations, revisions, deletions and renaming performed by the different task-teams in the considered period[10]. The table shows that, though all the typologies of page activities have been exercised by users, there exists a trend in their distribution: as expected, the most frequent one is the page revision activity, followed by page creation and deletion. By looking at the distribution of the activities per month (Figure 5a)[11], we found that: (i) the general trend of activity typologies on pages is also monthly confirmed; (ii) there have been peaks of work in November 2010, February 2011 and May 2011. These can be partially justified by users' autonomous ProDeMoKi training after the beginning of the modeling activities (October 2010) and the learning session (February 2011), and by project internal deadlines (May/June 2011).

---

[9] Collected data are available at
    https://dkm.fbk.eu/index.php/ProDeMoKi_Evaluation_Resources.
[10] The task-teams that are not required to use ProDeMoKi in this phase of the project are not reported in the table.
[11] The average number of daily activities on page is reported.

Table 1. Usage of ProDeMoKi by the different *task-teams* per type of activity on pages

task-team	Page Creations	Page Revisions	Page Deletions	Page Renaming
t1	171	542	71	0
t2	171	799	37	54
t3	135	534	13	33
t4	217	548	55	2
t5	384	1065	58	7
t6	103	316	16	0

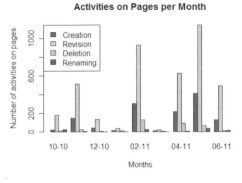

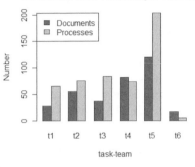

(a) Activities on pages per month   (b) Documents and processes per task-team

Fig. 5. ProDeMoKi usage

Currently, the two reference models contain overall 342 documents and 506 processes modeled with ProDeMoKi. Figure 5b shows their distribution per task-team. Most of the task-teams (4 out of 6) produced more processes than documents, thus remarking the importance of the procedural knowledge in domains such as the PA one. Furthermore, the number of documents used in processes' diagram (on average 4.2^{12}) and the number of processes in which a document is used (on average 1.32), also reveals the strength of the relationship between these two types of knowledge, thus confirming our intuition about the importance of providing adequate means for their integration.

Moreover, by looking at the usage of (the commands available in) the different views provided by ProDeMoKi, it is possible to have an insight of their suitability for specific purposes, as well as of users' preferences. In detail, Table 2a summarizes the results obtained by investigating the different views available for the model overview and navigation, while Table 2b for the creation of new documents and processes. While in case of documents, the number of model visualizations and document accesses from the unstructured list is more than double with respect to the corresponding numbers in the hierarchical view, for processes also the hierarchical visualization plays an important role. On the other hand, while the graphical representation of both documents and processes in BPMN diagrams has not been extensively exercised by users for the model navigation (probably because of its late introduction in the tool, in January 2011), it represents the most used way for creating new documents and processes: it in fact

---

[12] The average number decreases to 0.6 documents per process if we consider also atomic processes, i.e. with an empty flow.

Table 2. Use frequency of ProDeMoKi views

(a) Use frequency of model overview and navigation

Entity	Model Overview		Model Navigation			
	List	Hierarchy	List	Hierarchy	Diagram	Textual Search
Doc.s	503	239	341	155	11	-
Proc.s	939	626	758	478	55	-
Total	1042	865	1099	633	66	11

(b) Use frequency of entity creations

Document and Process Creation			
	Sidebar	Diagram	Hierarchy
Doc.s	113	275	61
Proc.s	90	580	-
Total	203	855	61

allows to introduce high level entities directly in the process modeling phase and to detail them later on. Both these results seem to confirm our intuition about the importance of abstraction layers in process modeling and visualization.

Finally, we investigated the usage of a special functionality offered by the tool (provided by MediaWiki), the log history. Although it has not been frequently used (52 times in total), the 40% of times it has been exercised by the most productive (120 documents and 203 processes) and among the most numerous (4 modelers) task-teams, thus remarking its usefulness in case of large models and of collaborative works.

5.2 Qualitative Evaluation Results

We analyzed the data collected by means of the on-line questionnaire proposed to a total of 14 ProDeMoKi users in order to investigate their perception about the tool. Table 3 reports the information about their background knowledge, as well as about the time spent and the approach followed to learn the ProDeMoKi use. All the participants are pc habitual users: they use the pc almost everyday, mainly for writing and reading documents, for job (using office suites) and for navigating the Internet. They also had frequent occasions to visit wiki pages, while only half of them had the opportunity to (on average rarely) edit wiki pages. 93% of the employees had previous experience in analyzing administrative documents and procedures, while 21% had never modeled diagrams before. On average, the time spent for learning ProDeMoKi has been 1-2 days and the preferred learning approach the autonomous training.

In the questionnaire, we investigated the users' evaluation with respect to the three main ProDeMoKi macro-functionalities, with a special focus on the different alternative approaches provided by ProDeMoKi for their realization.

Table 4 (left) reports the results related to the users' average ranking of the two alternative views provided by ProDeMoKi (where 1 and 2 denote respectively the first and the second position in the ranking) for the model overview with respect to ease of use and usefulness[13]. Similarly, Table 4 (right) reports the average value (and the standard deviation) of the ranking provided by subjects about the usefulness of the four alternative views for the model navigation, where 1 represents the first position in the ranking and 4 the last one. The table shows that the hierarchical view is perceived by users as the most useful for both the model overview and navigation, while the list is considered as the least useful. A similar result is obtained also in the case of the comparison between the ease of use of the list and the hierarchical views. Moreover,

---

[13] More details about the reliability of the agreement among users can be found in [5].

Table 3. Users' background knowledge and learning approach

Property	Value	%	Property	Value	%
	never	0		document reading and editing	27.66
	rarely	0		office suites	27.66
Pc use	sometimes	0	Pc use	Internet browsing	25.53
frequency	often	0	purpose	programming	17.02
	always	100		testing and customer care	2.13
	never	0		never	50
Wiki page	rarely	21.43	Wiki page	rarely	42.86
consultation	sometimes	28.57	editing	sometimes	7.14
frequency	often	14.29	frequency	often	0
	always	35.71		always	0
	none	7.14		domain analysis and textual documentation drafting	21.43
Experience in	bad	0	Type of	procedure analysis and diagram creation	21.43
domain analysis	medium	35.71	experience	document analysis and documentation production	0
and documentation	good	42.86		domain analysis as well as documentation and diagram production	57.14
	very good	14.29			
	< 1 day	36.36		autonomous training	72.73
Time spent	< 2 days	36.36	Learning	learning session	9.09
for learning	1 week	27.27	approach	tutorial	0
	10 days	0		talking with and asking colleagues	18.18
	> 10 days	0			

Table 4. Users' ranking of ProDeMoKi views for the model overview and navigation

Model Overview				Model Navigation			
Factor	View	Avg. rank	Std. dev.	Factor	View	Avg. rank	Std. dev.
Ease of use	Hierarchy	1.14	0.23	Usefulness	Hierarchy	2.00	0.74
	List	1.86	0.23		Diagram	2.14	0.78
Usefulness	Hierarchy	1.14	0.23		Textual Search	2.73	0.68
	List	1.86	0.23		List	3.14	0.92

the quality of the overall model navigability support provided by ProDeMoKi is also investigated: on average, it is perceived by users as more than reasonable.

Table 5a reports, instead, the subjective user perception, on a scale from 1 (very difficult) to 5 (intuitive), about the ease of performing modeling activities (i.e., creation, revision, deletion and renaming) on model entities. The table shows that on average users found entity creation easy, deletion between reasonable and easy, and revision and renaming more than reasonable.

Among the different activities, we focused in details on the entity creation, and we investigated the users' subjective ranking of ease of use and usefulness of the different alternatives for creating new documents/processes. Table 5b reports the average value and the standard deviation of the users' rankings, where 1, 2, and 3 are the three possible values provided by subjects for denoting the first, the second, and the third ranking position in case of documents, and 1 and 2 for processes. According to the subjective evaluation, the most useful and easy to use way for creating new documents and processes is by means of the sidebar commands, while the least useful and easy to use is the creation starting from the diagrammatical view.

Finally, Table 6 shows the results related to the subjective evaluation about the overall ease of use (resp. usefulness) of the tool reported by ProDeMoKi users on the base of a 5-point scale, where 1 means very difficult (resp. completely useless) and 5 intuitive (resp. absolutely useful). According to the users' answers, ProDeMoKi is easy to use and

Table 5. ProDeMoKi entity modeling

(a) Perceived ease of use of modeling activities

	Avg.	Std. dev.
Entity Creation	4	0.89
Entity Revision	3.45	0.69
Entity Deletion	3.55	0.69
Entity Renaming	3.27	0.65

(b) Users' ranking about ease of use and usefulness of alternative views for new entity creation

Document Creation				
View	Ease of use		Usefulness	
	Avg. rank	Std. dev.	Avg. rank	Std. dev.
Sidebar	1.68	0.46	1.68	0.6
Hierarchy	2.05	0.27	2.14	0.45
Diagram	2.27	0.41	2.18	0.6
Process Creation				
View	Ease of use		Usefulness	
	Avg. rank	Std. dev.	Avg. rank	Std. dev.
Sidebar	1.32	0.25	1.41	0.3
Diagram	1.68	0.25	1.59	0.3

Table 6. Perceived ease of use and usefulness of ProDeMoKi

	Average	Std. dev.	p-value
Ease of use	3.36	0.67	7.442e-06
Usefulness	3.36	0.81	3.739e-05

useful (the provided evaluation about ProDeMoKi is, on average, more than easy and more than fairly useful for the collaborative modeling). These positive results are also statistically relevant at 5% confidence level[14] (i.e., we have the 95% of confidence that the fact that the tool is perceived as easy to use/useful is not due to chance). Moreover, we found that there exists a strong positive correlation[15] between the subjective answers with respect to the ProDeMoKi usefulness for the collaborative work and the number of people involved in the team, thus strengthening the result related to usefulness of the tool for collaborative purposes.

5.3   Discussion and Lesson Learned

The quantitative data collected demonstrate a concrete and almost continual experience of actual users, distributed all over Italian territory, in the use of the tool, thus making us confident about the reliability of their answers.

By looking at the results obtained with respect to the ease of use of the creation, editing and deletion on entities and at the overall ease of use of the tool, we can state that the users perceive the tool as more than easy to use. This result is also strengthened by the amount of time spent and the approach exploited for learning how to use the tool: 72% of employees spent only less than two days to learn how to use ProDeMoKi, and the same percentage learned it autonomously. We can hence positively answer *RQ1*.

Moreover, we observed that users positively perceive the overall usefulness of the tool for the collaborative modeling of documents and processes. The validity of this result is also confirmed by the fact that such a usefulness is perceived more strongly by

---

[14] We applied a one-tailed Wilcoxon test and we obtained in both cases p-value$<$0.05.

[15] We performed a correlation analysis applying the Spearman's coefficient at 5 percent confidence level.

employees working in teams having more than two persons (on average 3.8 for teams with more than two persons versus 2.8 for those with less than three). There exists, in fact, a correlation between the size of the subject's team and his/her feedback about the ProDeMoKi usefulness for collaborative purposes. We can hence positively answer *RQ2* too.

With respect to the alternative views provided by the interface we found that, while the hierarchical one is perceived by users as the most useful (and the list as the least useful one) both for the model overview and navigation, their frequency of use shows that the list is actually the most used for both the macro-functionalities. Similarly, the creation of documents and processes starting from the process view, though perceived as difficult to use and less useful with respect to the commands available in other views (i.e., the classical ones on the sidebar and in the hierarchical view), has been extensively used. A possible explanation to this is that more mechanical actions (as the model overview through the list of its artifacts) are usually the first to be executed and only later, if not useful, they are replaced by more specific ones. While these results suggest to better investigate the usability of those views that, although perceived as more useful and easy to use, are less used in practice, they also reveal the importance of having all the available alternative views. In conclusion, for all the three sub-questions of *RQ3*, we can state that all the views have their own usefulness.

Finally, some interesting suggestions for improving ProDeMoKi and making it more suitable to users' needs came from the answers to the open questions. While quite satisfied with the functionalities related to the model overview and navigation, users found some space of improvement for the functionalities related to the modeling, in particular to the document modeling (almost 50% of the users found the document modeling a weakness of the tool). Some of the proposed suggestions have already been implemented in a new prototypical version of the tool, while we plan to implement others (e.g., a command for duplicating entities) in the next versions. However, the tool has been positively judged with respect to its process modeling capability (identified as the ProDeMoKi major strength), as well as to its support for collaboration and modeling of complex domains, and its broad accessibility (only a web browser is needed).

6   Concluding Remarks

Several works have focused on the application of Semantic Web technologies to the PA, for instance in the areas of semantic services [12], reference models [10], data integration [1], and collaborative knowledge sharing [7]. Similarly several efforts aim at using semantic wikis for the collaborative construction and visualization of conceptual models [8,6,2]. In this paper we report our experience in applying semantic-based wiki technology for the specific modeling needs of a complex PA domain, in which administrative procedures and related "entities" are tangled. The concrete use of the ProDeMoKi tool by real domain experts and their subjective evaluation revealed that the tool is easy to use and useful for the collaborative work, though still open to improvements. In the future, we plan to enhance ProDeMoKi by implementing the users' suggestions and further investigating how to improve the usage of the tool (e.g., strengthening the training, enhancing or adding functionalities). Moreover we aim at validating the overall MoKi customization approach by extending the MoKi customization to other specific domains.

References

1. Alani, H., Dupplaw, D., Sheridan, J., O'Hara, K., Darlington, J., Shadbolt, N., Tullo, C.: Unlocking the Potential of Public Sector Information with Semantic Web Technology. In: Aberer, K., Choi, K.-S., Noy, N., Allemang, D., Lee, K.-I., Nixon, L.J.B., Golbeck, J., Mika, P., Maynard, D., Mizoguchi, R., Schreiber, G., Cudré-Mauroux, P. (eds.) ASWC 2007 and ISWC 2007. LNCS, vol. 4825, pp. 708–721. Springer, Heidelberg (2007)
2. Auer, S., Dietzold, S., Riechert, T.: OntoWiki – A Tool for Social, Semantic Collaboration. In: Cruz, I., Decker, S., Allemang, D., Preist, C., Schwabe, D., Mika, P., Uschold, M., Aroyo, L.M. (eds.) ISWC 2006. LNCS, vol. 4273, pp. 736–749. Springer, Heidelberg (2006)
3. Decker, G., Overdick, H., Weske, M.: Oryx – An Open Modeling Platform for the BPM Community. In: Dumas, M., Reichert, M., Shan, M.-C. (eds.) BPM 2008. LNCS, vol. 5240, pp. 382–385. Springer, Heidelberg (2008)
4. Di Francescomarino, C., Ghidini, C., Rospocher, M., Serafini, L., Tonella, P.: Semantically-Aided Business Process Modeling. In: Bernstein, A., Karger, D.R., Heath, T., Feigenbaum, L., Maynard, D., Motta, E., Thirunarayan, K. (eds.) ISWC 2009. LNCS, vol. 5823, pp. 114–129. Springer, Heidelberg (2009)
5. Dragoni, M., Ghidini, C., Rospocher, M., Serafini, L., Di Francescomarino, C.: On the use and evaluation of a wiki-based tool. Technical report, FBK-IRST, Italy (2011), https://dkm.fbk.eu/index.php/ProDeMoKi_Evaluation_Resources
6. Ghidini, C., Rospocher, M., Serafini, L.: Moki: a wiki-based conceptual modeling tool. In: ISWC 2010 Posters & Demonstrations Track: Collected Abstracts, Shanghai, China. CEUR Workshop Proceedings (CEUR-WS.org), vol. 658, pp. 77–80 (2010)
7. Krabina, B.: A semantic wiki on cooperation in public administration. In: 7th International Semantic Web Conference (ISWC 2008), Poster & Demo session (2008)
8. Krotzsch, M., Vrandecic, D., Volkel, M.: Wikipedia and the semantic web - the missing links. In: Proc. of the 1st Int. Wikimedia Conference, Wikimania 2005 (2005)
9. OMG. Business process modeling notation, v1.1, http://www.omg.org/spec/BPMN/1.1/PDF
10. Peristeras, V., Tarabanis, K.: Reengineering the public administration modus operandi through the use of reference domain models and semantic web service technologies. In: AAAI Spring Symposium on The Semantic Web meets eGovernment (2006)
11. Smith, M.K., Welty, C., McGuinness, D.L.: Owl web ontology language guide. W3C Recommendation (February 10, 2004)
12. Vitvar, T., Kerrigan, M., van Overeem, A., Peristeras, V., Tarabanis, K.: Infrastructure for the semantic pan-european e-government services. In: AAAI Spring Symposium on The Semantic Web meets eGovernment (2006)

Linking Semantic Desktop Data to the Web of Data

Laura Drăgan[1], Renaud Delbru[1], Tudor Groza[2],
Siegfried Handschuh[1], and Stefan Decker[1]

[1] Digital Enterprise Research Institute (DERI),
National University of Ireland, Galway
firstname.lastname@deri.org
http://www.deri.ie
[2] School of ITEE, The University of Queensland, Australia
firstname.lastname@uq.edu.au
http://www.itee.uq.edu.au

Abstract. The goal of the Semantic Desktop is to enable better organization of the personal information on our computers, by applying semantic technologies on the desktop. However, information on our desktop is often incomplete, as it is based on our subjective view, or limited knowledge about an entity. On the other hand, the Web of Data contains information about virtually everything, generally from multiple sources. Connecting the desktop to the Web of Data would thus enrich and complement desktop information. Bringing in information from the Web of Data automatically would take the burden of searching for information off the user. In addition, connecting the two networks of data opens up the possibility of advanced personal services on the desktop.

Our solution tackles the problems raised above by using a semantic search engine for the Web of Data, such as Sindice, to find and retrieve a relevant subset of entities from the web. We present a matching framework, using a combination of configurable heuristics and rules to compare data graphs, that achieves a high degree of precision in the linking decision. We evaluate our methodology with real-world data; create a gold standard from relevance judgements by experts, and we measure the performance of our system against it. We show that it is possible to automatically link desktop data with web data in an effective way.

Keywords: Semantic Desktop, Semantic Web, Linked Data, Personal Information Management.

1 Introduction

The Semantic Desktop aims to enable better organization of the personal information on our computers, by applying semantic technologies on the desktop. Just like Linked Data connects distributed data on the web, creating a network of interlinked information, the Semantic Desktop connects personal data across

L. Aroyo et al. (Eds.): ISWC 2011, Part II, LNCS 7032, pp. 33–48, 2011.
© Springer-Verlag Berlin Heidelberg 2011

application boundaries on the desktop, creating a network of personal information. However, information on our desktop is often incomplete, as it is based on our subjective view, or limited knowledge about an entity.

On the other hand, the Web of Data contains information about virtually everything, generated by multiple sources, and theoretically unlimited. Connecting the desktop to the Web of Data would thus enrich and complement desktop information. Bringing in information from the Web of Data automatically would release the user from the burden of searching for information.

Connecting the two networks of information opens up the possibility of personal services on the desktop which use external data, but in the personal context of the user, highly connected to his personal data and focused on his interests. One such example is a service that finds implicit links between the publications that the user has on the desktop, and provides recommendations to other publications on the same topics, by the same authors, or related in another way. Another desktop service could use information from the Web of Data to notify the user of new concert dates in his area, based on the latest or most popular artists played on the desktop. web data can also be used as a point of reference when working collaboratively, e.g., documents linked by the user to people, projects, or other resources from his semantic desktop can be shared together with the annotations, which can be accessed and reused outside of the semantic desktop where they were generated.

From the perspective of interlinking information, and using the frameworks provided by the Semantic Desktop and the Web of Data, we have separate islands of knowledge, both containing similar data, related to the same topics of interest to the user, but disconnected from each other.

The disconnection appears in two forms:

- The data on the desktop, although similar to that on the Web of Data, is described using specific *desktop ontologies*, which are different from the ones found on the Web of Data. This schema mismatch makes interlinking data from the two datasets difficult.
- Identifiers (URIs) on the desktop are local to the desktop data space, they are not globally unique and cannot be dereferenced as normal Linked Data URIs are. Hence, it is impossible to access and connect to local data from the Web of Data.

To tackle this disconnection, it is necessary to create links between desktop identifiers and web identifiers that refer to the same real-world thing. This means we need to compare the data graph describing the entity on the desktop with the data graph of an entity on the web. Leaving aside the use of different terminology within the data, the Web of Data is large, billions of entities across hundreds of thousands of datasets. From this vast amount of information we must find and retrieve a relevant subset of entities, that are potential candidates with the desktop entity. Then we must decide if the candidates are similar enough with the desktop entity to create a link between the two. Because we wish to make the interlinking automatic, we must be able to decide with a high degree of precision which candidates among this subset are in fact referring to the same entity.

Our solution tackles the problems raised above by using a semantic search engine for the Web of Data, such as Sindice, to find and retrieve a relevant subset of entities from the web. We then present a matching framework, using a combination of configurable heuristics and rules to compare data graphs, that achieves a high degree of precision in the linking decision.

We evaluate our methodology with real-world data. We create a gold standard from relevance judgements by experts, and we measure the performance of our system against it.

Our solution proves that interlinking the two environments is feasible, and even more, it yields good results. Connecting desktop data with the web enables the system to bring web data to the user, instead of the user having to go find it by himself.

The paper is organised as follow. In Section 2, we start by presenting the Nepomuk Semantic Desktop, as it represents the framework on which we base our solution. We continue with the related work section. In Section 3 we describe the process for finding web aliases for desktop resources, and continue in Section 4 with the implementation of the process and a detailed description of the matching algorithm. We describe the set-up of the evaluation we performed and the results in Section 5. We discuss some of the results in Section 6, before concluding.

Throughout the paper we consider that all the data we are working with is represented as RDF — both on the desktop and on the web. When we mention the desktop, we always mean the Semantic Desktop, more specifically the Nepomuk instantiation of the Semantic Desktop. Similarly, when we mention web data, we refer to the Web of Data. In our implementation we only use Web of Data sources, which are freely available online. However, this is not a requirement of the system, since new data sources can be easily plugged in.

2   Background

In this section, we first provide an overview of the Nepomuk Semantic Desktop and the infrastructure it provides. Next, we review existing approaches for entity linking and entity identity management, and finally compare our entity matching framework with Silk, a linking discovery framework for the Semantic Web.

2.1   Semantic Desktop and Nepomuk

The Semantic Desktop aims to solve the problem of information interlinking and to help managing and organising in a better way our personal data by applying Semantic Web technologies on the desktop. The Semantic Desktop is gaining momentum by the adoption and integration of the Nepomuk framework [2] into mainstream desktop environments.

The Nepomuk Semantic Desktop defines and uses a set of ontologies[1], complemented by ontologies defined by the community, like Xesam[2].It also defines an

---

[1] http://www.semanticdesktop.org/ontologies/

[2] http://xesam.org/main/XesamOntology - is used in Nepomuk-KDE.

extension to RDF called Nepomuk Representational Language (NRL)[3], which adds Named Graphs and Graph Views to RDF/S and introduces the closed world assumption to the data.

The ontologies describe various aspects of desktop use cases for personal information management. The central ontology is the Personal Information Model (PIMO)[4]. According to its specification [13], "PIMO is based on the idea that users have a mental model to categorize their environment", and "each concept in the environment ... is represented as [a] Thing in the model". PIMO defines high level types like Person, Project, Event and Task. The desktop ontologies also include Nepomuk Annotation Ontology (NAO) which allows users to attach tags and ratings to the resources, Nepomuk Contact Ontology (NCO) which describes contact information for people and organizations, Task Model Ontology (TMO) which describes personal tasks and to-dos, etc. All the data is stored in a central repository that is accessible and shared across applications.

2.2   Related Work

The problem of entity linking is well known across various research communities with a variety of different names, such as record linkage [8], entity resolution [1], reference reconciliation [6] or object consolidation [9]. A wide variety of algorithms has been developed for resolving the coreference problem, but record linkage between distributed databases is still considered a difficult problem.

Recent initiatives within the Semantic Web community address the problem of linking entities across data sources. Jaffri et al. describe the phenomenon of proliferation of URIs and propose a Consistent Reference Service to manage URI equivalences [10]. The OKKAM project [4] proposes an infrastructure for assigning global identifiers at web scale. These approaches are more focussed towards the management of entity identity on the web, but do not provide easy means to create new links between data sources. Similar to our approach, Raimond et al. describe an algorithm and its implementation GNAT, for linking a personal music collection to corresponding MusicBrainz resources [11]. The approach measures recursively the similarity of the resource graphs from the two datasets, with the restriction that the same vocabularies are used in both. By contrast, using property paths in our mappings, we eliminate the need for recursion while still propagating the measures from connected resources. Silk is a framework to help linking multiple entities between two datasets [3]. It relies on user-defined rules and various string matching algorithms to measure the similarity between two entities. In this case it is necessary to know a priory which specific dataset to link to and to perform manual configuration of the matching algorithms, something that requires a high degree of expertise. Hogan et al. [9] and Saïs et al. [12] propose logical-based methodologies for merging identifiers of equivalent entities across multiples knowledge sources. While being precise, these techniques do not have a very good recall and are demanding in term of computation.

---

[3] http://www.semanticdesktop.org/ontologies/nrl
[4] http://www.semanticdesktop.org/ontologies/pimo/

The most relevant approach related to ours is the Silk framework. We provide a generic matching process that the user can configure based on its own expertise in order to get more precise results. However, our approach differs by the fact that the matching process is not restricted to link data between two predefined information sources. On the contrary, our approach gives the possibility to link desktop data with an arbitrary number of external data sources. This makes the problem harder since we are generally unaware of the data structure or schema of these data sources. We therefore need to first find potential entities of interest among a vast number of data sources, then retrieve a partial description of these entities and rely on more complex entity matching algorithms. This first step can be seen as a blocking pass [7] to reduce the information space before executing complex matching algorithms. The blocking step is implemented on top of the boolean query model for centralized search systems such as Sindice [14] and on top of the SPARQL query language for specific data sources providing a SPARQL endpoint.

3   The Process of Finding Web Aliases

The goal of the algorithm and system is to find web aliases for desktop resources. A web alias is a web entity identifier, i.e., URIs, that represents the same real-world thing as the desktop entity to which it was matched. To find web aliases, we use the information available on the desktop, like the contact information from the address book for people, or metadata of music files for songs, albums and artists. We also make use of knowledge about the desktop ontologies and the way data is organized and used on the desktop. The desktop data is used throughout the process, which consists of several steps:

1. Candidate Selection
   - Query and identify candidate entity URIs from various Web of Data sources
   - Retrieve data for each of the candidate from the appropriate Web of Data source.
2. Candidate Filtering
   - Compute similarity score based on the data of the entities.
   - Filter the candidates based on the similarity score.

The first step requires identifying a list of candidate entities and obtaining the data available about them. There are several options to do this: (i) through a small set of sources that we know have the data we need, and querying each of them independently for possible candidates, or (ii) through a search engine for the Web of Data, like Sindice [14], which indexes millions of documents containing semantic mark-ups. Each option has use cases where it is more suitable than the other. Querying specific sources is preferred for instance, if the desktop data we want to find aliases for is from a very specific domain, like cancer research, or when we are interested only in results from an organization's internal repository. Using a search engine is best when the information sources to query are not known a priori. It also has the advantage of covering a large number of information sources with only one query, and of selecting the most relevant data sources and candidates with respect to the query via the search engine ranking system. However, in the case of ambiguous entities, the latter option has the disadvantage of returning too many unrelated results, thus making the entity selection more difficult.

Once a list of candidates is available, we compute a similarity score for each of them with respect to the desktop entity. The algorithm checks first if the types of the candidate entities correspond to the type of the desktop entity, and discards the ones that do not. Only then, the data of the entities are examined and the properties and corresponding values are compared. If required, the algorithm looks at other related entities and their properties. The values of the properties are compared using either exact string matching or string similarity techniques.

4   Implementing the Process

We implemented the process described above, in a desktop daemon that finds web aliases for desktop entities. It sequentially searches for aliases for all resources that have no alias listed, and for the resources that changed since the last time aliases were determined for them. In the case when a resource is revisited, the previously found aliases are discarded and new ones are determined.

New links are created on the desktop between the local and the web resources, once the aliases are found. They can be used to enhance the available desktop data about the entities, or as entry points to access further information about them online.

The tool has two major components, each handling one step of the matching process. A query component that initiates the search and identifies the candidates, and a matching component that filters the candidates based on similarity measures.

4.1   The Query Modules

The query component can use either generic search engines or specific data sources. Therefore, we chose to make the query component plugin-based, thus allowing various new sources to be connected if needed. The query modules are responsible for finding the initial list of candidates, as well as for retrieving the data for each candidate. The maximum number of candidates to retrieve from a data source can be set as a parameter in the configuration. We allow three types of plugins:

SWSE — connect to semantic search engines, through their APIs. We provide a plugin of this type for Sindice.

Sparql — connect to sources that provide a SPARQL endpoint. We provide plugins of this type for DBpedia and the Semantic Web Conference Server.

Custom — connect to other sources, possibly ones that do not expose any data as RDF (e.g., relational databases or third-party APIs like last.fm).

Both DBpedia and SWC are indexed by Sindice, therefore the Sindice plugin is the only one enabled by default.

In the Sindice module, the initial query, which determines the list of candidates, is constructed using all the value properties of the desktop entity, combined using the boolean conjunction operator "OR". Multiple word terms are also tokenised and the tokens are added to the query. We rely on the search engine to interpret the query and rank higher the results that match most of the terms. For the music album shown in Figure 3, the query constructed is:

Example 1. "Bee Gees" OR "One Night Only" OR "1998" OR "Bee" OR "Gees" OR "One" OR "Night" OR "Only"

4.2   The Matching Module

The matching module computes a similarity score for each pair (*desktop entity—web candidate entity*). The way the score is computed depends on a set of parameters:

String matching (SM) — If this parameter is set to `true`, the matching module will use string similarity measures where appropriate. Currently the system supports Monge Elkan and Chapman distances. If the value is set to `false`, the matching module uses exact matching of property values.

Weighted properties (WP) — If `true`, the matching module will use weights for the properties compared, otherwise, all properties contribute the same to the final score.

Multi-valued properties (MVP) — If `true`, properties that have more than one matching value will contribute to the score proportionally to the number of values.

The algorithm also uses a set of mappings from the desktop ontologies to some of the more popular web vocabularies, like FOAF. There are two kinds: type mappings (see Figure 1 for an example) and property mappings, each described in a separate file. The property mapping supports paths of properties. For example, you can express a path composed of the property `dbpedia:artist` and `foaf:name` as shown in Figure 2. The mappings are relatively static configurations of the system. We have created a set of mappings for the most common ontologies, which can be used out of the box by the end users. Power users can edit the mapping files according to their need.

The algorithm for computing the score works as follows. Considering e_d and e_w the pair of entities to be compared, it first determines the sets T_{e_d} and T_{e_w} of types for each entity, and the set $Map[T_{e_d}]$ of types to which the elements of T_{e_d} are mapped to. If no types are matching, i.e., $T_{e_w} \cap Map[T_{e_d}] = \phi$, it gives a score $score(e_w) = 0$, and stop the matching. Otherwise, it continues the process by evaluating the properties.

The evaluation of the properties is driven by the relations and properties of the desktop entity e_d. For each property p_{e_d}, the algorithm retrieves the list of values

```
"http://www.semanticdesktop.org/ontologies/2007/11/01/pimo#Person":{
   "mapping":[
      "http://xmlns.com/foaf/0.1/Person",
      "http://xmlns.com/foaf/0.1/Agent",
      "http://dbpedia.org/ontology/Person",
      "http://www.w3.org/2000/10/swap/pim/contact#Person"
      "http://rdf.data-vocabulary.org#Person" ]}
```

Fig. 1. Type mapping for pimo:Person

```
"http://www.semantcdesktop.org/ontologies/2009/02/19/nmm#performer##
      http://www.semanticdesktop.org/ontologies/2007/03/22/nco#fullname":{
   "mapping":[
      "http://dbpedia.org/property/artist",
      "http://xmlns.com/foaf/0.1/maker##http://xmlns.com/foaf/0.1/name",
      "http://dbpedia.org/ontology/artist##http://xmlns.com/foaf/0.1/name"
   ],
   "approx":"true",
   "thresholds":[
      "MongeElkan:0.7",
      "Chapman:0.8"
   ],
   "weight":"0.7" }
```

Fig. 2. Property mapping for nmm:performer

$V(p_{e_d}) = \{v \,:\, \{e_d\ p_{e_d}\ v\}\}$. Based on the list of property mappings $Map[p_{e_d}]$, it determines the set of values $V(p_{e_w} \cap Map[p_{e_d}])$ that the properties from $Map[p_{e_d}]$ have in common with e_w. If there is no value in common, i.e., $V(p_{e_d}) = \phi$ or $V(p_{e_w} \cap Map[p_{e_d}]) = \phi$, it skips the pair and there is nothing added to the score. Otherwise, it continues the process by measuring the similarity between values.

The evaluation of values is performed using string similarity between each pair of values $(v_d, v_w) \in V(p_{e_d}) \times V(p_{e_w} \cap Map[p_{e_d}])$. The algorithm creates a sparse matrix where the value of a cell contains a string similarity score between 0 and 1. Let $sum_{p_{e_d}}$ be the sum of the best score for each row of the matrix.

The final score is computed as follows:

$$score(e_w) = \frac{\sum_{p_{e_d}} (w_{p_{e_d}} * sum_{p_{e_d}})}{\sum_{p_{e_d}} (w_{p_{e_d}} * |V(p_{e_d)}|)}$$

where $w_{p_{e_d}}$ is the weight assigned to a certain property mapping. If the score is above 0.5^5, the entity is accepted as a web alias for the desktop entity.

5   Evaluation

To evaluate our system, we wanted to measure the accuracy of the matches, in a real-world set-up, with real data. For this purpose we created two entity corpora,

---

[5] We found that the threshold 0.5 was providing better results in our experiment.

one with desktop data and one with web data. To assess the results returned by our system we created first a baseline from relevance judgements made by human experts, on these corpora. Then, we ran our entity matching algorithm and we computed precision, NDCG and MAP to measure its performance.

5.1 Data Collection

We created two corpora for the evaluation, one containing desktop entities, and one containing possible matching entities from the Web of Data.

Desktop Data Entity Corpus. The desktop data used in the evaluation was collected from a real, in-use Nepomuk-KDE Semantic Desktop. It was generated by Nepomuk applications, and extracted from the desktop repository.

We restricted the entities selected to three types: (i) people — of type `nco:PersonContact`, (ii) publications — of type `nfo:PaginatedTextDocument`, and (iii) music albums — `nmo:MusicAlbum`. From each type we collected fifty different resources, resulting in a corpus of 150 seed desktop entities, and other entities related to them. Examples of auxiliary entities are the authors of publications, which may or may not be already in the corpus as contacts, the tracks of the albums and the artists. In total the desktop data corpus has 11.917 triples.

We used information from our desktops, therefore the people are colleagues or other researchers we collaborate with; the publications are related to our research interests, and generally related to semantics and information extraction. The music albums data was gathered from several colleagues, for variety of genres.

The contact data is extracted by Nepomuk from the default KDE address book, and we made no changes to it. The correct way to use the `nco:Person-Contact` resources extracted automatically, is to link each of them to a corresponding `pimo:Person` representing the person that has the contact information. However, the current tools do not make the distinction, therefore we also used the "raw" `nco:PersonContact` resources, for simplicity. The algorithm makes no distinction between types, so it would yield identical results if we would have used the "proper" `pimo:Person`.

The information related to music albums is extracted automatically by Nepomuk from the ID3 tags of music files.

For publications we used existing tools to perform shallow metadata extraction from files to obtain the title and the authors of the publications, when the metadata of the documents was not set.

Web of Data Entity Corpus. We used the Sindice query module of our system to generate the second corpus, containing Web of Data entities. For each desktop entity we retrieved the first twenty results returned by Sindice, thus making a total of 3.000 URIs. The queries used in Sindice were constructed as presented in Section 4.1, a combination based on the properties of each desktop entity. For each URI we obtained all the triples extracted by Sindice — explicit and implicit. In total this corpus has 1.530.686 triples.

In this dataset we did not explicitly retrieve Sindice data for the auxiliary entities related to the result URIs. We assumed this data will be available when/if required — in the relevance judgements by experts, and in the matching process by the algorithm.

5.2 Relevance Judgements from Experts

We collected the relevance judgements from experts through an online experiment, in which we asked participants to decide if pairs of desktop and web URIs identify the same real-world object or person. We evaluated in this way all 3.000 pairs from the two corpora. Each pair was judged by three different experts. Eighteen people participated in the experiment, all researchers in the area of Semantic Web.

Fig. 3. The web interface of the experiment for collecting relevance judgements

To simplify the task, we presented the two entities side by side, with all the information which was available about them in the corpora (see Figure 3). The desktop entity is shown on the left, and the web entity on the right. On the web side we included hyperlinks to the related entities, for further exploration in the case when the information available was not enough to make the decision. For convenience, on the web side, we have separated and brought to the top the triples which partially matched any of the values from the desktop side.

There were only two decisions possible: *Yes* or *No*, with a *Skip* option, in case of uncertainty. Once a pair was judged or skipped, another one was shown to the participant. The pairs were randomly chosen from the remaining set. To make the experiment feel like a game, we kept count of the number of pairs judged by each participant, and displayed it on the page. We found that even such a small addition generated ad-hoc competition and made the dull task more interesting.

The results of the experiment show an average agreement and its standard deviation, computed with Fleiss's κ, of 0.638 ± 0.214, over all three types of entities,

Table 1. Inter-annotator agreement measures

	κ	σ	Avg
All	0.638	0.214	92.252
People	0.661	0.257	88.2
Publications	0.786	0.127	98.067
Albums	0.442	0.233	90.523

suggesting substantial agreement between annotators. Table 1 shows the Fleiss's κ and its standard deviation σ per type, as well as the average pairwise percent agreement. We observed that for music albums, there was only moderate agreement between annotators, visibly lower than the average, while for publications it is visibly higher. We believe the difference is caused by the fact that the data about publications is generated and curated by experts in the field — even more so, as the publications were largely from the domain of Semantic Web —, while the music data comes from much more heterogeneous sources.

5.3 Evaluation Results of the Matching Algorithm

To evaluate the performance of the algorithm, we evaluate each of the matching modules separately and using a combination of them, against a baseline which is the matching framework without any matching modules activated. In the following, the String Matching module is denoted by SM, the Weigthed Properties by WP and the Multi-Valued Properties by MVP.

We used the *trec_eval* tool[6] to compute standard information retrieval measures. The precision at k ($P@k$) with k=1,2,3,4,5, mean average precision (MAP) and normalized discounted cumulative gain (NDCG) are reported in Table 2 for music albums, Table 3 for people and Table 4 for publications. We report also the interpolated precision at recall cut-off points when all matching modules are activated. The goal for the system is high precision, i.e., achieving a maximum at P@1. Recall is not a target, as it is generally impossible to determine the entire set of correct results available in the Web of Data.

In Table 2, we can observe that only the SM module is enhancing the results compared to the baseline. The baseline and the other two modules do not help the system at matching certain candidates. Also, in term of MAP and NDCG, the system achieves the lowest performance on the albums corpus. This can be explained by the fact that the album entities are mostly matching entities representing e-commerce products, which are not defined as a type of interests, and therefore rejected by the system. Whether or not such candidates should have been kept by the system is open to discussion and left for a future work.

In Table 3, we can observe that the baseline, the WP and the MVP modules are each one able to match good candidates with high precision at P@1, with

---
[6] http://trec.nist.gov/trec_eval/

Table 2. Evaluation results for albums, when varying configuration parameters

	MAP	NDCG	P@1	P@2	P@3	P@4	P@5
SM WP MVP	0.2464	0.5117	1	0.625	0.4167	0.3125	0.25
SM WP	0.2464	0.5117	1	0.625	0.4167	0.3125	0.25
SM MVP	0.2464	0.5117	1	0.625	0.4167	0.3125	0.25
WP MVP	0	0	0	0	0	0	0
SM	0.2464	0.5117	1	0.625	0.4167	0.3125	0.25
WP	0	0	0	0	0	0	0
MVP	0	0	0	0	0	0	0
Baseline	0	0	0	0	0	0	0

WP providing slightly better MAP and NDCG. However, the system does not get significant advantage by combining them. The SM module alone provides slightly lower precision at P@1 but significantly better MAP and NDCG. By combining the three modules, the system does not get significant advantage and it seems that the SM module prevails.

In Table 4, the baseline provides good results from the start. The system is not able to return any candidates when the MVP is activated by itself. However, when WP and MVP are combined, the system achieves much better results (in term of MAP and NDCG) than the baseline or than the WP module alone. When the system combines the SM module with the two previous ones, the system achieves a lower MAP and NDCG but an improved precision with a larger cut-off rank. While on the two previous types of entities, the SM module seemed to be the most important matching feature, this corpus shows that the WP and MVP are important matching features in certain cases.

Table 3. Evaluation results for people, when varying configuration parameters

	MAP	NDCG	P@1	P@2	P@3	P@4	P@5
SM WP MVP	0.4212	0.6354	0.9302	0.8953	0.7597	0.6337	0.5442
SM WP	0.4174	0.6321	0.9286	0.8929	0.746	0.6131	0.5286
SM MVP	0.4212	0.6354	0.9302	0.8953	0.7597	0.6337	0.5442
WP MVP	0.2916	0.5338	1	0.8243	0.6036	0.473	0.3838
SM	0.4212	0.6354	0.9302	0.8953	0.7597	0.6337	0.5442
WP	0.2916	0.5338	1	0.8243	0.6036	0.473	0.3838
MVP	0.2877	0.53	1	0.8243	0.6036	0.4662	0.3784
Baseline	0.2877	0.53	1	0.8243	0.6036	0.4662	0.3784

Table 4. Evaluation results for publications, when varying configuration parameters

	MAP	NDCG	P@1	P@2	P@3	P@4	P@5
SM WP MVP	0.7773	0.8651	1	0.625	0.4167	0.3125	0.25
SM WP	0.8032	0.8609	0.9062	0.5781	0.3958	0.3047	0.2438
SM MVP	0.7175	0.7986	0.9231	0.5769	0.3846	0.2885	0.2308
WP MVP	1	1	1	0.5	0.3333	0.25	0.2
SM	0.7265	0.7883	0.8235	0.5294	0.3627	0.2868	0.2294
WP	0.6893	0.7347	1	0.55	0.3667	0.275	0.22
MVP	0	0	0	0	0	0	0
Baseline	0.7175	0.7588	1	0.5455	0.3636	0.2727	0.2182

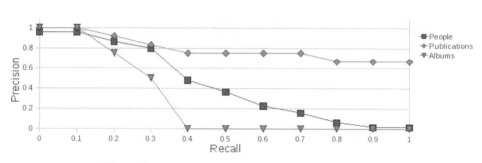

Fig. 4. Interpolated precision at recall cut-off points

Overall, the results are satisfying for our use cases where high precision prevails over recall. However, given the results shown in Figure 4, we can see that
the system could be configured to return more than one entity in order to achieve
a better recall while keeping a good precision. It can prove useful to implement
a semi-automatic system which presents the top n candidates to the user for
manual selection.

5.4 Performance

To determine the performance, we measure the time spent on each step of the
algorithm. To be noted that these results come from a prototypical implementation, still to be subject to technical optimisations. Table 5 shows the average
times overall, and for each resource type separately, when all three parameters
are active — SM, WP, MVP. We find only small variations in the measurements
when the parameter values are changed. We do not consider the time spent
on retrieving data from Sindice, as it depends on external factors, like network
speed and server availability.

The checking of types is the only value that in average does not depend on the
type of resource, as it must be performed for all pairs. The time spent in average
per property check is low, but it varies by type, and by the complexity of the

Table 5. Time performance (milliseconds)

	Overall	People	Publications	Albums
Pair total	375.04	52.19	977.87	53.18
Types check	0.23	0.26	0.21	0.23
Per property check	6.66	0.92	13.2	22.06
All properties	2026.22	7.17	5478.87	1963.88

properties (e.g. takes longer if several resources in the graph must be traversed, for long property paths like the name of the artist of an album). The "All properties" row shows the average time required for checking all the properties of an entity, and the computation of the final score[7]. These values depend on the type of resources as well, and on the complexity of the resource graph. We found that longer times correspond to very big graphs for online entities, e.g., the graph for `http://webconf.rkbexplorer.com/models/iswc-aswc-2007-complete.rdf`, which must be loaded for checking even if in most cases are not found to represent valid candidates.

6   Discussion and Future Work

The scope of the system presented here is limited to finding Web of Data aliases for desktop resources. We leave the use of the aliases found to future work, but the use cases include personalized desktop services like those described in Section 1 and enhancement of desktop information from online sources. We plan to develop a semi-automatic service that retrieves information from the web aliases and updates the local resources, while saving provenance information for the imported data and allowing synchronization when the web data changes.

There exist already web applications that provide similar services via specific APIs (e.g., last.fm). However this is not the goal of this work. Instead, we wish to leverage information across all public information sources accessible on the Web of Data. In addition, such third-party APIs are seen as an additional information sources on the web, and are supported by our system.

Within the system, we make use of existing semantic technologies, one of which are semantic search engines such as Sindice. In the process of determining the aliases we focus on selecting the most appropriate URI from the list of candidates returned by the search engine. In this case, the issues of data sources to trust is left to the search engine, that usually employs advanced techniques [5] for measuring the popularity of a data source. This is however not a requirement we impose on the users, who can choose to query other trusted data sources suitable for their use case.

---

[7] The "All properties" row has values higher that the "Pair total" row because the average time is computed only for those pairs who passed the type check, thus less in number, but with longer computation times.

The system we presented is automatic from the user's point of view, as there are no interactions required for it to work. Once set up it will find and save aliases to desktop resources. Power users can however tweak the settings to fit their specific needs by enabling/disabling modules, changing threshold values or managing mappings. Although the mappings were written manually, they are part of the system and do not need to be modified by end users. We envision for the future, a way of allowing power users to publish their own mappings and let other users install new mappings in a way similar to installing add-ons to web browsers.

7   Conclusion

In this paper, we have presented a framework to automatically link entities from the semantic desktop to the Web of Data. The framework uses existing technologies such as semantic search engines or SPARQL endpoints for retrieving a set of candidates. Each candidate is then evaluated more precisely based on a collection of matching components using string matching, heuristics and rule-based mechanisms. We evaluate qualitatively the system using real-world data retrieved from a Nepomuk Semantic Desktop and the Sindice search engine. The evaluation is based on relevance judgements from a group of experts. We show that the system in its current form provides satisfactory results in term of precision for automatic linking of entities.

Acknowledgments. The work presented in this paper was supported by the Líon-2 project funded by Science Foundation Ireland under Grant No. SFI/08/ CE/I1380, and by the European projects Digital.me (No. 257787) and LOD2 (No. 257943) under the Seventh Framework Program (FP7/2007- 2013).

References

1. Benjelloun, O., Garcia-Molina, H., Jonas, J., Su, Q., Widom, J.: Swoosh: A generic approach to entity resolution. Tech. rep., Stanford University (2006)
2. Bernardi, A., Decker, S., van Elst, L., Grimnes, G., Groza, T., Jazayeri, S.H.M., Mesnage, C., Moeller, K., Reif, G., Sintek, M.: The Social Semantic Desktop: A New Paradigm Towards Deploying the Semantic Web on the Desktop. IGI Global (2008)
3. Bizer, C., Volz, J., Kobilarov, G., Gaedke, M.: Silk - a link discovery framework for the web of data. In: Proceedings of the 18th International World Wide Web Conference (April 2009)
4. Bouquet, P., Stoermer, H., Giacomuzzi, D.: OKKAM: Enabling a web of entities. In: Proceedings of the WWW 2007 Workshop I3: Identity, Identifiers, Identification, Entity-Centric Approaches to Information and Knowledge Management on the Web (May 2007)
5. Delbru, R., Toupikov, N., Catasta, M., Tummarello, G., Decker, S.: Hierarchical Link Analysis for Ranking Web Data. In: Aroyo, L., Antoniou, G., Hyvönen, E., ten Teije, A., Stuckenschmidt, H., Cabral, L., Tudorache, T. (eds.) ESWC 2010. LNCS, vol. 6089, pp. 225–239. Springer, Heidelberg (2010)

6. Dong, X., Halevy, A.Y., Madhavan, J.: Reference reconciliation in complex information spaces. In: Özcan, F. (ed.) SIGMOD Conference, pp. 85–96. ACM (2005)
7. Elmagarmid, A.K., Ipeirotis, P.G., Verykios, V.S.: Duplicate record detection: A survey. IEEE Transactions on Knowledge and Data Engineering 19(1), 1–16 (2007)
8. Fellegi, I.P., Sunter, A.B.: A theory for record linkage. Journal of the American Statistical Association 64(328), 1183–1210 (1969)
9. Hogan, A., Harth, A., Decker, S.: Performing object consolidation on the semantic web data graph. In: Proceedings of the WWW 2007 Workshop I3: Identity, Identifiers, Identification, Entity-Centric Approaches to Information and Knowledge Management on the Web (May 2007)
10. Jaffri, A., Glaser, H., Millard, I.: URI Identity Management for Semantic Web Data Integration and Linkage. In: Meersman, R., Tari, Z., Herrero, P., et al. (eds.) OTM-WS 2007, Part II. LNCS, vol. 4806, pp. 1125–1134. Springer, Heidelberg (2007), http://eprints.ecs.soton.ac.uk/14361/
11. Raimond, Y., Sutton, C., Sandler, M.: Automatic interlinking of music datasets on the semantic web. In: Proceedings of the Linked Data on the Web workshop, LDOW 2008 (2008)
12. Saïs, F., Pernelle, N., Rousset, M.C.: L2r: a logical method for reference reconciliation. In: AAAI 2007: Proceedings of the 22nd National Conference on Artificial Intelligence, pp. 329–334. AAAI Press (2007)
13. Sauermann, L., Elst, L.V., Möller, K.: Personal Information Model (PIMO). Deliverable 1.1 (February 2009), http://www.semanticdesktop.org/ontologies/2007/11/01/pimo/v1.1/pimo_v1.1.pdf
14. Tummarello, G., Delbru, R., Oren, E.: Sindice.com: Weaving the open linked data. In: Aberer, K., Choi, K.-S., Noy, N., Allemang, D., Lee, K.-I., Nixon, L.J.B., Golbeck, J., Mika, P., Maynard, D., Mizoguchi, R., Schreiber, G., Cudré-Mauroux, P. (eds.) ASWC 2007 and ISWC 2007. LNCS, vol. 4825, pp. 552–565. Springer, Heidelberg (2007)

Linking Data across Universities:
An Integrated Video Lectures Dataset

Miriam Fernandez, Mathieu d'Aquin, and Enrico Motta

Knowledge Media Institute, Open University
Walton Hall, Milton Keynes, MK7 6AA, UK
{m.fernandez,m.daquin,e.motta}@open.ac.uk

Abstract. This paper presents our work and experience interlinking educational information across universities through the use of Linked Data principles and technologies. More specifically this paper is focused on selecting, extracting, structuring and interlinking information of video lectures produced by 27 different educational institutions. For this purpose, selected information from several websites and YouTube channels have been scraped and structured according to well-known vocabularies, like FOAF[1], or the W3C Ontology for Media Resources[2]. To integrate this information, the extracted videos have been categorized under a common classification space, the taxonomy defined by the Open Directory Project[3]. An evaluation of this categorization process has been conducted obtaining a 98% degree of coverage and 89% degree of correctness. As a result of this process a new Linked Data dataset has been released containing more than 14,000 video lectures from 27 different institutions and categorized under a common classification scheme.

Keywords: Linked Data, Education, Integration.

1 Introduction

Different educational institutions, and even different departments within the same institution, produce yearly large amounts of educational material (videos, slides, documents, etc.). However, when students and educational practitioners have to perform learning and investigation tasks, they generally: (i) not find the best available resources for the topic they aim to investigate or, (ii) spend large amounts of time browsing the websites of different institutions in order to collect and extract the key information about the topic. In this context, we believe that integrating the large amount of educational material produced by different institutions is a key requirement towards educational data sharing and exploitation. The fact that different institutions publish and describe their educational content using different formats, tags, categories and structure, makes this integration process a difficult and challenging problem.

---

[1] http://xmlns.com/foaf/spec/
[2] http://www.w3.org/TR/mediaont-10/
[3] http://www.dmoz.org/

L. Aroyo et al. (Eds.): ISWC 2011, Part II, LNCS 7032, pp. 49–64, 2011.
© Springer-Verlag Berlin Heidelberg 2011

The emergence of Linked Data (LD)[4] brings to this scenario a new dimension of possibilities under which educational material can be organized, integrated, archived and retrieved. LD refers to a set of principles to put raw data on the Web and making them Web addressable and linkable, so that they can be easily accessed, discovered, connected and reused. The number of universities, research organizations, publishers and funding agencies contributing to the LD cloud is constantly increasing. Universities such as The Open University[5], Southampton[6], Sheffield's Computer Science Department[7], or the University of Münster[8], among others, are embracing the LD principles and releasing educational resources as part of the LD cloud.

In this paper, we aim to expose our experience extracting, structuring and integrating video lectures material from 27 different educational institutions by exploiting LD principles. Since standardized practices for publishing and integrating educational LD across institutions are not yet in place, we expect that this work can contribute to reflect on the evolution of such practices.

The rest of the paper is structured as follows: Section 2 provides an overview of related work. Section 3 describes the processes of selecting, extracting and structuring video lectures from various information sources according to LD principles. Section 4 explains the data integration process, focused on the creation of a common searchable/browsable space for educational material. Section 5 shows the conducted evaluation for the data integration process. Conclusions are shown in Section 6.

2   Related Work

We are currently witnessing a substantial increase in universities adopting the Linked Data initiative. One of the currently strongest activities towards LD production and consumption within the context of education has been carried out by the Open University (OU), in the context of the Lucero project[9] (Linking University Content for Education and Research Online). This project performs OU data extraction, transformation, maintenance and exploitation [14]. At the time of writing, several datasets about publications, podcasts and course descriptions, among others, have been released and are accessible in an open way through online access at http://data.open.ac.uk. In addition, several applications[10] have been developed to show the potential of the OU linked data exposure. Although all these applications show several significant advantages of exploiting LD in the educational context, their coverage is currently limited to the OU. There is still no significant integration of educational data across universities that can be exploited by these applications.

Other examples of efforts towards the production and consumption of LD in the educational context are:

---

4   http://linkeddata.org/
5   http://data.open.ac.uk
6   http://data.southampton.ac.uk
7   http://data.dcs.shef.ac.uk/
8   http://lodum.de/about
9   http://lucero-project.info/
10 http://data.open.ac.uk/applications/

- The University of Sheffield's Department of Computer Science[7], which provides a LD service describing research groups, staff and publications, all semantically linked together [10].
- The University of Southampton, which has recently announced the release of their LD portal[6]. At the time of writing 26 different datasets including information about university buildings, educational videos, or university bus routes have been released.
- The University of Manchester's library catalogue. Its records can now be accessed in RDF format[11].
- The University of Edinburgh, where the university's buildings information is now generated as LD[12].
- The University of Münster, which recently announced LODUM[8], a project with the aim to release the university's research information as LD. This includes information related to people, projects, publications, prizes and patents.

Additionally to the initiatives of publishing educational content as LD, it is important to highlight some of the current works towards integrating library catalogs on a global scale. Some of these works are discussed in [4]. Examples include the American Library of Congress[13], the German National Library of Economics [8], and LIBRIS[14], the Swedish National Union Catalogue, which publish their subject heading taxonomies as LD. Similarly, the OpenLibrary[15], a collaborative effort to create "one Web page for every book ever published" has published its catalogue in RDF. Scholarly articles from journals and conferences are also well represented through community publishing efforts such as DBLP as LD[16], RKBexplorer[17], and the Semantic Web Dogfood Server [7].

We believe that the increase involvement of the library community in LD[18] will soon enhance the exchange and consumption of educational material, facilitating its search, exploration and comparison across institutions.

3 Generating RDF

Among the different types of educational resources (textual documents, slides, videos, etc.) this paper is focused on: (a) generating and (b) interlinking RDF descriptions from video lectures. In this section we will explain the RDF generation process including: (i) the processes of information selection and extraction, (ii) the process of vocabulary selection and, (iii) the process of information structuring according to the selected vocabularies.

---

[11] http://prism.talis.com/manchester-ac/
[12] http://ldfocus.blogs.edina.ac.uk/2011/03/03/
[13] http://id.loc.gov/authorities/about.html
[14] http://blog.libris.kb.se/semweb/?p=7
[15] http://openlibrary.org/
[16] http://dblp.l3s.de
[17] http://www.rkbexplorer.com/data/
[18] http://www.w3.org/2005/Incubator/lld/

3.1 Selecting and Extracting Educational Information from Various Sources

When selecting information sources, we have considered two of the currently most popular video lecture containers: YouTube[19] university channels and the videolectures.net website.

YouTube is a video-sharing website on which users can upload, share and view videos. In the context of education, YouTube has been used by several institutions to make their video lectures publicly available on the Web via YouTube channels. For the purpose of this work we have selected the YouTube channels of 25 different universities and research institutions including: Standford, Yale, Harvard, Oxford or Google Talks, among others. The complete list of YouTube channels used for this paper can be found in http://smartproducts1.kmi.open.ac.uk/web-linkeduniversities/index.htm. Video lectures information from YouTube channels is accessed and extracted via the YouTube data API[20]. Among the information that can be accessed through this API we have focused on: (i) video upload feeds and (ii) playlist feeds. Video upload feeds refer to all the videos uploaded by the same university channel. Video playlist feeds are collections of videos available via a particular university channel that may have been uploaded by the university or by other users/institutions. Figure 1 represents a summary of the common properties associated to video uploads and playlist feeds. When querying the YouTube data API, each element (video or playlist) is returned and

```
<entry gd:etag='W/"DkADSH47eCp7ImA9WhZWFEg."'>
    <id>tag:youtube.com,2008:video:zZCaHSW88Ts</id>
    <published>2011-02-18T11:41:08.000Z</published>
    <updated>2011-05-15T10:19:39.000Z</updated>
    <category scheme='http://gdata.youtube.com/schemas/2007/categories.cat'
             term='Education' label='Education'/>
    <category scheme='http://gdata.youtube.com/schemas/2007/keywords.cat'
             term='Dr Barry Cooper'/>
    <title>Intro to Professional Practice (Children & Families)</title>
    <author>
      <name>TheOpenUniversity</name>
      <uri>http://gdata.youtube.com/feeds/api/users/theopenuniversity</uri>
    </author>
  <media:description type='plain'>Free learning from The Open University
      http://www.open.ac.uk/openlearn/
      An introduction by Barry Cooper detailing the Postgraduate […]
  </media:description>
  <media:keywords>Dr Barry Cooper, postgraduate qualifications, social work,
      children and families, childcare worker, childcare practitioner,
healthcare
      practitioner, flexible pace of study, flexible award, online tutor panel,
      online classroom, ou_k14, ou_e70, open university
  </media:keywords>
  <media:thumbnail url='http://i.ytimg.com/vi/zZCaHSW88Ts/default.jpg'
      height='90' width='120' time='00:03:19.500' yt:name='default'/>
  <yt:duration seconds='399'/>
    <content type='application/x-shockwave-flash'
src='http://www.youtube.com/v/zZCaHSW88Ts?f=user_uploads&app=youtube_gdata'/>
    <gd:feedLink
href='http://gdata.youtube.com/feeds/api/videos/zZCaHSW88Ts/comments'
countHint='2'/>
```

Fig. 1. Summarized example of a YouTube upload video feed

---

[19] http://www.youtube.com/
[20] http://code.google.com/apis/youtube/getting_started.html#data_api

represented as an entry point with several associated properties. The complete list of properties can be found in the YouTube data API documentation[21].

Among these properties we have selected for the purpose of this work: the video ID, the publication date, the date at which it was updated, its duration, its title, its description, its authors, the link to the video content, the links to the associated thumbnails and the list of categories and keywords that describe it. Additionally, for videos extracted from a playlist, the playlist identifier is also extracted. When selecting the set of entities and properties on which to apply the LD principles, the goal we had in mind was to have an essential definition of the video lectures which would be reasonably independent from the original source.

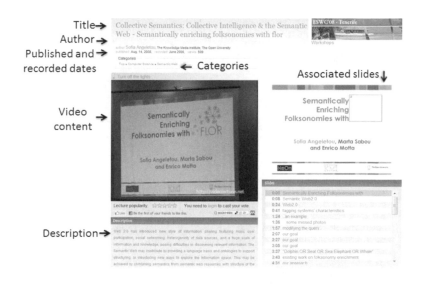

Fig. 2. Screenshot of `videolectures.net` describing one of its videos

`Videolectures.net` is a website for academic talks launched in 2007. It offers to the scientific, research, business and general public a large collection of video lectures that are enriched with slides. While the vast majority of talks belong to the subject of Computer Science, it also contains videos about Astronomy, Medicine or Philosophy among others. `Videolectures.net` does not provide any API for accessing its data so, for the purpose of this work, a tailor-made HTML scraper has been developed with the aim of extracting a selected set of information. In this case, the properties extracted for each particular video are: the video ID, the publication date, the recording date, its duration, its title, its description, its list of corresponding authors (including authors' names and affiliations), the link to the video content, the link to the associated slides when available, the links to the associated thumbnails and the list of categories used to describe it. A screenshot of the `videolectures.net` website where these properties are displayed for a particular video can be seen in Figure 2.

---

[21] `http://code.google.com/apis/youtube/2.0/`
`developers_guide_protocol_understanding_video_feeds.html`

Great-circle distance

http://data.open.ac.uk/podcast/218dce44a4ed17b36ada50d18b866b03

hasiTunesU	http://deimos.apple.com/WebObjects/Core.woa/Browse/itunes.open.ac.uk.2133244465
relatesToCourse	mu120
transcript	mu120.04showing04.pdf
depiction	mu120-showing-the-way_00359_std.jpg
download	mu120.04showing04.m4v
duration	00:03:39
description	Great-circle distances might be the shorte... always the easiest. We also look at how lin... that permission must be acquired to fly ov
isPart	58dbd5be4f01f4b1eec1df1e8f97eaad
published	2009-05-19T02:29:55+01:00
subject	Mathematics@en / Mathematics and Statistics@en
title	Great-circle distance

type	VideoPodcast
comment	Great-circle distances might be the shortest way to travel, but they are not always the easiest. We also look at how lines of latitude aren't great-circles and that permission must be acquired to fly over many countries.
label	Great-circle distance
collection	58dbd5be4f01f4b1eec1df1e8f97eaad
copyright	The Open University 2009
createDate	2009-05-19T02:29:55+01:00
description	Great-circle distances might be the shortest way to travel, but they are not always the easiest. We also look at how lines of latitude aren't great-circles and that permission must be acquired to fly over many countries.
duration	00:03:39
format	video/x-m4v
genre	Mathematics@en
language	en
locator	mu120.04showing04.m4v

Fig. 3. Example of an OU Podcast

Additionally to the two previously mentioned information sources, we have also added to the video lectures linking process an already LD structured video lectures dataset, the OU Podcasts[22]. OU Podcasts is a collection of Audio and Video material related to education and research at the Open University. This video and audio material has been remodeled using LD principles and is currently defined using a variety of ontologies[23]. Figure 3 shows an example of the information associated to the OU Podcast *"Great-circle distance"*, including: the video ID, the publication date, the creation date, its duration, its title, its description, its list of corresponding publishers, the link to the video content, the link to the video transcript, the links to the associated thumbnails and the list of categories used to describe it. Note that this information source is already structured according to the LD principles, so the process of RDF generation is not applied to it.

3.2 Reusing Vocabularies to Describe Educational Data

As described by Heath and Bizer [4], RDF provides a generic, abstract data model for describing resources using subject, predicate, object triples. However, it does not provide any domain-specific terms for describing classes of things in the world and how they relate to each other. This function is served by taxonomies, vocabularies and ontologies expressed in $SKOS^{24}$, $RDFS^{25}$ and OWL^{26}. In this context, and according to

---

[22] http://podcast.open.ac.uk/
[23] http://data.open.ac.uk/datasets/
[24] http://www.w3.org/TR/skos-reference/
[25] http://www.w3.org/TR/rdf-schema/
[26] http://www.w3.org/TR/owl-features/

linked data practices [4], "*if suitable terms can be found in existing vocabularies, these should be reused to describe data wherever possible, rather than reinvented*". Reuse of existing terms is highly desirable as it maximises the probability that data can be consumed by applications that may be tuned to well-known vocabularies, without requiring further pre-processing of the data or modification of the application.

   Following these guidelines, several vocabularies have been selected to describe the information extracted from the YouTube university channels and `videolectures.net`. The following list represents the chosen vocabularies:

- Dublin Core[27]: is a widely deployed vocabulary for representing provenance, particularly by the use of the `dcterms:creator` and `dcterms:published` predicates. It also provides descriptive predicates such as `dcterms:title`, `dcterms:description` or `dcterms:subject`.
- FOAF[28], the Friend Of A Friend vocabulary: defines terms for describing persons, their activities and their relations to other people and objects. The class `foaf:Person` and the predicates `foaf:name` and `foaf:homepage` are examples of reused elements to describe authors of video lectures.
- The W3C Ontology for Media Resources[29]: is both a core vocabulary (a set of properties describing media resources) and mappings to a set of metadata formats currently describing media resources published on the Web. Examples of reused elements include: `ma:publisher`, `ma:createData` or `ma:description`.
- The Media Vocabulary[30]: defines a minimal scheme for media content. Classes like `media:Recording`, to instantiate video lectures, as well as predicates like `media:download` or `media:depiction` have been reused to describe the video content and its associated thumbnails.
- The Nice Tag Ontology[31]: describes tags as generally as possible. Tags associated to videos from `videolectures.net` and YouTube channels have been modeled using the `nt:isRelatedTo` predicate.

Note that `dcterms`, `foaf`, `ma`, `media`, and `nt` are the corresponding prefixes for the vocabularies' associated namespaces.

3.3 Structuring Information According to the Previously Selected Vocabularies

When structuring information in RDF, one of the main discussions raised in the Linked Data community are the best practices to generate Unified Resource Identifiers (URIs). URIs should be representative as names for things (real-world entities or abstract concepts) and should be designed to be simple. The W3C Interest Group has generated "*Cool URIs for the Semantic Web*[32]", a guideline about good practices for URI generation. This guideline has been followed, when applicable, during the development of the present work.

---

[27] http://dublincore.org/documents/dcmi-terms/
[28] http://xmlns.com/foaf/spec/
[29] http://www.w3.org/TR/mediaont-10/
[30] http://payswarm.com/vocabs/media
[31] http://ns.inria.fr/nicetag/2010/09/09/voc.html
[32] http://www.w3.org/TR/cooluris/

The base URI, common to all elements of the dataset, is `http://linkeduniversities.org`.

Individuals of the class `media:Recording` have been generated to represent video lectures objects. The URIs for this type of objects are identified by five main elements: the base URI, the type of educational material (video, audio, text, etc.), the educational institution producing this material (Carnegie Mellon University, Open University, etc.), the storage/communication source used by the institution (YouTube, Podcasts, etc.) and the primary key, or video identifier within the storage source. The properties of each video are structured according to the set of vocabularies described in Section 3.2. An example of how to structure a video lecture, including its assigned URI and its list of associated properties is described in Table 1.

Table 1. Structure of a video lecture

Example of the structure associated to a video lecture

`http://linkeduniversities.org/video/CarnegieMellonU/youtube/B135229F3706D215/9949817F2FB77F0C`

`rdf:type`	`media:Recording`
`media:download`	`http://www.youtube.com/watch?v=TOTuStPIeFc&feature=youtube_gdata_player`
`dcterms:title`	`CMU Football Engineering Summer 2008 Video`
`rdfs:label`	`CMU Football Engineering Summer 2008 Video`
`dcterms:description`	`Football […]Summer 2008 Video`
`foaf:thumbnail`	`http://i.ytimg.com/vi/TOTuStPIeFc/3.jpg`
`foaf:thumbnail`	`http://i.ytimg.com/vi/TOTuStPIeFc/1.jpg`
`foaf:thumbnail`	`http://i.ytimg.com/vi/TOTuStPIeFc/2.jpg`
`foaf:thumbnail`	`http://i.ytimg.com/vi/TOTuStPIeFc/0.jpg`
`media:duration`	`155`
`dcterms:isPart`	`http://linkeduniversities.org/video/CarnegieMellonU/youtube/playlist/B135229F3706D215`
`ma:publisher`	`http://linkeduniversities.org/video/CarnegieMellonU/youtube/user/footballtracking`
`dcterms:published`	`2011-06-03T23:23:53.262Z`
`nt:isRelatedTo`	`http://linkeduniversities.org/video/CarnegieMellonU/tag/cmu`
`nt:isRelatedTo`	`http://linkeduniversities.org/video/CarnegieMellonU/tag/sports`
`nt:isRelatedTo`	`http://linkeduniversities.org/video/CarnegieMellonU/tag/football`
`nt:isRelatedTo`	`http://linkeduniversities.org/video/CarnegieMellonU/tag/engineering`
`dcterms:subject`	`http://dmoz.org/Society/People`
`dcterms:subject`	`http://linkeduniversities.org/video/CarnegieMellonU/dmoz/Society/People`
`dcterms:subject`	`http://dmoz.org/Sports/Football/Rugby_Union`
`dcterms:subject`	`http://linkeduniversities.org/video/CarnegieMellonU/dmoz/Sports/Football/Rugby_Union`

Looking at the table, it is important to highlight certain design decisions:

- The content of the title has been duplicated within the properties `dcterms:title` and `rdfs:label`. Since most current Semantic Web applications exploit the `rdfs:label` predicate as the main descriptive property of the object.
- The association of a video with a playlist is reflected using the `dcterms:isPart` predicate.
- The set of tags and categories describing the video are associated using the `nt:isRelatedTo` predicate. In addition, these tags are mapped to the base URI, i.e., the `linkeduniversities.org` domain.
- The use of the property `dcterms:subject` is extensively described in section 4. Basically it reflects the categorization of the video lecture within the unified searchable/browsable space. As we can see, the value of this property is also duplicated to maintain the URI of its original source, but also to add it as part of our base URI.

Individuals of the class `foaf:Person` have been generated to represent authors. The same URI elements used to represent individuals of the class `media:Recording` are used to represent this class (with the exception of the primary key, or author identifier, which is generated taking into account the author's name). An example of the structure of a video lecture, including its assigned URI and its list of associated properties, is described in Table 2.

Table 2. Structure of a Person/author

Example of a structure associated to an author

```
http://linkeduniversities.org/video/videolectures/michel_dumontier
```

rdf:type	foaf:Person	
foaf:name	Michel Dumontier	
foaf:homepage	http://videolectures.net/michel_dumontier	
vcard:organization-name	Carleton University	
<http://linkeduniversities.org/ video/videolectures/6593>	dcterms:contributor	<http://linkeduniversities.org/video/ videolectures/michel_dumontier>

As we can see, every identified author has at least an associated name, homepage, and organization. The last row of Table 2 describes how a video lecture is associated to its corresponding authors by the property `dcterms:contributor`.

4 Integrating Educational Information

YouTube channels, `videlectures.net` and OU Podcasts, use different classification schemes and systems. To unify the search and exploration tasks of all these educational material it is necessary to integrate the extracted videos under a common searchable/browsable space, in this case under a common topic hierarchy. To address this problem, three key issues should be tackled:

i) Select the most appropriate categorization scheme under which these video materials should be classified.

ii) Analyze the classifications assigned by each particular information source (YouTube channels, `videolectures.net`, OUPodcast) to determine how this information can be mapped to the common categorization scheme.

iii) Propose a categorization approach to classify every video lecture to the common categorization scheme.

4.1 Selecting a Common Categorization Scheme

When selecting the potential categorization schemes, three main requirements have been considered: (i) to be general, i.e., aiming to cover all subjects in "the universe of information", (ii) to be fully public and, (iii) to be available in RDF. Following these requirements, four potential categorization schemes have been selected:

• DMOZ[33], the Open Directory Project (ODP) topic hierarchy: the ODP is the largest, most comprehensive human-edited directory of the Web. It is constructed and maintained by a vast, global community of volunteer editors

---

[33] http://www.dmoz.org/docs/en/about.html

and is 100% free. RDF dumps of this topic hierarchy and its content are available for download[34]. ODP data powers the core directory services[35] for many of the Web's largest search engines and portals, including Netscape Search, AOL Search and Google. In addition, more than seventy-five languages are currently represented in this topic hierarchy and, at the time of writing, it claims to have over 1,007,233 categories.

- DBpedia categories[36]: The DBpedia project extracts various kinds of structured information from Wikipedia editions in 97 languages and combines this information into a large cross-domain knowledge base. The DBpedia knowledge base currently claims to describe more than 3.5 million things. It provides three different classification schemes for things: (i) Wikipedia categories, (ii) the YAGO Classification, derived from the Wikipedia category system using WordNet and, (iii) WordNet Synset Links, generated by manually relating Wikipedia infobox templates and Word Net synsets. Although DBPedia currently constitutes one of the main cores of the Web of Data, erroneous Wikipedia categories also cause the derivation of false facts [13].
- Library of Congress Subject Headings (LCSH)[37]: the LCSH comprises a thesaurus of subject headings, maintained by the United States Library of Congress, for use in bibliographic records.
- The International Press Telecommunications Council (IPTC) News Codes[38]: The IPTC creates and maintains sets of concepts to be assigned as metadata values to news objects like text, photographs, graphics, audio and video files and streams. Among this metadata they provide several taxonomies to describe the content of news items.

Although LCSH and IPTC are high quality classification schemes, developed and maintained by the library and journalism communities, to the best of our knowledge, they only support the English language. Considering that educational resources may be accessed and described in different languages we have opted for selecting a multilingual classification scheme, i.e., either DBPedia or ODP. DBPedia is currently considered the core of the LD cloud and there is a high level of activity towards its development. On the other hand, the ODP classification scheme has been longer established and there is a wide range of sites that are successfully exploiting it. Although both classification schemes seemed suitable for the task at a hand we have selected ODP because of its maturity and the availability of tools to exploit it in the context of classification tasks.

4.2 Analyzing the Categorization Schemes of Each Information Source

Mapping video lectures categorized under different schemes, to a common searchable/browsable space of topics is a challenging problem. YouTube videos,

---

[34] http://www.dmoz.org/rdf.html
[35] http://en.wikipedia.org/wiki/Open_Directory_Project
[36] http://dbpedia.org/About
[37] http://id.loc.gov/authorities/about.html
[38] http://www.iptc.org/site/NewsCodes/
NewsCodes_Retrieval_in_Different_Formats

for example, are categorized by YouTube categories, as well as by user's and developer's tags[39]:

- Each video can be associated with one predefined YouTube category, such as Comedy, News or Sports. A video's category is identified by the `<media:category>` and the `<category>` tags for which the value of the scheme attribute is `http://gdata.youtube.com/schemas/2007/categories.cat`.
- Each video can be associated with an arbitrary number of keywords, which are also known as tags. A video's tags are identified using the `<media:keywords>` tag in API requests and responses. Keyword tags are also identified by `<category>` tags for which the value of the scheme attribute is `http://gdata.youtube.com/schemas/2007/keywords.cat`.
- Each video can also be associated with an arbitrary number of developer tags. Video developer tags are identified in `<media:category>` and `<category>` tags for which the value of the scheme attribute is `http://gdata.youtube.com/schemas/2007/developertags.cat`.

`Videolectures.net` uses its own categorization scheme that contains 23 main root elements including: Architecture, Arts, Biology, Business, Chemistry or Computer Science among others. The categorization scheme is available through their website.
OU Podcasts are classified under three different categorization systems: OU specific subject headings, iTunes categories[40] and iTunes U categories[41].

In addition to the categorization information used by each individual source, properties such as title and description, available in all three sources, can be used as additional information to generate the corresponding mappings to the ODP categorization scheme.

4.3 The Categorization Approach

As mentioned in the previous section, we have three main different types of information to extract the most accurate ODP categories for each particular video lecture: (i) its source-dependent categories, (ii) its associated tags, and (iii) the text extracted from its title and description.

When mirroring this problem to current state of the art approaches, we found several interesting works that have attempt to: (i) generate mappings between category hierarchies [6, 9, 11], (ii) generate mappings from tag information spaces to category hierarchies [1, 12] and (iii) classify textual documents under category hierarchies [2, 5]. While the previously mentioned works are focused on using only one type of information, our purpose is to exploit simultaneously, tags, source-dependent categories and associated textual descriptions, to extract the most accurate ODP categorization for each video lecture.

For this purpose, among the available systems and techniques for information classification we have decided to reuse TextWise[42] software and services. The

---

[39] `http://code.google.com/apis/youtube/1.0/reference.html`
[40] `http://itunes.apple.com/us/genre`
[41] `http://deimos.apple.com/rsrc/doc/iTunesUAdministrationGuide/ iTunesUintheiTunesStore/chapter_13_section_3.html`
[42] `http://textwise.com/`

categorization service provided by Textwise identifies the main topic categories for an input text or URI using the ODP 2010 categorization[43]. According to Textwise, the categorization of content is performed by analyzing the dimensions and weights of the content's Semantic Signature[44], where a Semantic Signature represents the concepts in a text through a weighted vector entry of typically several thousand semantic dimensions.

Our proposal is therefore to exploit the three different types of information to generate the input text that the TextWise service needs to perform the categorization process. When generating this text, it is important to keep in mind that tags, and domain-dependent categories, are the key video classification properties. Although properties like title and description provide a more extended and coherent characterization of the video content, we have empirically observed that they tend to become ambiguous information elements when they are used to extract the key video topics. Based on these facts, the categorization approach is formulated as follows:

Let $S = \{s_1, s_2 ..., s_n\}$ be the list of different educational institutions, or information sources. Let $V_i = \{v_{i_1}, v_{i_2}, ..., v_{i_m}\}$ be the complete list of video lectures extracted from the educational institution S_i where each video lecture v_{ij} has associated: a set of tags, $T_{v_{ij}}$, a set of categories, $C_{v_{ij}}$, a title $Title_{v_{ij}}$ and a description, $Desc_{v_{ij}}$. Following these definitions, the pseudo-code of the proposed categorization approach is described in Table 3.

Table 3. Pseudo-code of the proposed video categorization approach

Categorization approach
For $i = 1$ to n
select information source s_i
For $j = 1$ to m
select the video lecture v_{ij}
extract $T_{v_{ij}}$,   extract $C_{v_{ij}}$, extract $Title_{v_{ij}}$,   extract $Desc_{v_{ij}}$
$HTMLD_{v_{ij}} = HTMLDoc\,(T_{v_{ij}}, C_{v_{ij}}, Title_{v_{ij}}, Desc_{v_{ij}})$
TextWise ($D_{v_{ij}}$, 2)

Basically, for each video lecture, v_{ij}, the approach extracts its set of tags, $T_{v_{ij}}$ and source-dependent categories, $C_{v_{ij}}$, its title, $Title_{v_{ij}}$ and its description, $Desc_{v_{ij}}$. With this information it generates an HTML document, $HTMLD_{v_{ij}}$, from which the Textwise service extracts up to two ODP classifications of v_{ij}. The generated HTML document contains the following structure:

---

[43] http://textwise.com/api_docs/labels/
2010-ODP-Topic-Category-Mapping.txt
[44] http://textwise.com/technology-0

```
<html>
    <head>
        <title> Title_{v_{ij}} </title>
        <meta name="keywords" content=" T_{v_{ij}} , C_{v_{ij}} ">
    </head>
    <body> <p> Desc_{v_{ij}} </p></body>
</html>
```

As we can see, the HTML page title and body correspond to the title and description of the video lecture respectively. Each tag and source-dependent category is added as a meta keyword element of the HTML document. Following this approach, tags and categories are emphasized within the HTML page. This emphasis is expected to produce a positive impact when using the TextWise categorization service, because it decreases the relevance of more ambiguous properties, such as title and description. To visualize the HTML page generation process, let's consider the video lecture presented in Figure 2:

Table 4. HTML page associated to the v_{ij} video lecture presented in Figure 2

v_{ij} **Information**	**HTMLDoc** $(T_{v_{ij}}, C_{v_{ij}}, Title_{v_{ij}}, Desc_{v_{ij}})$
s = videolectures.net, $T_{v_{ij}} = \varnothing$, there are no tags associated to this video. $C_{v_{ij}}$ ={Computer Science, Semantic Web} $Title_{v_{ij}}$ = "Collective Intelligence [...] enriching folksonomies with Flor" $Desc_{v_{ij}}$ = "Web 2.0 has introduced [...] with help of the Semantic Web"	`<html>` `<head>` `<title>` Collective Intelligence [...] enriching folksonomies with Flor `</title>` `<meta name = keywords content=` "Computer Science", "Semantic Web"> `</head>` `<body><p>` Web 2.0 has introduced [...] with help of the Semantic Web `</body></p>` `</html>`

For this generated HTML page, the TextWise service produces as response:

- Reference/Knowledge_Management (id=495), w=0.71
- Reference/Libraries/Library_and_Information_Science (id=497), w= 0.53

The TextWise service provides not only the ODP categorization label and its corresponding id, but also a weight which reflects the confidence of the service in the proposed classification. As we can see in the example, for the two proposed answers, the first one may be considered correct by most evaluators, but the correctness of the second one is more arguable. Following some empirical tests, we have decided to set up a threshold of 0.5 to accept the proposed categorization as valid.

5 Evaluating the Categorization Problem

Achieving a high degree of correctly categorized video lectures is a key requirement towards the generation of a high quality interlinked dataset of educational material. In this section we describe the evaluation conducted to assess the quality of the video lectures data integration process. The evaluation pursues three key goals: (i) measuring the *coverage* of the categorization process; i.e., how many video lectures have been assigned at least to one ODP category; (ii) measuring the *correctness* of the categorization process; i.e., which percentage of the assigned categories are considered correct and; (iii) measuring the *specialization* of the categorization process; i.e., are the assigned ODP classifications the most specialized ones or is it possible to find a more refined ODP category to describe the same video content?

Coverage: To evaluate the coverage of the categorization process we have analyzed the number and percentage of video lectures for which no ODP categories were assigned. From the total of 14,311 videos lectures extracted from the 27 different educational institutions, a total of 14,037 (**98%**) were successfully categorized using the approach presented in section 4. Additionally 55% of the videos were assigned a second ODP category. Over the remaining 274 video lectures we have performed an empirical analysis to find out the different reasons for their lack of classification. The most significant one is the use of different languages to represent the properties of the video lecture. As an illustration, consider the video lecture defined by the URI http://videolectures.net/inovativna_slovenija2010_golobic_kis/. This video lecture has its title described in Slovenian language *"Kdaj inovativna Slovenija?"* and its classification described in English language *" Top » Technology » Innovation"*. Other reasons include the simultaneous lack of video description, tags and categories.

Correctness and Specialization: To evaluate the correctness and specialization of the categorization approach, we have engaged 3 different evaluators in the campaign. Each of them has evaluated the categories assigned to 675 video lectures (25 randomly selected video lectures for each of the 27 information sources). Considering that 252 of the 675 selected videos where assigned two different ODP categories, the total number of video categorizations judged by each evaluator was 927. Note that, when randomly selecting the 25 video lectures for each information source, we have previously discarded those ones for which no ODP category was assigned. To judge the correctness and specialization of each video categorization, the evaluators were provided with: (i) all the available video information, (ii) its assigned categories and, (iii) the complete ODP hierarchical classification. Each video categorization was judged using a value from 0 to 2, where each number implies: (0) the classification is incorrect, (1), the classification is correct but a more specialized category could have been assigned, and (2), the classification is correct and the evaluator has not found any more specialized category in the ODP.

For each video categorization, given the three user's evaluations, the categorization was considered correct if at least two evaluators were rating it with values higher than 0, and it was considered specific, if at least two evaluators were rating it at level 2 and the remaining evaluation was not 0. There was a substantial agreement among users. Fleiss' kappa statistic [3] measuring user's agreement was k=0.71 (a value k=1 means complete agreement). Once the agreement results were established we found that over the 927 video categorizations 831 (**89%**) were considered correctly classified and 475 (**51%**) were considered specialized.

6 Conclusions and Discussion

This paper presents our work and experience interlinking educational information across universities through the use of LD principles and technologies. More specifically, this paper is focused on selecting, extracting, structuring and interlinking information of video lectures produced by 27 different educational institutions. For this purpose, selected information from several websites and YouTube channels have been scraped and structured according to several existing vocabularies. To integrate this information, the extracted videos have been categorized under a common searchable/browsable space, the taxonomy defined by the Open Directory Project. As a result of this process a new LD educational dataset has been released containing more than 14,000 video lectures from 27 different institutions. These videos have been categorized under a total of 569 different ODP categories. Among the most popular ones we can highlight: Science/Math, Science/Physics and Computers/Artificial_Intelligence.

High levels of coverage (**98%**) and accuracy (**89%**) have been achieved during the integration process. The complete dataset is available under `http://smartproducts1.kmi.open.ac.uk/web-linkeduniversities/index.htm`. Here, the reader can find a complete description of the dataset, including the RDF dumps for each institution, a SPARQL endpoint, and several SPARQL query examples.

Regarding our lessons learned we propose five main ingredients for a successful production and integration of educational content through the use of LD principles.

1. LD principles are simple. However, identifying available data, obtaining access to it and remodeling it is a high-cost process. Making educational institutions understand that it is worth doing it is a critical factor.
2. There is a need to agree on a set of collective vocabularies to model and structure educational information. Following a bottom up approach, those vocabularies should initially focus on modeling common elements across educational institutions like: educational material, courses or research staff.
3. There is a need to agree on common searchable/browsable spaces under which educational information can be explored and retrieve. Establishing a common space of topics under which educational material and courses can be classified could be a good starting point.
4. Establishing qualitative criteria and quantitative evaluation measures to assess these criteria are key requirements for the development of high quality educational LD.
5. Educational LD is not about a killer application, but is about multiple small things that are made easier (integrating information across university departments, enriching information with external sources, sharing educational content across institutions, etc.) Proposals should emerge about how to integrate the benefits of LD within the universities' practices and workflows.

To be truly effective, many of these improvements should be the results of community-wide efforts rather than advances at the level of individual research groups. We believe that this is an important time for the development of the education's Web of LD. Collaborative efforts to produce and integrate educational information are needed to achieve the envisioned data space from which, information across educational institutions will be search, explored, compared and retrieved in an homogenous way.

References

1. Cantador, I., Konstas, I., Jose, J.M.: Categorising Social Tags to Improve Folksonomy-based Recommendations. Journal of Web Semantics 9(1), 1–15 (2011)
2. Cesa-Bianchi, N., Conconi, A., Gentile, C.: Regret Bounds for Hierarchical Classification with Linear-Threshold Functions. In: Shawe-Taylor, J., Singer, Y. (eds.) COLT 2004. LNCS (LNAI), vol. 3120, pp. 93–108. Springer, Heidelberg (2004)
3. Fleiss, J.L., Cohen, J.: The equivalence of weighted kappa and the intraclass correlation coefficient as measures of reliability. Educational and Psychological Measurement 33, 613–619 (1973)
4. Heath, T., Bizer, C.: Linked Data: Evolving the Web into a Global Data Space. In: Synthesis Lectures on the Semantic Web: Theory and Technology, 1st edn., vol. 1(1), pp. 1–136. Morgan & Claypool (2011)
5. Hofmann, T., Cai, L., Ciaramita, M.: Learning with Taxonomies: Classifying Documents and Words. In: Syntax, Semantics and Statistics Workshop, NIPS (2003)
6. Kalfoglou, Y., Schorlemmer, M.: Ontology mapping: the state of the art. The Knowledge Engineering Review 18(1), 1–31 (2003)
7. Möller, K., Heath, T., Handschuh, S., Domingue, J.: Recipes for semantic web dog food — the ESWC and ISWC metadata projects. In: Aberer, K., Choi, K.-S., Noy, N., Allemang, D., Lee, K.-I., Nixon, L.J.B., Golbeck, J., Mika, P., Maynard, D., Mizoguchi, R., Schreiber, G., Cudré-Mauroux, P. (eds.) ASWC 2007 and ISWC 2007. LNCS, vol. 4825, pp. 802–815. Springer, Heidelberg (2007)
8. Neubert, J.: Bringing the "thesaurus for economics" on to the web of linked data. In: Workshop on Linked Data on the Web, WWW 2009 (2009)
9. Noy, N.: Semantic integration: a survey of ontology-based approaches. SIGMOD Rec. 33(4), 65–70 (2004)
10. Rowe, M.: Data.dcs: Converting legacy data into linked data. In: Workshop on Linked Data on the Web, WWW 2010 (2010)
11. Shvaiko, P., Euzenat, J.: A Survey of Schema-Based Matching Approaches. In: Spaccapietra, S. (ed.) Journal on Data Semantics IV. LNCS, vol. 3730, pp. 146–171. Springer, Heidelberg (2005)
12. Specia, L., Motta, E.: Integrating Folksonomies with the Semantic Web. In: Franconi, E., Kifer, M., May, W. (eds.) ESWC 2007. LNCS, vol. 4519, pp. 624–639. Springer, Heidelberg (2007)
13. Suchanek, F.M., Kasneci, G., Weikum, G.: Yago: a core of semantic knowledge. In: 16th International Conference on World Wide Web, WWW 2007, pp. 697–706. ACM, New York (2007)
14. Zablith, F., Fernandez, M., Rowe, M.: The OU Linked Open Data: Production and Consumption. In: eLearning Approaches for the Linked Data Age. Extended Semantic Web Conference, ESWC 2011 (2011)

Mind Your Metadata: Exploiting Semantics for Configuration, Adaptation, and Provenance in Scientific Workflows

Yolanda Gil[1], Pedro Szekely[1], Sandra Villamizar[2],
Thomas C. Harmon[2], Varun Ratnakar[1], Shubham Gupta[1],
Maria Muslea[1], Fabio Silva[1], Craig A. Knoblock[1]

[1] Information Sciences Institute, University of Southern California,
4676 Admiralty Way, Marina del Rey, CA 90292, USA
[2] School of Engineering, University of California Merced,
5200 North Lake Rd., Merced, CA 95343, USA
{gil,szekely,varunr,shubhamg,mariam,fabio,knoblock}@isi.edu
{tharmon,villamizar_amaya}@ucmerced.edu

Abstract. Scientific metadata containing semantic descriptions of scientific data is expensive to capture and is typically not used across entire data analytic processes. We present an approach where semantic metadata is generated as scientific data is being prepared, and then subsequently used to configure models and to customize them to the data. The metadata captured includes sensor descriptions, data characteristics, data types, and process documentation. This metadata is then used in a workflow system to select analytic models dynamically and to set up model parameters automatically. In addition, all aspects of data processing are documented, and the system is able to generate extensive provenance records for new data products based on the metadata. As a result, the system can dynamically select analytic models based on the metadata properties of the data it is processing, generating more accurate results. We show results in analyzing stream metabolism for watershed ecosystem management.

Keywords: Scientific metadata, semantic workflows, data integration.

1 Introduction

Despite significant advances in computational infrastructure and sensor network observatories, many environmental scientists are slowed down by the tasks required to set up their analyses as data comes in daily from their sensors. Data preparation is time-consuming: scientists gather data from multiple sources and sensors, they must first clean the data, normalize it so that data from different sources is represented using the same units and formats, and they must integrate it and configure it according to the requirements of their models and simulation software. Data analysis is also time consuming: scientists run different models and must make sure to provide each model the inputs it requires in the format it requires, and that the outputs of one model are compatible with the inputs of the next one.

L. Aroyo et al. (Eds.): ISWC 2011, Part II, LNCS 7032, pp. 65–80, 2011.
© Springer-Verlag Berlin Heidelberg 2011

An important aspect of data analysis is selecting and fine-tuning models according to the data characteristics. For example, for analyzing metabolism in a river some models are appropriate for high water flows, and others are best for low water flows. One way to do this is for the scientist to first prepare the data, and then based on the characteristics of the data select the appropriate models. This simple approach becomes cumbersome and time-consuming when scientists wish to run their analysis periodically (e.g., every day) to analyze data coming from sensors.

Despite best intentions and care, the execution of a model may fail, often because data violates a model assumption for which components do not explicitly check. Errors that surface in running a model may have been introduced in an earlier step. To understand and debug these problems, scientists need to trace back the provenance of the data.

Finally, to assess progress, scientists must be able to reproduce their analyses, to run new models on previous data, and to easily retrieve results of prior runs. Reproducing previous analyses becomes difficult if the process involved manual steps where scientists manually configured models, or interactively provided inputs. The process needs to be fully audited so it can be accurately reproduced. Results from many runs need to be found based on their properties. Inspecting prior results is often difficult because scientific analysis processes often generate vast amounts of data and files, and without explicit metadata and provenance information it is hard to understand where each piece of data came from and what it represents.

All these issues could be addressed if scientists invested the time and were thorough in creating and propagating metadata as they prepare and process data. However, the management of metadata often has to be done manually, so it becomes a burden and therefore it is seldom done.

The main contribution of our work is to show that by explicitly capturing the semantics of the data and their provenance, our tools enable scientists to focus on their science rather than on the mechanics of running their models. We show that capturing metadata and propagating it through the data preparation and analysis processes is useful to: 1) save manual effort in managing metadata and setting up and running analyses, 2) make all data and analytic results searchable, 3) make results understandable and interpretable, 4) share results with other scientists. We demonstrate two integrated systems for data preparation and analysis that capture and use metadata and provenance information as data flows through different steps of the process. Karma, our data preparation system [14] helps scientists extract, clean, normalize and integrate the data coming from sensors and third-party data sources. Karma uses a programming by example paradigm to enable scientists to perform these tasks by providing examples of how the data should be transformed. Karma infers general procedures from these examples that it can then apply to entire data sets. During this process, Karma also learns models of the data, aligning the data to a domain ontology and augmenting the data sets with metadata that records the learned models. This metadata is passed along with the data to Wings, a workflow system that uses the metadata to ensure that the workflow components fit together in a semantically meaningful way. Most importantly, this metadata enables Wings to dynamically select analytic models and parameters that are appropriate for the data being processed. Because all metadata is expressed according to a domain ontology, it is possible to query all workflow results in terms of the domain ontology.

Karma and Wings work together to make metadata management effective and accessible to scientists while saving them time throughout the data analysis processes. Karma generates the metadata that Wings uses to reason about the workflow almost as a side effect of preparing the data for the workflow. Much of the metadata is learned automatically from the data itself. Wings then propagates the metadata throughout the workflow to all intermediate data sets that each workflow component produces as it is executed. In the end, the provenance for all data sets is captured in the metadata, producing a complete audit trail of the workflow products.

This paper presents our approach in the context of a case study where scientists analyze stream metabolism of the Merced river in California's Central Valley. In the next section we describe the case study in more detail. In section 3 we present an overview of the approach, and in sections 4 and 5 we describe Karma and Wings in more detail. In section 6 we show the results of our case study, and in section 7 we present conclusions and directions for future work.

2   Motivation: Environmental Science

Despite tremendous advances in shared infrastructure, many daily tasks faced by scientists are disconnected from those capabilities. For environmental scientists and many of the observatory disciplines, the scientific method—hypothesize, observe, analyze, interpret—remains bogged down by myriad manual and routine data analysis processes aimed at separating environmental variability from the phenomena of interest. To truly enable transformative science, the time and effort required for these processes must be lowered in order to substantially compress the timeframe of observatory-scale analysis.

We motivate key requirements to support scientists with the problem of simulation of whole stream metabolism, where we use a model for estimating rates of aquatic photosynthesis known as gross primary production (GPP) and community respiration (CR_{24}) [1]. These estimates are useful for assessing the status of and changes in stream ecosystems in the context of a watershed management.

In our domain of interest, the hypothesize, observe, analyze, interpret cycle takes weeks to months longer than the timescale of observation, such that the best that scientists can achieve are *post-hoc* interpretations of river conditions. Furthermore, in uncontrolled (real) systems, they often learn only late in the observation or analysis parts of the cycle that an experiment has failed due to unexpected changes in river flow or water chemistry. These limitations make it difficult for researchers to discover the cause and effect links between different drivers (e.g., climate and land use change) and the aquatic ecosystem function on a timescale less than years, or even decades. By automating and compressing the cycle of data collection, integration, and analysis, we aim not only to enable the more rapid advancement of river science, but to advance the environmental science paradigm by enabling timely, practical resource management decisions [4]. We now describe the data preparation, integration, and processing steps in this cycle for water metabolism.

Data Preparation

Stream ecologists spend significant time collecting data in the field and preparing it to be useful for running computational models. **Metadata regarding data origins** is needed at collection time to annotate the station, location, type of sensor, and error rates. This metadata is important for selecting supplemental data and to determine which models to run. For example, the time and location are needed so that weather data from national weather sites is selected consistently with the sensor readings. Once the raw data is transmitted or brought back to the lab, it needs to be checked for consistency and anomalies. For example, data filtering is generally needed to remove noise and spurious data points, or sometimes sensor calibration drift necessitates systematic adjustment of the data. These quality control steps should be tracked, as they transform data in ways that are important to select models and to interpret results of future steps.

Data Integration

Beyond investigations of local, site-specific scope, a major need for environmental scientists is the integration of their data with the massive amounts of data and other resources that the national scale cyberobservatories are designed to provide. For example, river simulators require inputs such as: (1) the material properties (e.g., the soil type of the river bed in different parts of the model) and structure or geometry of the simulated domain (e.g., the bathymetry of the river bed), (2) the fluid properties, including dissolved chemical species, and (3) the initial and boundary conditions associated with the river (e.g., a constant flow condition on the upstream boundary would drive flow into the system). The data behind these parameters are populated from a variety of data sources and are in different formats, including spatial shapefiles, time series, and locally gridded data (as from robotic sensor platforms). It is important to capture **metadata about types and constraints** to represent the semantics of what the data means and what data each source provides. The origin metadata captured during earlier data preparation steps is useful here in order to select the appropriate supplemental data sources. In addition, **metadata about statistical properties** of the datasets provides extremely useful characteristics that drive model use and facilitate search. For example, extracting the average daily depth of a river based on hourly readings enables scientists to determine days of low flow and select models appropriately.

Data Processing

Once datasets are located, cleansed, and integrated, there are many possible data analysis processes performed using analytic software or simulation models. Analytic and simulation software may be developed in-house or by colleagues, or obtained from third parties such as government agencies or commercial vendors. Examples range from relatively sophisticated simulation engine codes for river flow, chemical fate, and transport modeling to statistical packages for time series analysis. These tools are used in diverse aspects of the modeling processes. For example, spatial interpolation routines are often used to prepare spatially distributed material properties or physical parameters for input to gridded numerical simulation models.

For our task of river metabolism analysis, there are a variety of model types available. Selecting the most appropriate models for analysis is important. In the current *post-hoc* modeling paradigm, the researcher needs to select an appropriate one on the basis of key observational parameters and knowledge of the field. For instance, some metabolism rate models work better when the river flow is relatively low while others are better suited to high flow conditions. Indeed, for a given location in a river, different models may be appropriate over different time periods because the amount of water may change drastically based on changes in conditions dictated by weather (e.g. heavy precipitation or snow melt) or human activities (e.g., reservoir releases). Therefore, the metadata about statistical properties captured earlier during data integration is useful to select models at this stage.

As if managing the data transfers across individual tools and models were not challenging and time consuming enough, metadata is often poorly managed and laborious to integrate into the analysis. Key metadata is often archived locally by key investigators but not moved along with the data throughout the analytic process steps. Metadata for analytic results is tracked manually and seldom published. This **process metadata** is key for documenting results, so that they can be interpreted appropriately, searched based on what processes were used to generate them, and so that they can be understood and used by other investigators. For example, the fact that a particular model was used to generate a result and what the parameter settings were matter tremendously if the result is to be integrated with other results.

In summary, scientists need integrated environments for managing end-to-end data preparation, integration, and analysis that offer a comprehensive treatment of metadata throughout the processes. In order to make scientific data analysis processes more efficient and useful, we must offer better support to capture metadata about: 1) the origins of raw data, 2) the types and relationships across datasets, 3) the statistical properties of datasets, and 4) the processes applied to the data.

3   Approach: Provenance-Aware Systems That Manage Metadata

Our approach is to develop **provenance-aware systems that create, propagate, and use metadata** as they contribute to scientific data analysis processes. Metadata can be extracted from original data sources, created during data integration and analysis, and propagated throughout the different steps of the analysis process so that the provenance of any result (whether intermediate or final) is well documented. All this metadata is useful throughout the process to integrate with new data sources, to select and setup analytic steps, and to understand analytic results.

To demonstrate our approach, we have developed two provenance-aware systems that address complementary steps in the scientific analysis process. Karma, our data preparation tool, carries out data preparation and integration steps [14]. Wings, our workflow system, carries our data processing steps through computations [6]. Both systems capture and use metadata as the data flows through different steps.

Figure 1 shows an overview of the interaction between Karma and Wings, which will be described in detail in the rest of the paper. The bottom-left part of the figure.

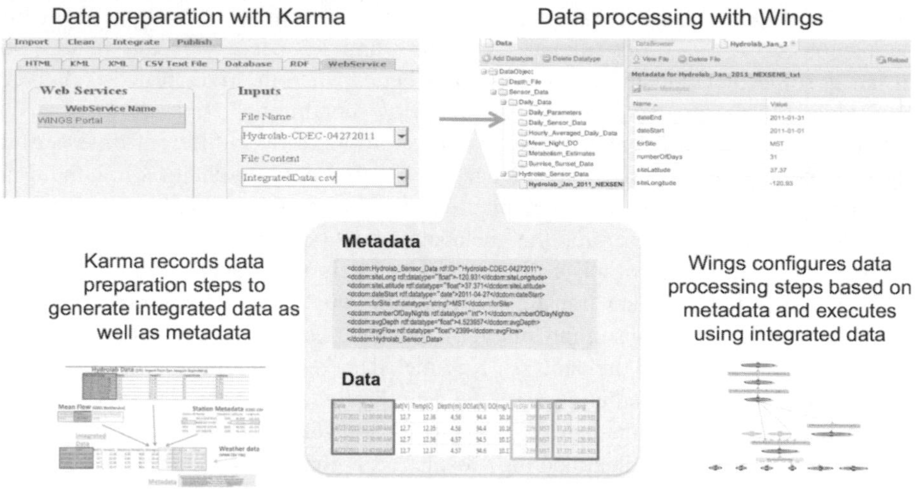

Data preparation with Karma   Data processing with Wings

Fig. 1. Overview of creation and use of metadata as the data is processed throughout our provenance-aware system

shows a sketch of the data preparation process where a scientist cleans, normalizes and integrates data from multiple sources. Different parts of the integrated dataset are color-coded to show the original sources where the data came from. Behind the scenes, Karma creates metadata for the dataset so that when a dataset is exported its metadata is exported with it, as shown in the blue bubble at the center bottom of the figure. Once a dataset is prepared, the scientist can upload the data set to Wings, as shown in the top-left part of the figure. The right part of the figure shows the data processing aspects of the system. The top-right part shows a Wings screen where users can review the data sets that have been uploaded from Karma for processing as well as the metadata associated with each dataset. The bottom-right part of the figure shows the workflow used to analyze the data. Wings propagates metadata received from Karma for the initial data inputs of the workflow so that newly generated results can be described appropriately. Wings also uses metadata to dynamically select models and set up their parameters. All the metadata is used to generate provenance.

The next sections describe how Karma and Wings create and use metadata in a synergistic manner, and how this integrated and comprehensive treatment of metadata benefits scientists.

4   Data Preparation and Integration with Karma: Metadata about Origins, Type, and Characteristics

Karma [14] is an information integration tool designed to enable users unfamiliar with databases, ontologies, scripting languages or any other programming concepts to extract, clean, normalize and integrate data.

Karma uses a programming-by-example paradigm where users provide examples of how these steps are carried out and Karma generalizes these examples into procedures that can be applied to entire datasets. User studies [14] showed that users were able to complete three information integration tasks about three times faster using Karma than using Dapper/Yahoo Pipes (a state of the art tool). These studies also revealed that the Karma users were able to complete the tasks without error. In contrast, 83% of Dapper/Yahoo Pipes users made at least one error in the first task, 45% in the second and 95% in the third. Karma is a visual tool that offers users a table representation of their data and commands to import, clean, integrate and publish their data. We present the Karma capabilities in the context of our stream metabolism case study.

The first data preparation step is to import the various data sets needed to drive the stream metabolism analysis. The first set of sources comes from the California Data Exchange Center (CDEC, water.ca.gov). We defined a Web service that provides programmatic access to the data published in this web site. After users select the appropriate web service from the library of web services registered in Karma, they can select the parameters of the service that are of interest. In the CDEC service, users can choose the station, sensor and date ranges. The data retrieved from the Web service is subsequently shown in a table where users can proceed with further data preparation steps. In our case study, users import data for multiple sensors obtaining a collection of five tables with data for the dates of interest. Our users also use data from their own water quality sensors, which comes from a comma-separated-value (CSV) file, and a metadata source from CDEC that records the geospatial coordinates of all the CDEC sensors, also a CSV file.

The next data preparation step is to integrate the data from all these tables into a single table that contains the sensor values for all the CDEC sensors, the water quality sensors, and the location of each sensor. In order to integrate the data, the date and time formats need to be normalized, and changed to the format required by the simulation software. The water quality dates are in the format "2010-03-10 00:15:00" and the CDEC dates formatted as "20100309" and "2300" in two separate columns.

Figure 2 illustrates Karma's by-example data normalization capabilities. To normalize the CDEC dates to the required format, users provide an example of how the data ought to be transformed. Karma generalizes the example and applies the general rule to all the values in the column. If the generalization is incorrect, users can provide additional examples. Using this procedure, users can quickly normalize all the date and time formats of the five tables imported from CDEC and the water quality tables to transform the data as shown in Figure 3.

Once the data is normalized, users must join the five CDEC tables and their water quality data table into a single table that contains the sensor values for all the sensors, as shown in Figure 3. To do so, they use the Karma 'Integrate" command on the consolidated table. Behind the scenes, Karma has analyzed the tables to automatically determine that they can be joined based on the Date and Time fields that are common to all tables and using this information it creates a menu of the columns from the CDEC and water quality tables that are appropriate to add to the consolidated table using database join operations. Users can successively choose from this menu the columns that they want to add to the table, unaware of the database join operations that Karma is performing to appropriately align the joined values based on Date and Time. Similarly, users can integrate the geospatial coordinates of the sensors from the CDEC metadata source.

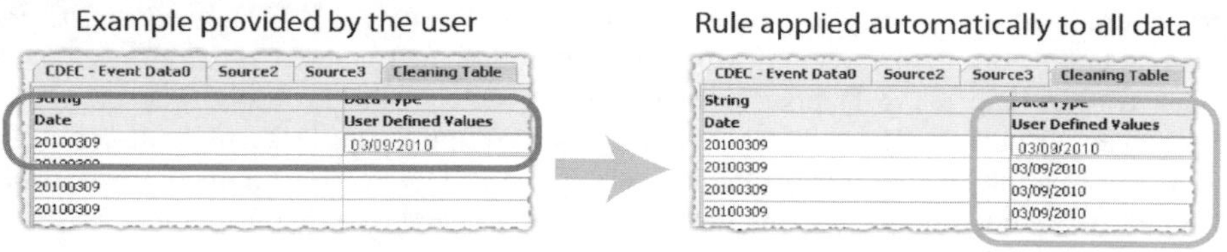

Fig. 2. Normalizing the date format: the user provides one example, and the system learns a rule and applies it automatically to the entire dataset

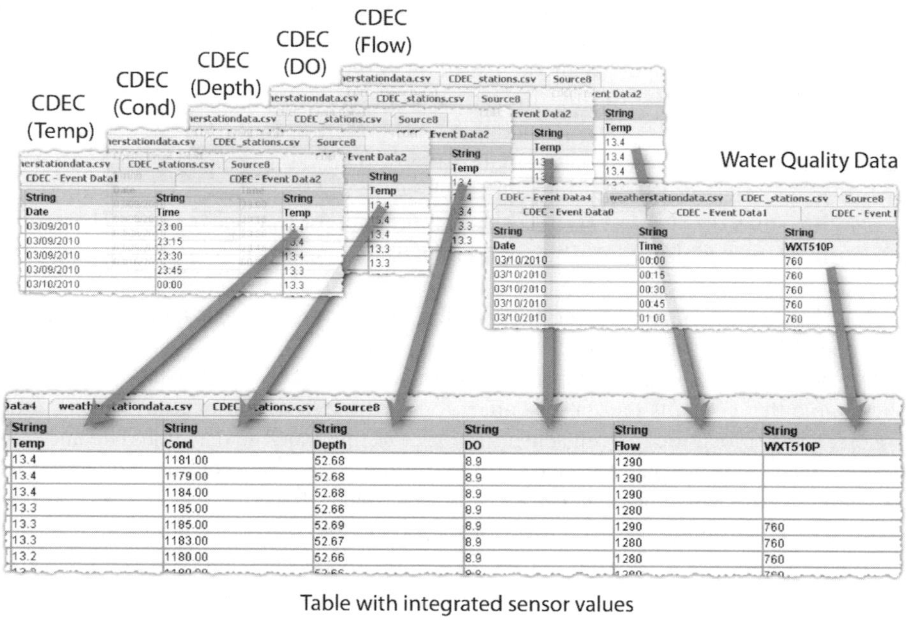

Table with integrated sensor values

Fig. 3. Integration of sensor sources into a consolidated dataset

The next step in the data preparation phase is to build the metadata for the consolidated table so that in the final step, when the table is deployed to Wings, it carries the metadata used for workflow processing. To do so, users invoke the ontology alignment capability in Karma that enables them to map each column of the consolidated table to the classes defined in the domain ontology, which is also used by Wings and may contain community ontologies for the domain. To map a column to the ontology, users click on the grey cells above the column headings and choose from the menu that appears the appropriate ontology class. Using the information in the ontology, Karma generates standard metadata for the source. For numeric fields and date fields Karma will generate metadata with the minimum, maximum and average values. It is also possible to associate with classes in the ontology custom computations that compute additional metadata. For example, we defined custom computations to compute a velocity metadata field. Karma represents the metadata as RDF. An example of the metadata for a daily dataset is:

```
<dcdom:Daily_Sensor_Data rdf:ID="DailyData-04272011">
<dcdom:siteLong rdf:datatype="float">-120.931</dcdom:siteLongitude>
<dcdom:siteLatitude rdf:datatype="float">37.371</dcdom:siteLatitude>
<dcdom:dateStart rdf:datatype="date">2011-04-27</dcdom:dateStart>
<dcdom:forSite rdf:datatype="string">MST</dcdom:forSite>
<dcdom:numOfDayNights rdf:datatype="int">1</dcdom:numOfDayNights>
<dcdom:avgDepth rdf:datatype="float">4.523957</dcdom:avgDepth>
<dcdom:avgFlow rdf:datatype="float">2399</dcdom:avgFlow>
</dcdom:Daily_Sensor_Data>
```

The final data preparation step is to deploy the table and its metadata to Wings. This is done using the "Publish" command in Karma that supports publishing the data in a variety of formats, as HTML pages that visualize the data, to a table in a database, as CSV files, or to a Web service. In our case, Wings uses a Web service to deploy data, so users will publish their data as a Web service. Karma can also publish the data and the metadata as RDF aligned to the user's ontology. This capability enables scientists to contribute the metadata, and the data if they so desire as Linked Open Data aligned to the domain ontology. Because the published data is aligned to an ontology it is much easier to link it to other data that uses the same ontology.

A dataset needs to be created for each day, as required by the simulation software used in later steps in the workflow. The data preparation procedure for each dataset is the same. To accommodate this, Karma allows users to save all the data preparation steps for one data set as a script. Then they can parameterize the script with respect to the dates and replay the script for the desired days. These steps are explicitly recorded as metadata to capture the process provenance for each dataset created.

5   Data Processing with Wings: Metadata about Analysis Processes

In our work, we use the Wings workflow system [6]. Wings is unique in that it uses semantic workflow representations to describe the kinds of data and computational steps in the workflow. Wings can reason about the constraints of the workflow components (steps) and the characteristics of the data and propagate them through the workflow structure. In contrast, most workflow systems focus either on execution management or on including extensive libraries of analytic tools [10; 3; 12]. Semantic reasoning in Wings is used to provide interactive assistance and automation in many aspects of workflow design and configuration. In [5], we show details of the interaction of a user with Wings through its web-based interface.

Wings uses OWL2 to represent ontologies of workflows, components, and data. Metadata is represented as RDF assertions that refer to those ontologies. Some constraints are represented as rules. A set of rules are associated with particular workflow component to express constraints on the applicability of the component, how to set up component parameters, or what the metadata of its outputs should be, given metadata of its inputs. We show examples of these rules and their use later.

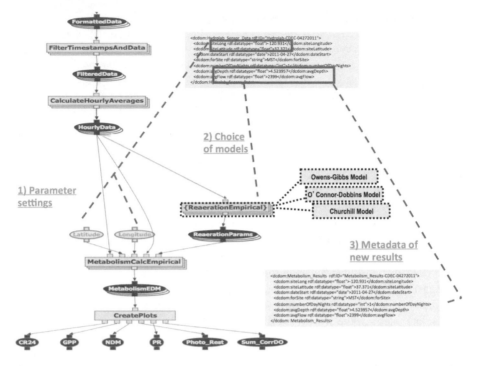

Fig. 4. A Wings workflow template for estimating water metabolism rates, illustrating how metadata created by Karma is used to: 1) choose a simulation model appropriate for the daily water quality data, 2) set up parameters of the models, 3) create metadata for new data generated by the workflow and provide provenance of any new results

Figure 4 shows an example workflow to carry out a daily metabolism calculation in a river site for a given time period, which needs as input the daily reaeration rates calculated in the initial steps of the workflow. The metabolism calculation also uses as input hydrological data collected by in-situ sensors for that day. The representation of this workflow exploits several key features in Wings, highlighted in the figure.

First, for water metabolism analysis, there are three different empirical reaeration models that must be selected based on the morphometry and flow conditions of the river at the site of interest [11; 9; 2]. Wings can represent abstract workflow steps as classes of components. In this example, Wings represents ReaerationEmpirical as a class of components, which has three instances (one per model).

Second, the workflow needs to be run for a time period of n days, while the data for each day is stored in a separate file. This means that a collection of hydrological data for the last n days is required as input to the workflow. Wings can reason about data collections and execute as many metabolism calculations as there are days in the time period.

Third, in the semantic workflows used in Wings, every dataset and step is associated with a variable. Variables can have associated semantic constraints. Figure 4 shows these constraints for the input variable Formatted_Data in a small box at the top, they are not readable but they are the metadata generated by Karma shown in the last section. These semantic constraints can be used to represent metadata properties of input datasets, such as the type of sensor and the collection date of a dataset of water flow rates. Wings checks that the metadata properties of the input datasets are consistent with the constraints of the workflow variable Formatted_Data.

Several important capabilities of Wings are used in this application and described in the rest of this section.

Dynamic Selection of Models Based on Metadata

An important capability is the ability to represent constraints associated with inputs to a component to express its requirements. In our case, each reaeration method is appropriate for certain ranges of flow conditions. For example, the O'Connor-Dobbins model is only valid when the depth is either greater than 0.61m or greater than 3.45 times the velocity to the 2.5 power (velocity in m/s). These constraints are expressed as rules, which invalidate that component for a workflow that uses data outside of those ranges. The above example is expressed as:

```
# ODM not valid for Depth <= 3.45 * pow(velocity, 2.5)
[ODMInvalidity2:
 (?c rdf:type pcdom:ReaerationODM)
 (?c pc:hasInput ?idv) (?idv pc:hasArgumentID "HourlyData")
 (?idv dcdom:depth ?depth) (?idv dcdom:velocity ?vel)
 pow(?vel, 2.5, ?vpow) product(?vpow, 3.45, ?condn) le(?depth, ?condn)
    -> (?c pc:isInvalid "true"^^xsd:boolean)]
```

Wings takes the abstract workflow step ReaerationEmpirical and specializes it to the three models, creating three possible workflow candidates for a given input dataset. Next, it applies the rule above to each of the workflows. The metadata of HourlyData has to contain a depth and velocity that conform with the requirement of this model to be greater than $3.45 \mathrm{x} \mathrm{V}^{2.5}$ (velocity in m/s), otherwise the candidate workflow that uses this model is invalidated and rejected. Similar rules exist for the other two models. Each component has its own set of rules. This approach is more modular than representing such constraints as a conditional branch of the wofklow, as is done in other workflow systems.

Automatic Parameter Set Up Based on Metadata

Wings can set up input parameters of the models selected based on characteristics of input datasets. In our workflow example, the latitude/longitude of the location are parameters to the metabolism estimation model. In Wings, they are set automatically by the system based on the location of the sensor that was used to collect the initial data. This is done with a rule for that workflow component:

```
[SetValuesLatLong:
    (?c rdf:type pcdom:MetabolismCalcEmpiricalClass)
    (?c pc:hasInput ?idv) (?idv pc:hasArgumentID "HourlyData")
    (?c pc:hasInput ?ipv1) (?ipv1 pc:hasArgumentID "Latitude")
    (?c pc:hasInput ?ipv2) (?ipv2 pc:hasArgumentID "Longitude")
    (?idv dcdom:siteLatitude ?lat) (?idv dcdom:siteLongitude ?long)
        -> (?ipv1 pc:hasValue ?lat) (?ipv2 pc:hasValue ?long)]
```

Note however that the HourlyData input to the metabolism calculation is not an input to the workflow, so it does not have any metadata. That is, while we know what the characteristics are for the input water quality datasets generated by Karma (FormattedData), we do not know what the characteristics are for other datasets in the workflow. Wings has the ability to create metadata for HourlyData by propagating metadata throughout the workflow, as we explain below.

Automatic Generation of Metadata for New Results
Wings generates metadata for all new workflow data products, and we already discussed how this metadata is used by the two types of rules that we just described.

Wings uses rules for components that express what the output metadata properties are based on input metadata properties. This is handled through metadata propagation rules associated with each component. An example rule for the first workflow step is:

```
[FwdPropFilter:
    (?c rdf:type pcdom:FilterTimestampsAndDataClass)
    (?c pc:hasInput ?idv) (?idv pc:hasArgumentID "InputSensorData")
    (?c pc:hasOutput ?odv) (?odv pc:hasArgumentID "outputSensorData")
    (?idv ?p ?val) (?p rdfs:subPropertyOf dc:hasDataMetrics)
        -> (?odv ?p ?val)]
```

Here, the property hasDataMetrics is a superclass of all the metadata properties that must be propagated to the outputs of a component, otherwise they are assumed to be different for the output dataset created by the component. Other rules express how metadata of the outputs will be different based on the metadata of the inputs, what the computation is about, and what the parameter settings are. These metadata propagation rules are used to automatically create metadata for all workflow data products. As a result, any resulting data can be queried based on their properties.

Workflow data products also have detailed provenance metadata that records what workflow was used to generate them. All workflow executions with their provenance can be exported as Linked Data using the Open Provenance Model (http://openprovenance.org). Unlike other provenance systems, Wings provenance records include semantic metadata associated to datasets, which can be used to query about workflow results. We show examples of such queries in the next section.

6   Results

An important contribution of our work is that our provenance-aware system automatically chooses a model each day based on metadata characteristics that are created and propagated by the system about the daily data. The results of the data analysis are more meaningful from the point of view of the scientific application.

Figure 5 shows plots of the calculated reaeration rates (K2) for the cases (a) when one model is used for every day of the period of analysis and (b) when each model is used only for the conditions of flow for which it applies. Notice how the models predict roughly the same values during the highest flows but diverge significantly as flow decreases.

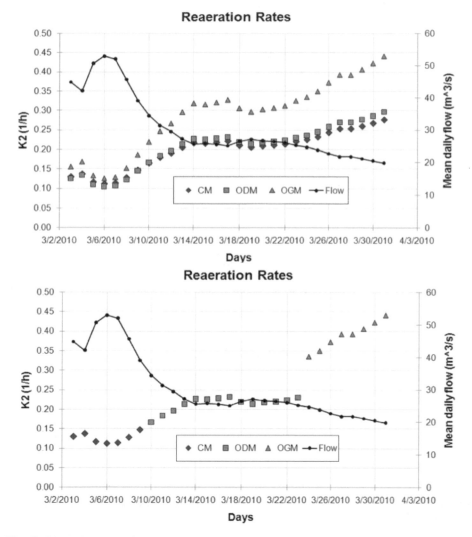

Fig. 5. Reaeration rates plotted against river flow a) Inaccurate results when a single model is used throughout the time period, b) Accurate results when different models are used for each day depending on the flow conditions

Figure 6 shows the plots of the net daily metabolism for a given monthly time period during Spring when the flow of the river has high fluctuations. The Wings workflow system selects dynamically the models each day based on flow conditions represented as metadata and captured in Karma. For the first few days, the Churchill model is best and is the one selected by the system. The O'Connor-Dobbins model is close, but the Owens-Gibbs model would not be appropriate. In the later part of the month the Owens-Gibbs model is significantly better, and is the one selected by our system. All models roughly agree for the dates around the middle of the month, which happen to have intermediate flow conditions.

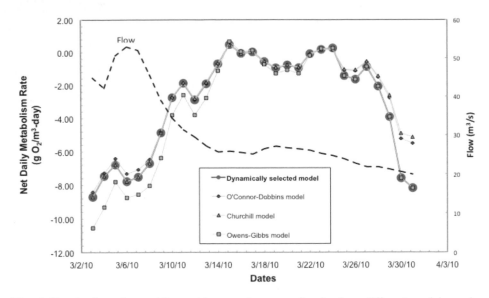

Fig. 6. Results from the workflow with reaeration correction by three different models, each optimal at different flow conditions (smaller symbols). The workflow system automatically selects the models (larger red circles) based on flow conditions.

Metadata for Metabolism_SMN_2010_03_03Z_ODM	
Name ˅	Value
velocity	0.66163415
usedAlgorithm	dcdom:ODM
slope	1.0E-4
siteLongitude	-120.97618
siteLatitude	37.347214
forSite	SMN
forDate	2010-03-03Z
flow	1581.6842
depth	1.0403947
barpress	760.0

```
PREFIX wflow:
  <http://www.isi.edu/2007/08/workflow.owl#>
PREFIX dcdom:
  <http://www.isi.edu/dc/Water/ontology.owl#>

SELECT ?url
WHERE {
?data dcdom:usedAlgorithm dcdom:ODM .
  ?data rdf:type dcdom:Metabolism_Estimates .
  ?data wflow:hasLocation ?url
}
```

Fig. 7. All workflow data products have metadata (left), and can be queried by their metadata properties (right)

The results demonstrate immediate scientific relevance, with the different reaeration models producing a significant divergence in the estimates (roughly 30-35% differences) for the highest and lowest flows.

Another important contribution of our work is the ability of our provenance-aware system to associate metadata with any analytic data products, so it can be meaningfully queried and interpreted. Figure 7 shows the metadata for one of the metabolism datasets that resulted from workflow execution on the left. On the right it shows an example of a query over the provenance of the results. The query retrieves all data products that were obtained with workflows that used the ODM model as a component and were metabolism data.

7   Related Work

Although there are many tools for scientific data preparation and integration, Karma provides a unique approach that learns from user-provided examples.   Many scientists still use spreadsheets for these tasks, and Karma retains that paradigm and adds to it novel capabilities for capturing metadata.   There has been recent related work on integrating large data sets using a spreadsheet paradigm in a project called Google Fusion Tables [7; 8]. In this effort they developed an online service that allows users to integrate large datasets, annotate and collaborate on the data, and visualize the integrated results. The focus of the work is primarily on the scalability and visualization of the integrated data instead of on the problems of extracting, modeling, cleaning, and integrating the data. The ideas in Google Fusion Tables are relevant to Karma and would apply directly to deal with the problems of scaling up to large datasets and visualizing the integrated results.   In contrast, Karma is able to generate valuable metadata for daily datasets that it exports together with the integrated data.   This is a novel capability in a spreadsheet paradigm for data manipulation.

Scientific workflow systems generate provenance of new data products (see [13] for an overview), many using the community-developed Open Provenance Model (http://www.openprovenance.org).   However, other systems focus on provenance concerning execution details, such as the specific invocations of the components and the execution times and other details.   Wings is the only workflow system that uses semantic constraints and rules to generate metadata, as well as to represent abstract steps as classes of components.

8   Conclusions

We have presented two integrated complementary systems that create, propagate, and use metadata in different stages of scientific data analysis processes.   They are provenance-aware systems that capture metadata about: 1) the origins of raw data, 2) the types and relationships across datasets, 3) the statistical properties of datasets, and 4) the processes applied during data analysis.   Our approach makes the management of metadata more efficient for scientists, and at the same time the metadata captured reduces the amount of manual work by using the metadata to dynamically select models, set up their parameters, and generate provenance metadata.   These capabilities are crucial in environmental sciences, where sensor networks report daily on datasets that cannot be analyzed manually in a timely manner.   We used our approach for water metabolism analysis, showing significant improvements in accuracy as the system dynamically analyzes daily water quality data.

We are currently setting up the system to produce water metabolism results in a timely manner as data comes in daily from sensors in the observation site. We are also using the system to analyze data for the last five years, in order to produce more accurate historical models of the metabolism in the observation site.

In future work, we want to use metadata to control the sensor collection and transmittal system. By making the sensor system provenance aware and integrating it with the data preparation and processing systems that we already have, we can have a feedback loop to guide the sensors to produce data that is most useful for scientists.

References

1. Bott, T.L.: Primary productivity and community respiration. In: Hauer, F.R., Lamberti, G. (eds.) Methods in Stream Ecology, 2nd edn., pp. 663–690. Academic Press (2007)
2. Churchill, M.A., Elmore, H.L., Buckingham, R.A.: The prediction of stream reaeration rates. Am. Soc. Civil Engineers Journ. 88(SA-4), 1–46 (1962)
3. Deelman, E., Singh, G., Su, M., Blythe, J., Gil, Y., Kesselman, C., Kim, J., Mehta, G., Vahi, K., Berriman, G.B., Good, J., Laity, A., Jacob, J.C., Katz, D.S.: Pegasus: A Framework for Mapping Complex Scientific Workflows onto Distributed Systems. Scientific Programming 13(3) (2005)
4. Dozier, J., Gail, W.B.: The emerging science of environmental applications. In: Hey, T., Tansley, S., Toll, K. (eds.) The Fourth Paradigm: Data-Intensive Scientific Discovery. Microsoft Research (2009)
5. Gil, Y., Ratnakar, V., Fritz, C.: Assisting Scientists with Complex Data Analysis Tasks through Semantic Workflows. In: Proceedings of the AAAI Fall Symposium on Proactive Assistant Agents, Arlington, VA (2010)
6. Gil, Y., Gonzalez-Calero, P.A., Kim, J., Moody, J., Ratnakar, V.: A Semantic Framework for Automatic Generation of Computational Workflows Using Distributed Data and Component Catalogs. To appear in the Journal of Experimental and Theoretical Artificial Intelligence (2011)
7. Gonzalez, H., Halevy, A.Y., Jensen, C.S., Langen, A., Madhavan, J., Shapley, R., Shen, W.: Google fusion tables: data management, integration and collaboration in the cloud. In: Proceedings of the First Symposium on Cloud Computing, Industrial Track, pp. 175–180 (2010)
8. Gonzalez, H., Halevy, A.Y., Jensen, C.S., Langen, A., Madhavan, J., Shapley, R., Shen, W., Goldberg-Kidon, J.: Google fusion tables: web-centered data management and collaboration. In: Proceedings of SIGMOD, Industrial Track, pp. 1061–1066 (2010)
9. O'Connor, D.J., Dobbins, W.E.: Mechanisms of reaeration in natural streams. Am. Soc. Civil Engineers Trans. 123, 641–684 (1958)
10. Oinn, T., Greenwood, M., Addis, M., Alpdemir, N., Ferris, J., Glover, K., Goble, C., Goderis, A., Hull, D., Marvin, D., Li, P., Lord, P., Pocock, M., Senger, M., Stevens, R., Wipat, A., Wroe, C.: Taverna: lessons in creating a workflow environment for the life sciences. Concurrency and Computation: Practice and Experience 18(10) (2006)
11. Owens, M., Edwards, R.W., Gibbs, J.W.: Some reaeration studies in streams. Int. Jour. Air and Water Pollution 8, 469–486 (1964)
12. Reich, M., Liefeld, T., Gould, J., Lerner, J., Tamayo, P., Mesirov, J.P.: GenePattern 2.0. Nature Genetics 38(5) (2006), doi:10.1038/ng0506-500
13. Taylor, I., Deelman, E., Gannon, D., Shields, M. (eds.): Workflows for e-Science. Springer, Heidelberg (2007)
14. Tuchinda, R., Knoblock, C.A., Szekely, P.: Building Mashups by Demonstration. To appear in ACM Transactions on the Web, TWEB (2011)

Using Semantic Web Technologies to Build a Community-Driven Knowledge Curation Platform for the Skeletal Dysplasia Domain

Tudor Groza[1], Andreas Zankl[2,3], Yuan-Fang Li[4], and Jane Hunter[1]

[1] School of ITEE, The University of Queensland, Australia
tudor.groza@uq.edu.au, jane@itee.uq.edu.au
[2] Bone Dysplasia Research Group
UQ Centre for Clinical Research (UQCCR)
The University of Queensland, Australia
[3] Genetic Health Queensland,
Royal Brisbane and Women's Hospital, Herston, Australia
a.zankl@uq.edu.au
[4] Monash University, Melbourne, Australia
yuanfang.li@monash.edu

Abstract. In this paper we report on our on-going efforts in building SKELETOME – a community-driven knowledge curation platform for the skeletal dysplasia domain. SKELETOME introduces an ontology-driven knowledge engineering cycle that supports the continuous evolution of the domain knowledge. Newly submitted, undiagnosed patient cases undergo a collaborative diagnosis process that transforms them into well-structured case studies, classified, linked and discoverable based on their likely diagnosis(es). The paper presents the community requirements driving the design of the platform, the underlying implementation details and the results of a preliminary usability study. Because SKELE-TOME is built on Drupal 7, we discuss the limitations of some of its embedded Semantic Web components and describe a set of new modules, developed to handle these limitations (which will soon be released as open source to the community).

1 Introduction

Skeletal dysplasias are a heterogeneous group of genetic disorders affecting human skeletal development. Currently, there are over 440 recognized types, categorized into 40 groups. Patients with skeletal dysplasias have complex medical issues, such as short stature, degenerative joint disease, scoliosis or neurological complications. Since most skeletal dysplasias are very rare (<1:10,000 births), data on clinical presentation, natural history and best management is sparse. The lack of data makes existing patient cases a precious resource for biomedical research because they enable scientists to study, among other things, the effects of single genes on human bone and cartilage development and function. The resulting insights may lead to a better understanding of the pathogenesis of more common connective tissue disorders, such as arthritis or osteoporosis.

L. Aroyo et al. (Eds.): ISWC 2011, Part II, LNCS 7032, pp. 81–96, 2011.
© Springer-Verlag Berlin Heidelberg 2011

Unfortunately, due to the intrinsic complexity of dysplasias, correct diagnosis is often difficult. At the same time, only a few centres worldwide have the necessary expertise in diagnosis and management of these disorders. On the other hand, the identification of many skeletal dysplasia-causing genes and subsequent studies of their functions and interactions have led to an explosion in the knowledge of bone and cartilage biology. The biomedical literature now contains a large amount of information about individual genes and gene interactions, but it is often difficult to grasp how these interactions work together in a broader context (such as skeletal dysplasias). In turn, the focus on specific patient cases or genes makes it difficult to identify etiological relationships between skeletal dysplasias, or to recognise clinical or radiological characteristics that are indicative of defects within a specific molecular pathway.

The International Skeletal Dysplasia Society (ISDS)[1] has attempted to address some of these problems with its Nosology of Genetic Skeletal Disorders [1]. Since 1972, the ISDS Nosology lists all recognised skeletal dysplasias and tries to group them by common clinical-radiographic characteristics and/or molecular disease mechanisms. The ISDS Nosology is revised every 4 years by an expert committee and the updated version is published in a medical journal, being widely accepted as the "official" nomenclature for skeletal dysplasias within the biomedical community. While the content is invaluable, the format of the Nosology has several short-comings, including: (i) an inflexible classification scheme – each disorder being listed in one group based either on its clinical radiographic appearance or on its underlying molecular genetic mechanism; (ii) limited amount of cross-referenced information – each entry contains only the Online Mendelian Inheritance in Man (OMIM) number [2], the chromosome locus and the gene name, without being linked to widely used semantic data repositories, like the Gene Ontology [3] or UniProt [4], which would allow users to study further up-to-date relevant information; and most importantly, (iii) the lack of a shorter publishing cycle – the content quickly becomes outdated, as genes or disorders discovered after the publication date can no longer be included until the next revision (4 years later).

In addition to the above-mentioned Nosology issues, collaboration among experts is also adversely affected by a lack of an appropriate tool support. Currently, the community uses the ESDN (European Skeletal Dysplasia Network) Case manager[2] and Google mailing lists to share information and to exchange and discuss patient cases. Neither of these provides an ideal collaboration environment. While ESDN provides a structured (form-based) discussion forum to support the diagnosis process, mailing lists are merely long threads of free text. Leaving aside the complete lack of any formal representation or semantics, a major issue is the inability to transfer knowledge or provide links between the rich pool of patient reports and the ISDS Nosology.

In this paper we report on the efforts of the SKELETOME project[3], which aims to develop a community-driven knowledge curation platform for the

---

[1] http://www.isds.ch/

[2] https://cm.esdn.org/

[3] http://itee.uq.edu.au/~eresearch/projects/skeletome/index.html

skeletal dysplasia domain. The SKELETOME platform[4] introduces an ontology-driven knowledge engineering cycle that supports the continuous evolution of the knowledge captured in the ISDS Nosology from existing patient studies, thus transforming into a living knowledge base. Concurrently, this knowledge informs the collaborative decision making process associated with newly arriving cases. Moreover, the underlying SKELETOME ontologies represent a foundational building block for linking to external resources and a mechanism for facilitating knowledge extraction and reasoning. SKELETOME is being developed by extending Drupal 7[5] with additional Semantic Web components to enable seamless and semantic-aware collaborative input, sharing and re-use of data and information among the experts in the field. The knowledge engineering cycle, together with the set of new Semantic Web Drupal modules (and some lessons learned from the existing ones) represent the main contributions of this paper.

The remainder of the paper is organized as follows. Section 2 describes the representational and functional requirements supporting the SKELETOME platform. Section 3 provides a detailed overview of the SKELETOME components and information flow. In Section 4 we discuss the preliminary evaluation, and before concluding in Section 6, we analyze some of the existing related efforts in Section 5.

2 Requirements

Since genetic disorders are typically quite rare, a global network of patients, clinicians and researchers is necessary to accumulate the critical mass of data and knowledge needed to address some of the greatest challenges in medical genetics, i.e., the development of evidence-based clinical management guidelines, the study of genotype-phenotype[6] correlations and the identification of disease modifier genes. Skeletal dysplasias are an ideal topic for a global medical collaboration network as the number of medical conditions is relatively small and well defined and there is an existing, tightly-knit and motivated community of clinicians and scientists willing to contribute, share and exchange case studies, data, diagnoses and clinical information.

Recognition of this opportunity, led to the establishment of the SKELETOME project – a collaboration between information scientists, Semantic Web researchers and clinical geneticists, led by the University of Queensland. In addition to a Web-based framework for enabling and encouraging the international skeletal dysplasia community (researchers, experts, clinicians) to contribute content, the most important requirements for the project (which emerged from direct discussions with the community) are the following:

Common Terminology. The diagnosis and management of skeletal dysplasias depends on highly specialised domain knowledge across a number of disciplines

---

[4] http://skeletome.metadata.net/skeletome
[5] http://drupal.org/drupal-7.0
[6] *Genotype* refers to the genetic information of an individual, while *phenotype* describes the actual observed properties of an individual, such as morphology or development.

(radiography, genetics, orthopaedics), which is not easily comprehensible to individual communities or hospitals. In order to enable the exchange of knowledge between experts (across languages and disciplines), a common terminology is required, hence leading to a shared conceptualisation of the domain.

Data Integration. Large datasets containing rich information on molecules (genes, proteins) already exist and the information relevant to skeletal dysplasias needs to be extracted and cross-referenced with the clinical data and knowledge produced by SKELETOME. The data cross-reference requires integration both at conceptual level, as well as, at actual data and instance level.

Privacy and Access Control. Actual patient studies and reports need to be visible only to the experts participating in the decision making process. Moreover, sensitive patient data (e.g., name, address, relatives) should only be accessible to the case initiator.

Knowledge Transfer and Sustainable Knowledge Evolution. The knowledge collectively acquired from the anonymized pool of patients represents a valuable asset from the conceptual perspective of the domain (materialized in the ISDS Nosology). Consequently, a seamless transfer of this knowledge is required to enable the dynamic and continuous evolution of the conceptual domain.

Capturing Provenance and Expertise. The contributed content may take several forms, ranging from personal observations to scientific publications. Independently of the form, SKELETOME requires a mechanism to keep track of the provenance of the data and knowledge, in order to ensure proper privacy and access control. It also needs to provide a measure of certainty of derived data and to leverage expertise from the content and to streamline the delivery of the most relevant information to the most appropriate person.

In order to support the above requirements, the SKELETOME platform provides the following services: (i) a collaboration environment for experts to exchange knowledge and patient cases and to build a repository of patient case studies linked to related evidence and Web resources (e.g., publications, radiographic data, gene databases, etc); (ii) a set of ontologies that capture the domain knowledge and underpin the platform; (iii) semantically enhanced content annotation and integration services; (iv) ontology-driven text processing of publications leading to rich semantic annotations; (v) enhanced image search and retrieval via ontology-based annotation; (vi) reasoning on anonymised patient data for semi-automated decision making.

3   The SKELETOME Platform

The innovative aspect of the SKELETOME platform is the ontology-driven knowledge engineering cycle, introduced to bridge the current knowledge about the domain (partly captured in the ISDS Nosology) to the continuously growing pool of patient cases. The engineering cycle consists of two concurrent phases: (1) semantic annotation of patient instance data, and (2) ontology learning from patient instance data.

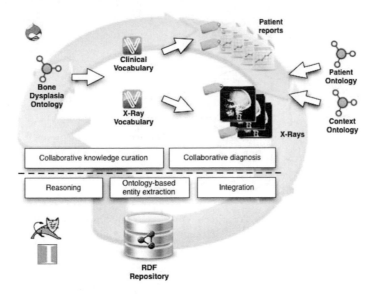

Fig. 1. The high level architecture of the SKELETOME platform

We developed the Bone Dysplasia ontology[7] to overcome the short-comings of the ISDS Nosology and to describe the relations between bone dysplasias and the genotype and phenotype characteristics. The ontology is used to semantically-enrich patient reports and the associated X-Ray imagery. Additionally, in conjunction to two auxiliary ontologies (the Patient[8] and Context[9] ontologies), which capture patient and provenance information, we use the Bone Dysplasia ontology to enhance the collaborative diagnosis process. The resulting (RDF) instance data is then used in the reasoning process to propose novel genotype and phenotype characteristics to be associated to bone dysplasias, and hence support the collaborative knowledge curation and the evolution of the conceptual knowledge of the domain.

Fig. 1 depicts a high level overview on the SKELETOME architecture. The upper part of the architecture, including also the front-end, is developed using Drupal 7 and contains two main components (implemented via several Drupal modules): (i) the collaborative knowledge curation component, responsible for generating Drupal pages associated to ontology concepts, in addition to generating tagging vocabularies from the underlying ontologies, and (ii) the collaborative diagnosis component, responsible for capturing the information exchanged by the experts in the diagnosis process (i.e., diagnosis creation and rating or open discussions on diagnoses). The lower part of the architecture consists of the ontology-driven services, developed via a set of servlets hosted in Tomcat and using OpenRDF Sesame as RDF triple store. The Integration service bridges the Drupal world to the RDF back-end by managing several Drupal hooks on

---

[7] http://purl.org/skeletome/bonedysplasia
[8] http://purl.org/skeletome/patient
[9] http://purl.org/skeletome/context

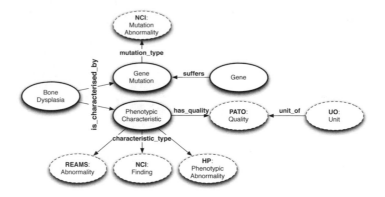

Fig. 2. A snippet of the Bone Dysplasia Ontology. The upper part of the ontology describes the genotypic information of bone dysplasias. The lower part relates bone dysplasias to phenotypic characteristics.

certain content types (or pages), e.g., Bone Dysplasia or Patient. Its role is to keep the RDF triple store in sync with the Drupal data, in addition to ensuring that no sensitive patient data is stored in the back-end. The other two services, i.e., Ontology-based entity extraction and Reasoning, have self-explanatory roles.

In the following sections we describe the underlying mechanisms used to develop the knowledge engineering cycle by means of the requirements introduced in the previous section. From a technical perspective we also identify some of the short-comings of the current Drupal 7 RDF support.

3.1 Common Terminology and Data Integration

As mentioned in Section 1 the ISDS Nosology has a rigid structure and only partially covers the genotype information of the domain. More concretely, it merely lists the skeletal dysplasias, the genes responsible for the diseases and their locus, which leads to a poor description of the domain. Elements such as the gene mutation information and the radiographic or phenotypic characteristics are unfortunately ignored. For example, if we consider the *Stickler syndrome*, the ISDS Nosology only lists *COL2A1* as the responsible gene, and it does not mention that it might be caused by a *Missense mutation* in the gene (leading to a Glycine substitution with Arginine on position 219), or that some of the phenotypic characteristics are *Myopia* and *Cleft palate*, or that radiographically it can be characterized by *Dolichocephaly*.

To overcome these issues, and to extend the existing common terminology used by the community, we developed the Bone Dysplasia ontology (that defines more than 1200 concepts) to capture all the relevant knowledge by integrating and re-using well known ontologies, such as NCI Thesaurus [5], Human Phenotype Ontology (HP) [6] or the REAMS ontology[10] – describing radiographic features. Fig. 2 depicts a snippet of the ontology, showing the relation between

---

[10] http://d-reams.org/?page_id=84

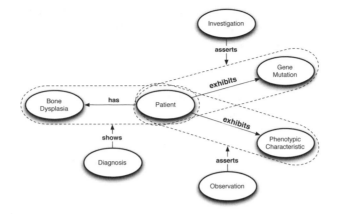

Fig. 3. A snippet of the Patient Ontology. The direct relations between Patient and the other three concepts are reified in order to capture the context in which their are materialized.

the root **Bone Dysplasia** class (further sub-classed by 40 bone dysplasia groups and then by specific skeletal dysplasias) and the **Gene Mutation** and **Phenotypic Characteristic** classes. As opposed to the ISDS Nosology, our **Gene Mutation** class provides the ground for encoding richer information about the characteristics of the mutation, e.g., type, position, original and mutated content. Similarly, the **Gene** class is linked (via annotation properties) to OMIM, MesH, UMLS and UniProt. These are only two examples where the ontology accommodates extended domain knowledge when compared to the ISDS Nosology.

In addition to the Bone Dysplasia ontology we have also developed a Patient ontology and a Context ontology. The Patient ontology captures knowledge about specific patient reports, hence describing "instances" of genotypic, phenotypic and radiographic characteristics of bone dysplasias in particular patients. As can be observed in Fig. 3 a **Patient** may *exhibit* diverse **Gene Mutations** or **Phenotypic Characteristics** which are *asserted* by **Investigations** or **Observations**. Similarly a **Diagnosis** *shows* that a **Patient** may *have* a particular **Bone Dysplasia**. The Context ontology is used to model the provenance of the patient information, including, for example, who suggested an **Investigation** or made an **Observation**, who and where a **Diagnosis** is documented, or even when a **Patient** exhibited certain **Phenotypic Characteristics**.

3.2   Knowledge Transfer and Sustainable Knowledge Evolution

The knowledge engineering cycle briefly introduced in the beginning of Section 3 was specifically designed to support this requirement. The first phase of the cycle uses the above-described ontologies to enrich the content created by the experts and to support the collective diagnosis process. This phase has, in reality, two sub-phases: (1) a sub-phase dealing with the evolution of the generic domain knowledge, i.e., classification of bone dysplasias and their descriptions, and (2) a second sub-phase covering the actual use of ontologies for semantic annotation. The second sub-phase processes the patient instance data (i.e., semantically-annotated reports and diagnoses) to propose novel findings about bone dysplasias.

(1) Domain knowledge maintenance and evolution
From a functional perspective the experts need to keep the domain knowledge up-to-date. SKELETOME publishes all the bone dysplasias and associated information (e.g., Genes) as Web pages (via specific Drupal content types). Hence each bone dysplasia has its own publicly available Web page, similar to the way in which Wiki systems work (see, for example, `http://skeletome.metadata.net/ skeletome/bonedysplasia/achondroplasia`). The Bone Dysplasia ontology acts as a backbone for this set of pages, as they are automatically generated from (and are in sync with) the concepts defined in the ontology. The page generation is realized via a Drupal module that we have developed, in conjunction with the Integration service from the RDF backend. This module will be released as open source and may be useful to anyone who wants to build and maintain an ontology-driven content management site, starting from an existing ontology.

Currently, the Drupal RDF extensions allow one to map existing content types to ontological concepts and/or properties. Creating pages of those content types will result in Drupal creating the associated concept instances (via `rdf:type`). We were, unfortunately, unable to use this support for two reasons. Firstly, while the generic Bone Dysplasia content type could have been mapped to the corresponding ontological class, we required all its instance pages to represent classes themselves, and not instances (i.e., via `rdfs:subClassOf`). For example, the Web page about the **Stickler syndrome** should be mapped to the **Stickler syndrome** class in the ontology, and not to an instance of the **Bone Dysplasia** class. At the same time, mapping manually over 1000 concepts, currently present in the ontology, is neither feasible, not sustainable. Secondly, **Bone Dysplasias** are related to **Genes**, also modelled as content type instances. Instantiating node reference property values between custom content types is currently not supported by the RDF extensions. The need to provide reasoning support, which intrinsically requires explicit relations between instances, forced us to maintain all the RDF in the RDF back-end and develop specific Drupal hooks to keep the store constantly updated, via the Integration service.

In order to support maintenance and evolution of the knowledge base, SKELETOME provides support for adding, renaming or removing dysplasia groups (which are direct subclasses of the **Bone Dysplasia** class), moving bone dysplasias (which are direct subclasses of the groups) between groups, adding newly discovered gene mutations or phenotypic characteristics, hence manipulating the structure and content of the Bone Dysplasia ontology, without the experts being aware of it. Hiding the underlying ontological concepts and details was an easy decision, because the vast majority of the experts are simple computer users.

To ensure quality control over the content of the Bone Dysplasia pages, we have imposed an editorial process. Each bone dysplasia has an associated editor, responsible for keeping the explicit information and related knowledge

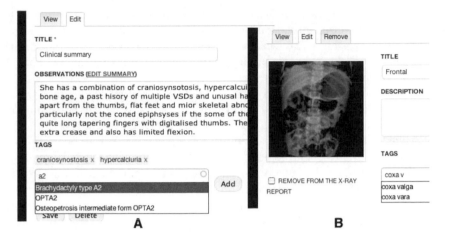

Fig. 4. [A] Semantic annotation of clinical summaries using a mixture of terms from Clinical Summary Vocabulary describing phenotypic characteristics; [B] X-Ray imagery tagging using terms from the X-Ray Vocabulary describing radiographic characteristics.

up-to-date, by reviewing input from the community. In addition to scientific publications, this community input takes the form of statements asserted about the disease on its page, which can then be discussed and commented on by the entire community (thus enabling a *"wisdom of the crowds"* approach). On a periodic basis, the editor will incorporate in the main disease description those statements that were accepted by the community, and have reasonable support though scientific evidence.

In practice, the statements (or micro-contributions) carry a dual role: (i) they enable the transfer of knowledge from patient cases to the conceptual domain knowledge, as typically they would report on novel findings about the dysplasia from a particular pool of patients, and (ii) they allow us to create and maintain an expertise profile of the authors, which will lead to an authorship reputation system, similar to the one in WikiGenes [7]. The reputation of an author-expert is calculated based on the acceptance of her statements by the community, and hence the extent to which her contribution impacts on and advances the field.

(2) Semantic annotation of patient cases
The Bone Dysplasia ontology provides not only the backbone for the evolution of the domain knowledge, but also the means for enriching patient cases with semantic annotations. Our goal is to provide experts with the mechanism for annotating both clinical summaries, as well as X-Ray imagery with domain concepts. In order to realize the annotation, we implemented a Drupal module that transforms a given ontology into a Drupal taxonomy (i.e., vocabulary) that enables tagging. The vocabulary import may be invoked from particular root concepts and can traverse the ontology up to a specified level or the leaf nodes. The actual tags are created by looking at the literal values of specified properties. For example, one may choose to create tags from

rdfs:label, but also from *skos:altLabel*. One significant aspect of this module is that the generated tags retain a relation to the URI of the originating concept. Hence, when an expert tags an X-Ray with a particular tag, s/he actually annotates the image with the ontological concept supporting the tag. This module will also be released as open source later this year.

Within the context of SKELETOME, this module is used to generate two vocabularies, as depicted in Fig. 1: a vocabulary for annotating clinical summaries (from the Bone Dysplasia ontology, HP, NCI and PATO, Phenotype and Trait Ontology [8], ontologies) and a vocabulary for annotating X-Ray imagery (from the HP and the REAMS ontologies). This distinction was specifically requested by the community in order to support their current terminological practice. The annotation of clinical summaries can be done manually or semi-automatically. The semi-automatic annotation is implemented by integrating the NCBO annotator [9] for entity extraction (via the Ontology-based entity extraction component of the backend).

The annotation of patient resources is transformed in the backend, via Drupal hooks, into relations between the patient instance and the corresponding concept instances. For example, consider the annotation depicted in Fig. 4: the patient instance being examined would be related via the *exhibits* relation to **HP:cranyosynostosis** and to **REAMS:coxa-valga**.

The collaborative diagnosis process works in a similar manner as the annotation of clinical summaries. By adding a diagnosis to a patient, in reality, the expert annotates the patient case with the corresponding Bone Dysplasia concept. In the backend, this translates into a reification of the relation between the patient instance and a particular Bone Dysplasia instance (see Fig. 3), which is then related to a freshly created **Diagnosis** instance that has context information attached to it. This context information is generated from the discussion among experts and includes the votes cast on the diagnosis.

Reasoning and Searching on Patient Data

Semantically annotated patient cases create a wealth of knowledge that represents a perfect application for reasoning. Leaving aside the access control aspects (detailed in the next part of the section), our goal is to close the knowledge engineering cycle by supporting both the evolution of the conceptual domain, as well as the collective decision making process in two ways. Firstly, we want to apply reasoning across current cases to propose diagnoses on newly published cases. Secondly, we want to infer novel findings on the conceptual domain, by focusing on the similarities between phenotypic, radiographic and genotypic characteristics in patient cases sharing the same diagnosis. With respect to this latter goal, we want to avoid discovering the obvious, e.g., that all patients diagnosed with *Achondroplasia* have a mutation in the *FGFR3* gene – this information is already in the domain knowledge.

The complexity of the two tasks requires a thorough investigation of the most suitable mechanisms to support them. Initially we considered reasoning across the instance data using SWRL rules for both tasks, however, we quickly realized that the rigidity of rule-based inferencing will not help us in fulfilling our goals.

The SWRL rules had especially negative consequences on our second goal – the identification of features that are not present in the vast majority of cases (features that you would expect to be inferred via reasoning). Both diagnoses and the presence of phenotypic characteristics cannot be stated with a 100% certainty. As a result, the collaborative voting mechanism currently featured in SKELETOME does not record a simple Yes/No, but uses a 5-star rating, hence allowing the experts to associate a level of uncertainty with their opinion. This rating cannot be converted into rigid/strict rules. Consequently, we are currently investigating ways of encoding the diagnosis information using fuzzy rules, in addition to using uncertainty and/or statistical reasoning techniques.

Besides reasoning (currently under investigation), the use of ontologies enables SKELETOME to provide semantic search functionality, dynamically related items and faceted browsing. This last aspect is particularly important for expert users as it allows them to quickly filter search results based on criteria, such as: patient ids, phenotypic or genotypic characteristics. Additionally, for an even richer browsing experience, we have integrated dynamic links to related knowledge items in some of the views (e.g., the dysplasia descriptions and patient clinical summaries). For example, a bone dysplasia description might have suggested links to patients diagnosed with this dysplasia or to related phenotypic characteristics. From a technical perspective, this is realized by following the relations in the ontology for the instance under scrutiny, and secondly, by analyzing the textual content (where possible) and extracting and linking domain concepts present in the knowledge base.

3.3   Privacy and Access Control

The information captured in SKELETOME is accessible via four layers of privacy and access control policies. The generic conceptual knowledge of the domain (i.e., the bone dysplasia pages and associated resources) are publicly available. The rest of the knowledge is private and accessible via group and individual-based access controls. Different groups of experts are registered within the platform and act as sub-communities within the greater community. The reason behind this group division is the need to share patient information only with a specific set of experts. Experts can, nevertheless, be members of multiple groups, and hence share their information and knowledge across all of them.

Sensitive patient information, such as name or address, is accessible only to the case initiator (individual-based access control). The purpose of exchanging patient cases is to foster advances in the field and to take advantage of the community-driven diagnosis process. However, sensitive patient information is not relevant for the diagnosis process, and hence is maintained only for provenance purposes. In reality, the so-called "participation" of the patients in these community exercises is acknowledged by the patients via written consents (also maintained within the platform, and included within the patient information).

As described in the previous sections, each patient's semantically annotated clinical data (including annotated X-Ray imagery) is stored and processed in the RDF backend. In order to enforce the individual-based access control over sensitive information, yet take advantage of the wealth of knowledge present in the entire pool of patient cases across all groups, we followed the principle of separation of concerns [10] (the fourth layer of access control). Drupal hooks were implemented on the patient content type to filter the fields that are transformed into RDF instance data via the Patient and Context ontologies, and then stored via the Integration backend service. As a result, the RDF triples will model strictly phenotypic, radiographic or genotypic information, while the rest of the information (including the sensitive data) remains stored only in the Drupal database, and is subject to the three access restrictions described above. This allows us to perform reasoning on the entire set of patient clinical data, hence taking advantage of both the quantity and quality of the knowledge created by experts – whilst still restricting access to sensitive data.

3.4 Capturing Provenance and Expertise

The previous sub-sections have already provided an insight into the mechanisms implemented by SKELETOME to capture provenance and expertise. The Context ontology is used to capture provenance information, ranging from the author names and dates of assertions to timestamps on diagnoses or phenotypic characteristics. This information is then used to generate expertise profiles from asserted statements, forum discussions and collaborative diagnoses.

Currently, our expertise modeling is based on the mining of micro-contributions, based on a bag-of-concepts approach by aggregating concepts extracted via the NCBO annotator from any micro-contributions. Following this approach leads to several issues, especially when dealing with qualities of the phenotypic characteristics. For the future, our plan is to filter the output of the annotated entities by organizing them into tensors expressing quality – phenotypic characteristic associations and then use the tensors to compute the weights of the domain concepts in the context of both local and global contributions of particular individuals. Moreover, we also plan to take into account the processes that the micro-contributions undergo during their lifespan, e.g., how many times they were altered, the extent of alteration and whether or not they were incorporated in the main disease description.

4 Preliminary Evaluation

We performed a preliminary usability study of the SKELETOME platform with a small group of eight experts from the community. The goal of the study was to compare the usability of SKELETOME against the two other "systems" currently used by the community, i.e., ESDN and Google mailing lists. At the same time we also wanted to understand how easy it is for the experts to adapt to using SKELETOME.

The evaluation consisted of two parts, with no training provided: (1) performing a series of operations on both SKELETOME and ESDN or Google mailing lists, and (2) completing a questionnaire about the usability of SKELETOME.

The tasks required to be performed for the first part were the following: (1) **Search & Browse:** search for a particular patient based on a given set

Table 1. System usability questionnaire

Question
I found SKELETOME easy to use
I found SKELETOME to be unnecessarily complex
I think I require technical support to be able to use SKELETOME
I found the various features of SKELETOME to be well integrated
I think most colleagues would learn SKELETOME quickly
I felt very confident using SKELETOME
I needed to learn a lot about SKELETOME before I could effectively use it

of phenotypic characteristics; (2) **Patient case manipulation:** upload and annotate a new patient case (i.e., clinical summary + 5 X-Ray reports); (3) **Collaborative diagnosis:** participate in the collaborative diagnosis process on a given patient; (4) **Domain knowledge manipulation:** modify the description of a bone dysplasia and add statements about it. This operation had to be performed only on SKELETOME (as the other systems do not have support for it). The experts were asked to pick one system in each category and motivate the choice by highlighting (via free text input) both the positive and the negative aspects.

The questionnaire required for the second part contained seven questions (see Table 1) with answers on a 5-point Likert scale ranging from "Strongly disagree" to "Strongly agree". These questions were adapted from the evaluation of the iCAT system [11] and from the System Usability Scale (SUS) [12].

The results of the evaluation were very positive. In the first part (choosing the "best" system), SKELETOME outperformed its opponents. In the **Search & Browse** task, the dynamically related items and the ontology-based faceted browsing and filtering of the search results were found to be very useful by all experts, and judged as critical missing aspects from the other systems. The **Patient case manipulation** task was the highlight of the evaluation, as all eight experts were particularly impressed by the possibility of annotating X-Rays and clinical summaries with domain concepts, and by the drag'n'drop functionality for uploading X-Rays. The third and last common task was found to be very similar to the ESDN functionality (50% voted for ESDN and 50% for SKELETOME), although the 5-star rating system was regarded as a positive addition by six out of eight experts. Finally, the fourth task (with no competition) was regarded as extremely useful to support their collaborative effort of maintaining and evolving the domain knowledge.

The questionnaire results were also positive. SKELETOME was found to be easy to use and well integrated by 87% of experts (7 out of 8), although 75% of experts did fell that they would require some time to learn to use it effectively. Functionalities, such as drag'n'drop or related items, made all the experts confident in using the platform. Some technical support was required with some users (37% of experts), however, it was only at the very beginning and did not influence the positive results of the questionnaire.

Overall, SKELETOME performed as we have expected, and indicated that Semantic Web technologies have the potential to make a real positive difference when applied in the right context and seamlessly built into familiar user interaction components.

5   Related Work

The SKELETOME platform is built on the work performed by the initiators and developers of the Drupal RDF extensions (described initially for Drupal 6 in [13]). Their continuous efforts, that resulted in support for RDF in core Drupal 7, are highly appreciated. However, as discussed earlier in the paper (see Sect. 3.2), in order to support a dynamically evolving skeletal dysplasia knowledge base, we required a different approach than the current top-down RDF mappings. Hence, we used the Drupal RDF core support, but developed our own set of modules, to be openly released to the community later this year.

The literature contains numerous descriptions of related systems, many using Wikis as their backbone. Semantic MediaWiki [14] was a pioneer in the area, being one of the first Wikis to embed semantic capabilities, packaged as extensions of MediaWiki. This led to its wide-spread adoption and application in many diverse domains. [15] is one example that adopted and applied the principles of Semantic MediaWiki. IkeWiki [16] on the other hand, is a stand-alone Semantic Wiki, providing similar functionality to Semantic MediaWiki and developed entirely in Java and AJAX.

Focusing on the biomedical domain, we identified BOWiki [17] and ConceptWiki[11] [18] as the most relevant with respect to our platform. BOWiki is a Semantic Wiki, designed for expert database curation, providing users with automated reasoning capabilities to verify the consistency of continuously added content to the knowledge base. ConceptWiki, or more concretely the WikiProteins part of it, is based on MediaWiki and enables experts to collaboratively curate knowledge about proteins. It incorporates several large knowledge bases, such as Gene Ontology, UMLS or Swissprot [19], to be used for the annotation and definition of terms, however, without providing a strict formalization of the knowledge or any reasoning support. The skeletal dysplasia curation aspect of the SKELETOME platform is reasonably similar to these approaches. However, while the generic goal of such Wikis is ontology engineering and population, SKELETOME extends this goal via the knowledge engineering cycle to learn new knowledge both from the growing pool of patient studies, as well as from the collaborative decision making process.

Another system of particular relevance is WikiGenes [7]. As opposed to typical Wikis, WikiGenes shifts the focus from creating knowledge to capturing the context of the knowledge via scientific artefacts (e.g., hypotheses or claims), in addition to fine-grained provenance. WikiGenes is special because it supports a reputation system for authors of the scientific artefacts based on their contributions to the field and their rating from other researchers. We adopted this

---

[11] http://conceptwiki.org/

concept and implemented it via the statements that can be added in conjunction to Bone Dysplasia concepts. However, in our case, these are more than mere conjectures, as they are (usually) supported by evidence emerging from the patient studies thus enabling one to track their evolution from the original patient data to the hypothesis at the generic conceptual level.

Outside of Wikis, the most notable and recent related effort is the custom WebProtégé [20] system built by the Stanford Center for Biomedical Informatics Research to help develop the 11th revision of the International Classification of Diseases (ICD-11) [11]. As with all the other tools in the Protégé suite, this system is specifically tailored towards efficient collaborative ontology engineering. As a result, in this respect it provides superior functionalities when compared to the Bone Dysplasia engineering aspect of SKELETOME. Nevertheless, we don't regard this as a negative point since it represents only one of the steps in the platform's knowledge engineering cycle. Additionally, in its current shape, SKELETOME serves its purpose of hiding the actual ontology evolution from the experts yet providing the mechanisms for keeping the knowledge up-to-date and enabling a shorter publishing cycle for the ISDS Nosology.

6   Conclusions

This paper describes the results to-date of our on-going effort in building a community-driven knowledge curation platform for the skeletal dysplasia domain. SKELETOME deploys an ontology-driven knowledge engineering cycle, aimed to support the evolution of the domain knowledge through the semantic enrichment of patient cases and via reasoning support that enables faster discovery of new knowledge and relationships. As the evaluation has shown, SKELETOME generates many benefits to the community and improves collaboration and knowledge exchange among the experts in the field. From a technical perspective, we believe that one of SKELETOME's main contributions is advancing the work started by Corlosquet et al. [13] in integrating Semantic Web technologies in the widely adopted Drupal CMS.

Future work on the platform will focus on developing novel mechanisms to support expertise modeling from micro-contributions and reasoning using fuzzy statements. From a functional perspective, we intend to integrate a notification mechanism, with personalized triggers on user-defined actions (e.g., notify me if anyone uploads a patient case that has these clinical attributes). In addition, to enhance the extent and ease of user interaction with the system, we plan to develop an email-based and iPhone app that enables the upload of clinical summaries and X-Ray reports.

Acknowledgments. The work presented in this paper is supported by the Australian Research Council (ARC) under the Linkage grant SKELETOME – LP100100156. The authors would like to thank Hasti Ziamatin and Razan Paul for their implementation support. Special thanks go to Tania Tudorache for her comprehensive and useful feed-back.

References

1. Warman, M.L., et al.: Nosology and Classification of Genetic Skeletal Disorders: 2010 revision. American Journal of Medical Genetics Part A 155(5), 943–968 (2011)
2. Hamosh, A., et al.: Online Mendelian Inheritance in Man (OMIM), a knowledge base of human genes and genetic disorders. Nucl. Acids Res. 33(1), 514–517 (2005)
3. Ashburner, M., et al.: Gene Ontology: Tool for the Unification of Biology. Nature Genetics 25(1), 25–29 (2000)
4. Bairoch, A., et al.: The Universal Protein Resource (UniProt). Nucleic Acids Research 33(1), 154–159 (2005)
5. Hartel, F.W., et al.: Modeling a description logic vocabulary for cancer research. Journal of Biomedical Informatics 38(2), 114–129 (2005)
6. Mabee, P.M., et al.: Phenotype ontologies: the bridge between genomics and evolution. Trends in Ecology and Evolution 22(7), 345–350 (2007)
7. Hoffmann, R.: A wiki for the life sciences where authorship matters. Nature Genetics 40, 1047–1051 (2008)
8. Gkoutos, G.V., et al.: Entity/Quality-Based Logical Definitions for the Human Skeletal Phenome using PATO. In: Proc. of the 31st Annual International Conference of the IEEE EMBS, Minneapolis, Minnesota, USA, pp. 7069–7072 (2009)
9. Jonquet, C., et al.: The open biomedical annotator. In: Proc. of the 2010 AMIA Summit of Translational Bioinformatics, San Francisco, California, US, pp. 56–60 (2010)
10. Dijkstra, E.W.: Selected Writings on Computing: A Personal Perspective. Springer, Heidelberg (1982)
11. Tudorache, T., et al.: Will Semantic Web Technologies Work for the Development of ICD-11? In: Patel-Schneider, P.F., Pan, Y., Hitzler, P., Mika, P., Zhang, L., Pan, J.Z., Horrocks, I., Glimm, B. (eds.) ISWC 2010, Part II. LNCS, vol. 6497, pp. 257–272. Springer, Heidelberg (2010)
12. Brooke, J.: SUS: a "quick and dirty" usability scale. In: Jordan, P.W., Thomas, B., Weerdmeester, B.A., McClelland, A.L. (eds.) Usability Evaluation in Industry, pp. 184–194. Taylor and Francis, London (1996)
13. Corlosquet, S., et al.: Produce and Consume Linked Data with Drupal! In: Bernstein, A., Karger, D.R., Heath, T., Feigenbaum, L., Maynard, D., Motta, E., Thirunarayan, K. (eds.) ISWC 2009. LNCS, vol. 5823, pp. 763–778. Springer, Heidelberg (2009)
14. Kroetzsch, M., et al.: Semantic Wikipedia. Journal of Web Semantics 5(4), 251–261 (2007)
15. He, S., et al.: Collaborative Authoring of Biomedical Terminologies Using A Semantic Wiki. In: Proc. of AMIA 2009 Symposium, San Francisco, California, US, pp. 234–238 (2009)
16. Schaffert, S.: IkeWiki: A Semantic Wiki for Collaborative Knowledge Management. In: Proc. of the 15th IEEE International Workshops on Enabling Technologies: Infrastructure for Collaborative Enterprises, Manchester, UK (2006)
17. Hoehndorf, R., et al.: BOWiki: an ontology-based wiki for annotation of data and integration of knowledge in biology. BMC Bioinformatics 10(S-5) (2009)
18. Giles, J.: Key biology databases go wiki. Nature 445, 691 (2007)
19. Boeckmann, B., et al.: The SWISS-PROT protein knowledgebase and its supplement TrEMBL in 2003. Nucleic Acids Res 31(1), 365–370 (2003)
20. Tudorache, T., et al.: Supporting Collaborative Ontology Development in Protégé. In: Sheth, A.P., Staab, S., Dean, M., Paolucci, M., Maynard, D., Finin, T., Thirunarayan, K. (eds.) ISWC 2008. LNCS, vol. 5318, pp. 17–32. Springer, Heidelberg (2008)

Cyber Scientific Test Language

Peter Haglich[1], Robert Grimshaw[2], Steven Wilder[2],
Marian Nodine[3], and Bryan Lyles[4]

[1] Lockheed Martin Advanced Technology Laboratories, Virginia Beach, VA
peter.haglich@lmco.com
[2] Lockheed Martin Advanced Technology Laboratories, Cherry Hill, NJ
{robert.s.grimshaw,steven.m.wilder}@lmco.com
[3] Telcordia Technologies, Austin, TX
nodine@research.telcordia.com
[4] Telcordia Technologies, Piscataway, NJ
jblyles@acm.org

Abstract. The Cyber Scientific Method (CSM) formalizes experimentation on computer systems, hardware, software, and networks on the U.S. National Cyber Range. This formalism provides rigor to cyber tests to ensure knowledge can be shared and experiment results can be viewed with confidence, knowing exactly what was tested under what conditions. Cyber Scientific Test Language (CSTL) is an ontology-based language for CSM experiments. CSTL describes test objectives, statistical experiment design, test network composition, sensor placement, and data analysis and visualization. CSTL represents CSM experiments throughout their lifecycle, from test design through detailed test network description, instrumentation and control network augmentation, testbed buildout, data collection, and analysis. The representation of this information in a formal ontology has several benefits. It enables use of general-purpose reasoners to query and recombine test specifications for rapidly building an experiment network and testbed on the range. Additionally, it facilitates knowledge management and retrieval of test procedures and results.

Keywords: Ontology, Computer Hardware and Software, Experimentation, Statistical Design of Experiments.

1 Introduction

This paper presents the ontology-based Cyber Scientific Test Language (CSTL) for the Cyber Scientific Method (CSM). The CSM was developed in conjunction with the Lockheed Martin team's implementation of the U.S. Defense Advanced Research Projects Agency (DARPA) National Cyber Range (NCR) program. CSM provides a discipline for rigorous experimental design in the performance of tests of computer hardware, software, systems, and networks. The CSM and CSTL enable the formal description of the system under test (SUT), the test instrumentation and environment, experiment assumptions and objectives, and test results. CSTL serves as a lingua franca for cyber test representation throughout the entire experiment planning, execution, and analysis cycle. During experiment planning, CSTL helps the Test Scientist

L. Aroyo et al. (Eds.): ISWC 2011, Part II, LNCS 7032, pp. 97–111, 2011.
© Springer-Verlag Berlin Heidelberg 2011

to develop a statistical design for the experiment. As the experiment is executed, range software uses the CSTL test specification to automatically build the testbed and to perform the experiment via the insertion of test events. After the test has been conducted, experiment results expressed in CSTL are processed and are available for later review.

CSTL consists of a modular ontology library expressed in the OWL 2 Web Ontology Language [1]. We chose OWL 2 for several reasons, including our team's familiarity with OWL and OWL editing tools. We also wanted to leverage OWL 2 language features such as property chains. Finally, because NCR is a multi-year project we wanted to start with the newest version of OWL.

Our design approach for CSTL was driven by the overall system-engineering plan for NCR. Our ontology development was grounded in the need to express information related to test design, execution, and analysis and testbed buildout and sanitization. We validated our design through the representation of examples in CSTL, reviewed by subject matter experts and NCR component developers. Later in the ontology development phase we instituted a configuration control board to review and approve proposed changes to CSTL.

The major components of CSTL cover:

- **Experimental Design:** This segment of CSTL describes ideas used in designing an experiment, including the statistical design. Topics covered here include objectives, hypotheses, experiment type, treatments, factors, and response variables.

- **Test Planning and Administration:** A small portion of CSTL provides terms to describe some of the administrative details of planning an experiment, such as facility scheduling and experiment sponsorship.

- **Network Descriptions:** An important module within CSTL describes computer networks in terms of their topology. These terms include links, nodes, and subnets.

- **Asset Descriptions:** Terms for representing hardware and software assets form a major coherent and independent module within CSTL. This asset description ontology is explained more fully in [2].

- **Test Execution:** CSTL supports the automated transition of test specifications to a testbed during experiment buildout via transformations to the Network Simulator (ns-2) language [3]. Furthermore, CSTL describes automatic network traffic generation and test event execution subsystems for use during the performance of the experiment.

- **Data Analysis and Visualization:** A segment of CSTL is used to represent the instrumentation of the testbed with sensors and the associated metrics to be computed. This segment also describes the data archiving and analysis process.

- **Templates:** In cyber testing it is often desirable to define the SUT using multiple clones of identical individuals. CSTL provides a method for defining and instantiating templates to provide this capability. Templates are also used in incorporating predefined control structures such as traffic generation into the testbed.

We will provide more detail on each of these topic areas.

2   Experiment Design and Planning in CSTL

A major goal for the CSM was the unification of test and evaluation methodology with computer and network system descriptions. Statistical experiment design formalizes the definition of an experiment through the identification of response variables, treatment variables, and other experimental factors. In the CSTL ontology library these concepts are captured in a separable design of experiments (DoEx) ontology module, *designofexperiments.owl*, which is applicable to other types of test and evaluation. Note that this segment of CSTL can be used to describe general statistical experiment design, even outside of the cyber domain.

In this section we will describe the major concepts in this module.

2.1   Experiment Types

The DoEx framework in CSTL includes ontology classes for four experiment types. The most general is the Discovery Experiment, in which testers observe the system under a variety of conditions. Discovery experiments often include humans in the loop. A second type of experiment is the Diagnostic Experiment, typically used for troubleshooting the behavior of a malfunctioning system or network. A third type of experiment is the Benchmark Experiment, whereby network or system performance characteristics are measured. The fourth type of experiment is the Statistically Designed Experiment. The remainder of this section will summarize the major concepts related to this type of experiment. The representation and process described herein is based on a Job Aid developed by Sandia National Laboratories [4] to assist researchers with DoEx.

2.2   Objectives and Hypotheses

At the start of experiment planning, it is good to state the objective or purpose of the experiment. Along with the objective, statistically designed experiments address a quantitative hypothesis based on a test statistic, which is a function of the experiment observations. In DoEx this hypothesis is expressed in terms of a null hypothesis and an alternative hypothesis. Most commonly, the null hypothesis represents the status quo and the alternative hypothesis represents the effect that is being investigated. For example, in an experiment to examine the effectiveness of an update to an anti-virus program, the null hypothesis could be that the new version has the same probability of allowing infection as the old version. An alternative hypothesis could be that the new program is 90% more effective than the old program.

The DoEx ontology includes classes for Objective, Hypothesis, and Test Statistic.

2.3   Treatments, Factors, and Response Variables

DoEx is designed to support the efficient planning of a set of experiment runs for data collection to determine the degree of evidence to reject the null hypothesis. This is based on the value of a test statistic computed from observations of response variable values during repetitions of the experiment, or runs. For the anti-virus example, the test statistic might be the number of infections observed divided by the number of tests run on each version. Statistical tests can use the values of the test statistic for the original version and the updated version in order to recommend rejection of the null hypothesis in favor of the alternative hypothesis.

The response variables are assumed to have a dependency on one or more input variables or experimental factors. In fact, the objective of the experiment is usually related to determining the response of the system to a subset of the input variables, known as treatments. The other input variables are also important, and need to be controlled or measured for later analysis.

In the cyber domain, these input variables represent system or network characteristics (e.g., available bandwidth), capacity (e.g., number of connected hosts), or configurations (e.g., which variant of Windows is installed on experiment hosts). In our anti-virus example, the fundamental treatment is the version of the anti-virus program used. An example of a non-treatment experimental factor to be controlled in this scenario is the service pack level of the underlying host operating system.

2.4 Designs

At an early stage of designing the experiment, the test scientist(s) may use a brainstorming process to identify potential treatment and factor variables for inclusion in the experiment, along with response variables. In CSTL we refer to this identification of variables as a proto-design. After further consideration the scientist will establish a final set of response, treatment, and factor variables in the experiment design. CSTL allows the scientist to tag the variables considered during brainstorming as being accepted for inclusion in the final design or as being excluded. This allows a later review of the test to determine which variables were considered but not included. For each accepted treatment and controlled factor variable, he will also specify the potential setting values. In our example there are only two setting values for the anti-virus treatment {'New Version', 'Old Version'}. If the underlying operating system is Windows XP there might be four setting values for service pack level {'SP0', 'SP1', 'SP2', 'SP3'}. A set of treatments or controlled factors with their setting values is called a treatment combination [6]. For example, combining the new version of the anti-virus software with Windows XP SP1 would produce one of the eight possible unique treatment combinations. Each run in the experiment is associated with a treatment combination. In the DoEx ontology the Experiment Design individual is linked to its response, input, and factor variables and its treatment combinations.

The top-level concepts in the DoEx ontology are shown in Fig. 1.[1]

---

[1] The figures in this paper are concept maps of ontology classes (rounded rectangular nodes), properties (labeled links), and individuals (rectangular nodes). These concept maps were developed using CMap Tools [5].

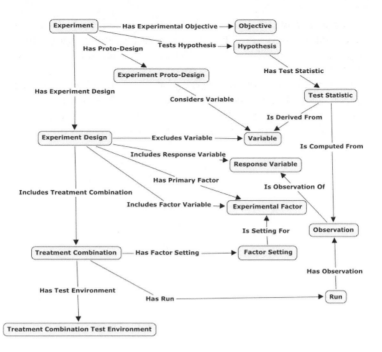

Fig. 1. DoEx Ontology

3   Test Environment

Each treatment combination defines a unique (up to isomorphism) test environment that is used for all runs associated with that treatment combination. This environment is represented in the CSTL ontology by the Treatment Combination Test Environment class. This class serves as a bridge between the statistical experiment design structure of the experiment and the system under test as embedded in the test. Individuals of the Treatment Combination Test Environment class are linked to two major individuals that further describe the test environment. These are members of the Network and Cyber Scientific Test Procedure ontology classes, each of which will be described in this section.

3.1   Network Topology Concepts

The *networktopology.owl* module provides extensions to the basic CSTL network concepts. The fundamental network concept is the Network Topology, which can describe either the logical topology or the physical topology of the network. The basic model for the network topology is a graph defined by a set of links and nodes. In most cases these links are symmetric but the network topology ontology does include terms to model asymmetric networks. Links are associated with nodes via the "Has Endpoint" property. Nodes may be directly asserted to be members of the network topology, but it is preferable to infer that the nodes belong to the topology using the property chain **Has Link** composed with **Has Endpoint** implies **Has Node**. Additional link and node properties such as bandwidth, latency, addressing, and subnet masks are also defined in the network topology ontology. The current version of this ontology focuses on IPv4 terminology that is compatible with ns, but does provide for extensions to other addressing protocols.

The nodes and links provide the topological view of the network. Each node may also be associated with a device such as a computer host or a routing device. The network topology ontology provides object properties to make these associations.

A simple view of the network topology ontology is shown in Fig. 2.

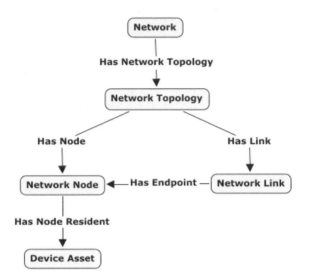

Fig. 2. Simple Network Topology

3.2 Test Procedures

The test environment is also linked to an individual from the Cyber Scientific Test Procedure class. This class serves as a collector for test instructions and event execution scripts. Test instructions can include written instructions for those aspects of the test that are not automated, or ground rules for human-in-the-loop experiments. Event execution scripts are described in Section 5.

4 Asset Description

A major portion of the CSTL ontology library is the collection of concepts that describe hardware and software assets. The asset description ontology included in CSTL is described at length in [2]. In this section we summarize the use of asset descriptions in CSTL.

4.1 Hardware and Software Assets

Repeatable experiment specification requires precise definition of the system under test and the testbed articles. For computers and other devices, a full description includes identification of hardware components and installed software. Assets are described nominally (make and model) and according to their functions. Software asset function descriptions also specify installation and runtime dependencies and compatibility with other software and hardware.

4.2   Recipes, Images, and Asset Controllers

A goal of the NCR is automation of cyber experiments so that they can be performed quickly and exactly. This is important for efficiency in large-scale experiments involving large numbers of nodes and links, treatment and factor variables, or runs per treatment combination. Leveraging previous experience with Emulab [7], CSTL includes terms to describe the recipes used to build and configure host assets using disk images and asset controllers. Asset controllers are auxiliary software assets that can install and configure software on hosts. Disk images and partitions are prebuilt and copied to the experiment host. Host recipes coordinate and specify the building of host assets from images and asset controllers. The recipe optionally starts with a disk image loader which identifies the image that is to be copied onto the host and which partition is to be booted. The recipe then specifies an ordered list of recipe steps, each of which points to an asset controller and provides parameter value assignments for items such as license keys or installation directories. During testbed buildout, experiment control software builds the hosts according to the information contained in the host recipe.

Figure 3 illustrates the classes and object properties that describe the building of host assets from recipes via images and asset controllers.

Given a recipe, a reasoner can determine what software asset types are installed on each host in the experiment by following property chains using the object properties shown above. There are two property chains that provide this information, represented by paths from Host Asset to Software Asset or Software Asset Type respectively. Absent a reasoner, this information can also be computed by joining SPARQL queries over an RDF graph.

5   Test Execution

CSTL includes terms to describe the control structures and events involved during the performance of experiment runs. These include execution scripts, event coordination, and production of network traffic.

5.1   Event Execution Language

The Lockheed Martin implementation of NCR features a scripting language for the execution of test events. This Event Execution Language (EEL) is not directly ontology-based. Rather, CSTL encapsulates EEL scripts and can express script arguments. Each test procedure may refer to a set of EEL scripts for testbed buildout, validation, test execution, data archiving, post-test analysis, and testbed sanitization associated with it. Additionally, EEL scripts may be passed argument values in a manner analogous to the manner of providing parameters for asset controllers described in Section 4.2. So, while EEL is mostly opaque to CSTL, CSTL still provides ways to link to appropriate EEL content.

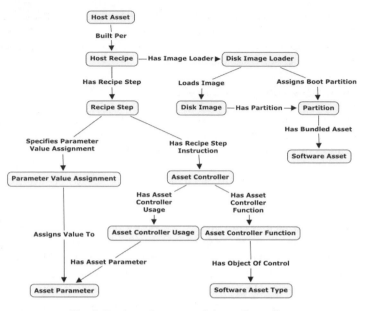

Fig. 3. Recipes, Images, and Asset Controllers

5.2 Test Execution Control Infrastructure

To ensure that test events are triggered automatically, in a synchronized fashion, the testbed needs to provide control infrastructure that interacts with the system under test but is not part of the system under test. The test designer does not usually need to specify the details for the control network at test specification. However, CSTL does need to be able to describe the control infrastructure for two reasons. First, elements of the system under test need to be identified as connection points for the control infrastructure. Second, after the test has been performed it is important to record all of the elements of the testbed and the system under test for provenance and experiment repeatability.

The Lockheed Martin implementation of NCR includes deployment of a test event timing execution server (TET Server) and test management control/view application (TMVC) for test directors. CSTL provides descriptions for these as well as test event execution software added to the hosts in the system under test.

5.3 Artificial Network Traffic Generation

In order to run a cyber experiment that represents real network usage without requiring humans in the loop, it is necessary to generate network traffic artificially. As used by Lockheed Martin in NCR, artificial traffic generation is managed by a node in the control network, built according to a pre-defined recipe expressed in CSTL. Hosts in the system under test are also provided with additional instrumentation software to support the generation of web browsing or email traffic.

Similarly to TET and TMVC, the traffic generation control node and additional experiment host software are described in CSTL and included with the full description of the experiment after execution.

6 Data Analysis and Visualization

Collecting the appropriate data, analyzing it, and visualizing it are important aspects of experimentation. The nature of cyber experimentation is such that large-scale experiments with a myriad of potential data elements are possible. In order to keep up with the pace of the experiment and to ensure all of the desired data is collected, it is imperative that the test specification include descriptions for the sensors and data to be collected.

6.1 Sensors and Metrics

Many of the sensors that are used to collect experiment data are themselves software applications, resident on either the hosts in the system under test or on the control network. When software sensors are installed on experiment hosts there is a slight potential for interference with the experiment itself. For example, a CPU load-monitoring sensor contributes in a small way to the load on the CPU. Therefore, to facilitate scientific rigor and provenance, it is important to capture details about the sensors and their operation during the experiment. Additionally, the sensors need to be controlled by the test event subsystem using the EEL language. For this reason the CSTL representation for sensors includes descriptions of the EEL statements used to control the sensor. The sensor description also includes asset descriptions for the sensor assets as discussed in Section 4.1.

The sensors provide the data observations. Normally, the raw observations are only useful to the extent that they can be used to compute metrics describing the performance of the system under test. These metrics are computed using the data analysis subsystem described in Section 6.2. The test scientist selects the metrics. CSTL describes these metrics in terms of the sensors from which they receive input and other parameters.

6.2 Data Archiving and Analysis Subsystem

To support data archiving and analysis, the experiment control network includes a Data Analytics Supercomputer (DAS) developed by Lexis-Nexis [8]. As part of experiment definition the test scientist also needs to provide some details for DAS operation, including data archiving location and frequency and metric computation intervals. For provenance reasons this is also encoded in CSTL and saved with the experiment description.

7 Templates

There are two major cases where cyber experiment descriptions require the description of individuals that are identical modulo a small set of data property values or object property linkages. The first major case occurs most frequently during scalability testing, where a large number of identical hosts are added to the system under test. The second major case occurs with the addition of control network infrastructure to the testbed. To support these major cases, CSTL provides for the creation, application, and instantiation of templates. The idea behind a template is that it provides the instructions needed to create a set of individuals that can be linked into the Resource Description Framework (RDF) graph of ontology individuals that describes the test specification. In some cases, the graph of the template looks like the graph of the individuals generated from it, with Internationalized Resource Identifiers (IRIs) assigned and data property values provided. The specification of a template includes object parameters, data parameters, invariant parameters, and multiplicity values.

The framework for templates that we developed for CSTL is not limited to use in cyber test specification. Thus, the same approach would work in other domains. A detailed exposition of our template approach will appear in a future paper. In the remainder of this section we will discuss the main ways templates are used in CSTL.

7.1 Templates for Cloning

One common scenario in cyber testing is the connection of multiple hosts to a switch or router in a star network topology as part of the system under test. The hosts may be identical except for host name and IP address assignment. That is, they may all be built from the same recipe or disk image. The experiment may vary the number of hosts for purposes of scalability testing. The template for this network would feature:

- Object parameters for the hosts, network nodes, and network links.
- Invariant (singleton) object parameters for the host recipe, the switch and its network node.
- Data parameters for switch, host, node, and link characteristics.
- A multiplicity parameter for the number of hosts desired, expressed as a data property value on the host node object parameter individual.

The template is illustrated in Fig. 4. An instantiation of the template is shown in Fig. 5.

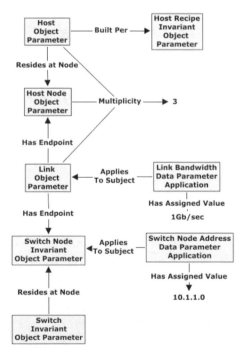

Fig. 4. Example Star Topology Template

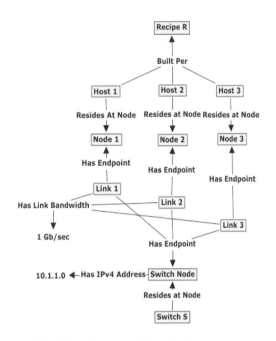

Fig. 5. Star Topology Template Instantiated

The instantiation of the template produced 12 new individuals, 11 object property assertions linking them, and four data property assertions. The resulting network has three host nodes connected to a single switch node. Different values for the multiplicity parameter would have produced different numbers of hosts, host nodes, and links. Tying this back to the design of experiment considerations discussed in Section 2.3, one can envision an experiment design with two factors: the number of hosts in the network and the software recipe used to build each of the hosts. For example, one of the treatment combinations might use seven hosts and recipe "B". Another might use five hosts and recipe "A". The setting values would be used as parameter assignments by the template to quickly define the system under test for a given run.

7.2 Templates for Test Infrastructure

The second major motivating case for using templates is the use of predefined test support assets in the experiment. For example, artificial traffic generation requires the installation of client software on hosts in the system under test. In Fig. 6 we see an example of such a template. Traffic generation requires software added to the experiment hosts to control Thunderbird for email and Internet Explorer for web browsing. The template is applied to the system under test by passing IRIs for the hosts and their recipes as arguments. The recipe extension individual is created and linked to the recipe steps and the original recipe for the host via appropriate object properties.

The Lockheed Martin specific portion of the CSTL ontology library includes similar templates for traffic generation, test execution, test management view, and sensors.

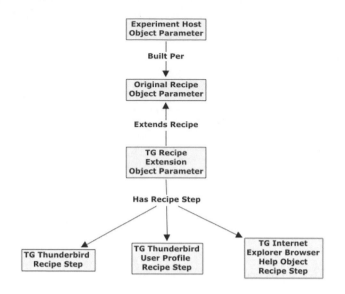

Fig. 6. Traffic Generation Support Template

8   Sparse Descriptions

In cyber testing it is not desirable to require a test scientist to provide exact descriptions of all aspects of the elements of the system under test. For example, the test scientist may not care about the exact clock speed of host CPUs above some minimum threshold. Furthermore, range scheduling is more flexible when reasonable substitute assets can be used in experiments. For this reason CSTL has adopted the notion of sparse description of test assets. A sparse description defines a class expression on the set of available asset types. It may be associated with a query on the inventory of assets at the range to determine a set of acceptable assets for use in the experiment. This allows the test scientist to focus on the characteristics of greatest importance while being flexible in other characteristics. It also allows the range facility operator to apply additional business rules in scheduling range assets while meeting the needs of the test scientists.

As we studied the situations where sparse descriptions were desired we realized that there are two fundamental kinds of sparseness. The first kind we refer to as *categorical sparseness*. The simplest form of categorical sparseness is a named ontology class. (For example, "x86 Host Asset Type".) More complicated categorical sparseness might include existential restrictions on object properties defined on a base class. (Example: "Software providing the NTP Client Function"). The second kind of sparseness we refer to as *range-constrained sparseness*. This kind of sparseness focuses on values for data properties and whether they fall in a particular range. An example might be "Processor with clock speed no less than 2.0 GHz".

Sparse descriptions correspond naturally with queries against the knowledge base to list the individuals that fit the sparse descriptions. For example, in the SPARQL [9] query language, data property value range constraints correspond to FILTER statements in queries.

Please note that while the idea of a sparse description was motivated by some of the challenges of cyber test design and planning, the idea can be generalized to other domains.

9 Transforms

While CSTL can be used to represent most of the details of a cyber experiment, there are some components of the NCR that require different input formats. Also, there are elements of the experiment lifecycle that require code to perform transformations on the test specification. For this reason the Lockheed Martin team has developed a set of transformations that operate on CSTL to produce either modified CSTL or other output formats. These transformations are performed by custom Java code using the Jena OWL Microreasoner [10]. Some of the CSTL transforms are:

- **NS Translator:** This transform takes a CSTL network description in OWL and produces a network description in ns format, suitable for use by Emulab.
- **Treatment Combination Expansion Transform:** This transform creates treatment combination test environments from a basic test environment and the collection of factor settings from the experiment design. The input and output are both expressed in CSTL.
- **Recipe Infrastructure Transform:** This transform takes host recipes as inputs and extends them to add several pieces of software needed for test instrumentation. The input and output of the transform are expressed in CSTL.
- **Template Instantiator:** As the name suggests, this transform takes a template specification in CSTL along with parameter value assignments and produces the resulting individuals in a similar manner to that discussed in Sections 7.1 and 7.2. The output format is CSTL.

10 Query Library

In addition to the CSTL ontology library, we have developed a library of knowledge base queries in the SPARQL language. These queries are used by components of NCR to retrieve germane information. For example, tools to automatically build the system under test in the testbed perform queries against the test specification to determine which disk images and installers to use. The query library contains many queries that can be used by any tool that can execute SPARQL queries. An additional use of the query library is in regression testing. A collection of sample test specifications serves as an adjunct extension to the main CSTL ontology library. Whenever changes are made to CSTL we can run the query library and compare it with a baseline set of results to determine the impact of the ontology changes. We can examine this impact from the standpoint of APIs for the NCR software components that use the query. We also have a set of sanity check queries that allow us to look for missing or correct information that might result from an ontology change.

11 Challenges

The development of the CSTL ontology library presented a number of challenges. These are categorized and discussed below.

11.1 Reasoner Support and OWL 2 Features

The CSTL ontology library is fairly complex. We wished to leverage OWL 2 features such as property chains to express natural domain axioms. We mentioned some of these earlier in the paper. Unfortunately, the available reasoners that could compute property chains were unable to reason on the whole of the CSTL ontology library in a satisfactory time. Therefore we had to suspend the use of property chains and require tools to make assertions of statements that would have been derivable using property chains. We are hopeful that as reasoners get more capable we will be able to return to using these advanced features in the future.

11.2 Second-Order Reasoning and Punning

There were a few cases where the design of CSTL had the potential to require second order reasoning. One of these areas, involving assets, asset types, and classes of asset types, is discussed in [2]. The second case arose in the development of templates, specifically in the application of data parameters. The description of how a template data parameter is to be applied needs to identify the appropriate data property in the ontology that provides the predicate for the assertion (subject, predicate, assigned data value). We did not want to use an object property that took the data property as a value, as this would cause difficulty in reasoning. To work around this, we use the IRI for the data property as a data value, which provides enough information to the template instantiator.

12 Summary

This paper provided an overview of the Cyber Scientific Test Language (CSTL) ontology library as used by the Lockheed Martin team for the National Cyber Range (NCR). CSTL can be used to describe experiment design, the computer systems and networks under test, test instrumentation, test planning and administration details, and data analysis. Test specification and reports expressed in CSTL facilitate knowledge management and provenance of results. CSTL includes approaches for the use of templates and sparse descriptions that are broadly applicable outside of the cyber domain.

The Lockheed Martin implementation of NCR includes custom code for transforming CSTL, with output to CSTL or other languages such as ns. An extensive query library is used as an API for NCR components and for regression testing. The implementation of a knowledge base for NCR needed to overcome or work around some challenges that limited our ability to use the full expressivity of OWL 2. We hope to overcome this in the future.

Acknowledgment. The United States Defense Advanced Research Projects Agency (DARPA) funded this research and development activity under contract HR0011-09-C-0042. The views expressed are those of the author and do not reflect the official policy or position of the Department of Defense or the U.S. Government. This is in accordance with DoDI 5230.29, January 8, 2009. Distribution Statement "A" (Approved for Public Release, Distribution Unlimited).

References

1. OWL 2 Web Ontology Language Document Overview (2009), `http://www.w3.org/TR/owl2-overview/` (last accessed on August 19, 2011)
2. Nodine, M., Grimshaw, R., Haglich, P., Wilder, S., Lyles, B.: Computational Asset Description for Cyber Experiment Support using OWL. Submitted to International Conference on Semantic Computing 2011 (2011)
3. The Network Simulator – ns-2 (2010), `http://www.isi.edu/nsnam/ns/` (last accessed on August 19, 2011)
4. Halbleib, L.L., Crowder, S.V.: Teaching Design of Experiments Using a Job Aid and Minitab, `http://www.minitab.com/en-US/uploadedFiles/Shared_Resources/Documents/Articles/Crowder.pdf` (last accessed on August 19, 2011)
5. Eskridge, T., Hayes, P., Hoffman, R., Warren, M.: Formalizing the Informal: A Confluence of Concept Mapping and the Semantic Web. In: Cañas, A.J., Novak, J.D. (eds.) Concept Maps: Theory, Methodology, Technology. Proceedings of the Second International Conference on Concept Mapping, vol. 1. Universidad de Costa Rica, San Jose (2006)
6. NIST/SEMATECH e-Handbook of Statistical Methods (2010), `http://www.itl.nist.gov/div898/handbook/pri/section1/pri1.htm` (last accessed on August 19, 2011)
7. Emulab.Net – Emulab – Network Emulation Testbed Home (2011), `http://www.emulab.net/` (last accessed on August 19, 2011)
8. Data Analytics Supercomputer – LexisNexis (2011), `http://www.lexisnexis.com/government/solutions/data-analytics/supercomputer.aspx` (last accessed on August 19, 2011)
9. W3C. SPARQL Query Language for RDF (2008), `http://www.w3.org/TR/rdf-sparql-query/` (last accessed on August 19, 2011)
10. Reynolds, D.: Jena 2 Inference Support (2010), `http://jena.sourceforge.net/inference/#owl` (last accessed on August 19, 2011)

How to "Make a Bridge to the New Town" Using OntoAccess

Matthias Hert, Giacomo Ghezzi, Michael Würsch, and Harald C. Gall

s.e.a.l. – software architecture and evolution lab
Department of Informatics
University of Zurich, Switzerland
{hert,ghezzi,wuersch,gall}@ifi.uzh.ch

Abstract. Business-critical legacy applications often rely on relational databases to sustain daily operations. Introducing Semantic Web technology in newly developed systems is often difficult, as these systems need to run in tandem with their predecessors and cooperatively read and update existing data.

A common pattern is to incrementally migrate data from a legacy system to its successor by running the new system in parallel, with a data bridge in between. Existing approaches that can be deployed as a data bridge in theory, restrict Semantic Web-enabled applications to read legacy data in practice, disallowing update operations completely.

This paper explains how our RDB-to-RDF platform ONTOACCESS can be used to transition legacy systems into Semantic Web-enabled applications. By means of a case study, we exemplify how we successfully made a bridge between one of our own large-scale legacy systems and its long-term replacement. We elaborate on challenges we faced during the migration process and how we were able to overcome them.

1 Introduction

The field of software engineering is in a constant state of flux. New paradigms, programming languages, frameworks, and tools gain tremendous momentum all of a sudden – and then they sink into oblivion as quickly as they have emerged. Short time-to-marked intervals are therefore critical for the success of new tools, be it in an industrial context or in research.

In contrast to short-lived tools, the body of acquired knowledge of a company usually evolves less rapidly and sometimes even remains relevant for decades, stored as data in different, mostly relational databases (RDB). This inevitably leads to challenges, when different generations of applications have to operate on this data.

There are legacy systems relying on a relational view of the database—these applications can not easily be upgraded or simply taken offline and thrown away when requirements change. Legacy systems are often crucial for daily operations and therefore need to be highly available. They are inherently valuable to many organizations but bear typical problems: Maintenance and especially further development have become difficult and costly.

L. Aroyo et al. (Eds.): ISWC 2011, Part II, LNCS 7032, pp. 112–127, 2011.
© Springer-Verlag Berlin Heidelberg 2011

These circumstances and new business opportunities emerging with the advent of paradigms, such as Service-Oriented Architectures and the Semantic Web, lead to the development of next-generation systems. While new development opens the door for incorporating recent best-practices and state-of-the-art technologies, the newly developed applications usually will run in tandem with legacy systems and still need to access legacy databases.

In such scenarios, it is common to *make a bridge to the new town*, that is, to incrementally migrate data from a legacy system by running the new system in parallel, with a data bridge in between [8]. Tools such as D2R Server [4] and OpenLink Virtuoso [9] serve RDF views on relational databases. However, they restrict Semantic Web-enabled applications to read legacy data, disallowing update operations completely.

In this paper, we describe how our RDB-to-RDF platform ONTOACCESS [19] can be used to facilitate the transition from legacy systems to Semantic Web-enabled applications in practice. ONTOACCESS provides a semantic layer on top of existing relational databases. It enables RDF-based read and write access to relational data. Based on mappings that bridge the conceptual gap between RDF and the relational model, a mediator translates Semantic Web requests on-the-fly to SQL. This enables relational and RDF-based applications to cooperate on the same data and to further exploit the advantages of the well established database technology such as query performance, scalability, transaction support, and security.

The contribution of this paper is a case study on how we successfully used ONTOACCESS to advance our Eclipse-based software evolution analysis framework EVOLIZER [13] to SOFAS [17], a service-oriented, distributed, and collaborative software analysis platform. We describe use cases where existing RDB-to-RDF approaches are insufficient and an approach such as ONTOACCESS is needed.

The remainder of this paper is structured as follows. Section 2 gives a brief introduction to the two systems between which ONTOACCESS acts as a data-bridge: EVOLIZER and SOFAS. ONTOACCESS itself bridges the conceptual gap between the relational model and RDF. It is described in Section 3. In Section 4, we present the case study on how we successfully used ONTOACCESS to advance EVOLIZER to SOFAS. Related work in the context of RDB-to-RDF mapping is reviewed in Section 5. Section 6 concludes this paper with a summary.

2   Background

In this section, we describe our two platforms for software analysis that run in tandem and are able to share data thanks to ONTOACCESS. The first platform, EVOLIZER, is considered to be a legacy system, whereas SOFAS represents our latest ambitions in providing a scalable, distributed means to analyze the evolution of a software system.

2.1 Evolizer

In the past, we have developed EVOLIZER [13] – a plug-in-based software evolution analysis and research platform, tightly woven into the Eclipse IDE.

At its core, EVOLIZER is based on the idea of a *Release History Database (RHDB)* [11]. It is implemented as a set of Eclipse plug-ins and integrates information originating from different software repositories, such as version control, issue tracking, mailing lists, etc. The combination of this diverse, yet interconnected data allows one to uncover and analyze the many different facets of the evolution of a software system and its parts. Examples are the system's fine-grained change history or bug-proneness over time, as well as a complete source code model.

EVOLIZER has become a typical legacy system over time: While the platform is still in active use, it becomes harder to adapt it to new requirements and recent advances in technology. The tight coupling to Eclipse makes it hard to adapt and re-use EVOLIZER's tools and algorithms in new environments such as in a service-oriented context. Further, the RHDB is based on classical relational database technology. It is therefore difficult to interlink information stored in the RHDB with other external data sources, because the relations that we store are local – not universal – and our entities lack unique resource identifiers that can be dereferenced over the Internet. Synergies with related approaches are therefore difficult to exploit. EVOLIZER's models also lack explicit semantics, such as cardinality, transitivity, symmetry, and so on. Bringing EVOLIZER and its RHDB to the Semantic Web would therefore be desirable.

EVOLIZER is still very valuable and our RHDB contains data about the software life-cycle of hundreds of systems. Re-importing this vast amount of data from version control and bug tracking systems would take months, and some of these repositories might not even be available online anymore.

To overcome these limitations, we are in need of a gradual migration path from EVOLIZER to the next generation of software evolution analysis platforms, allowing us to run the existing platform together with its replacement for the years to come.

2.2 SOFAS

Evolizer allowed us to combine and analyze different aspects of a software's evolution and its development. However, we realized that a big potential lies in having analyses easily accessible and composable, without platform and language limitations, and not having to install and configure particular tools. Based on these premises, we introduced the concept of "Software Analysis as a Service" [16]: getting easy access to different analyses from various tools and providers using Web services. We implemented that concept into a lightweight and flexible platform called SOFAS (SOFtware Analysis Services) [17].

SOFAS follows the principles of a RESTful architecture [10] and allows for a simple yet effective provisioning and use of software analyses based upon the principles of Representational State Transfer around resources on the Web. Its architecture is made up by three main constituents: Software Analysis Web Services, a Software Analysis Broker, and Software Analysis Ontologies. The services expose their functionalities and data through standard RESTful Web service interfaces. The Software Analysis Broker acts as the service manager and provides the interface between the services and the users. It contains a catalog of all the registered analysis services with respect to a specific software analysis taxonomy. As such, the domain of analysis services is described in a semantic way, enabling users to browse and search for their analysis service of interest. The ontologies – we call them SEON (*cf.* Section 4.2) – are used to define and represent the RDF data consumed and produced by the different services.

3   OntoAccess as a Bridge to the New Town

ONTOACCESS is a RDB-to-RDF mediation platform that enables Semantic Web-based applications to operate on relational data. It provides a semantic layer on top of existing relational databases to enable RDF-based read and write access to the relational data. Semantic Web requests, *i.e.*, query and update requests, are translated on-the-fly to SQL for execution in the database. ONTOACCESS therefore eliminates the need for mirroring and synchronizing the relational data and the RDF representation – both data models always operate on the same state of the data. This results in a cooperative use of the data in RDF-based as well as in relational applications. In addition, mediation allows one to further exploit the advantages of the well-established database technology such as query performance, scalability, transaction support, and security. The existing, read-only RDB-to-RDF mapping approaches are limited to data warehouse-like applications where the data can be queried and analyzed but not modified. In comparison, ONTOACCESS puts relational databases on par with native RDF triple stores by allowing read and write access to the data. This facilitates the transition from RDB-based legacy systems to Semantic Web-enabled applications in practice.

3.1   Architectural Principles of OntoAccess

ONTOACCESS is designed and implemented as an extensible platform. It encapsulates the RDB-to-RDF translation logic in the core layer which provides the foundation for an extensible set of data access interfaces in the interface layer. The RDB-to-RDF core is responsible for the translation of RDF-based request to SQL and interacts with the database system. The interface layer exposes the functionality of the ONTOACCESS core to RDF-based applications via different data access approaches. It translates the interface-specific operations to the basic ONTOACCESS operations, and results back into the interface-specific format. This facilitates the development of additional data access interfaces because the main RDB-to-RDF translation work is performed in the core layer. Currently, the ONTOACCESS platform supports data access via SPARQL [23], SPARQL/Update [24], Linked Data [2], and various Semantic Web Frameworks,

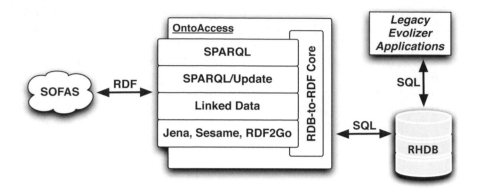

Fig. 1. OntoAccess Architecture Overview

such as Jena, Sesame, and RDF2Go.[1] The data access interfaces are accessible via a HTTP service endpoint if deployed as a stand-alone server. Alternatively, ONTOACCESS can be integrated into other applications as a library, in which case the data access interfaces are exposed via specific APIs. Figure 1 presents an overview of the overall architecture of ONTOACCESS and exemplifies how it can be used as a data-bridge in the context of SOFAS and the EVOLIZER RHDB. Next, before we discuss this example in detail in Section 4, we elaborate on the mapping principles of ONTOACCESS.

3.2   Mapping Principles of OntoAccess

Mediation requires a mapping from concepts in a relational database schema to terms defined in an ontology. For ONTOACCESS, we developed R3M [20] as a bidirectional RDB-to-RDF mapping language that incorporates the requirements of RDF-based write access to relational databases. Existing mapping languages developed for read-only use cases are unsuitable for write access as shown in [15] and by D2R/Update.[2]

R3M extends the mapping approach described in [3]. Tables of the database schema are mapped to classes in an ontology and the attributes of those tables to properties. Special support is provided for *link tables* that are used to represent M:N relationships in the relational model. As such helper constructs are not needed in RDF, link tables are mapped to properties instead of classes. In addition, R3M mappings contain information about datatypes, as well as integrity constraints of the database schema. This results in a mapping language that is not as expressive as the existing, read-only languages (*cf.* [21]) but it is sufficient to cover many application scenarios, including the one presented in this paper. In general, R3M is targeted at mapping highly normalized relational database schemata such as the ones generated by object-relational mappers

---

[1] http://openjena.org/, http://openrdf.org/, http://rdf2go.semweb4j.org/
[2] http://d2rqupdate.cs.technion.ac.il/

Listing 1.1. Example R3M Mappings

```
1   a)   ex:revision      a    r3m:TableMap;
2           r3m:hasTableName    "Revision";
3           r3m:mapsToClass     ver:Version;
4           r3m:uriPattern      "http://.../revision_%%number%%";
5           r3m:hasAttribute    ex:revision_number, ....
6
7   b)   ex:revision_number    a   r3m:AttributeMap;
8           r3m:hasAttributeName      "number";
9           r3m:mapsToObjectProperty  ver:hasID;
10          r3m:dbType                [ a   r3m:VarChar;
11                                        r3m:length 255 ];
12          r3m:hasConstraint         [ a   r3m:NotNull ].
13
14  c)   ex:release_revision    a   r3m:LinkTableMap;
15          r3m:hasTableName          "Release_Revision";
16          r3m:mapsToObjectProperty  ver:comprises;
17          r3m:hasSubjectAttribute   ex:rr_release;
18          r3m:hasObjectAttribute    ex:rr_revision.
```

(*e.g.,* Hibernate[3]), and at the so-called *direct mapping* where an equivalent RDF representation of the relational data is needed (*e.g.,* for use cases as described in [12]).

Listing 1.1 presents examples of the three main mapping constructs in R3M. The namespace prefixes used in the examples are defined as follows: `r3m` represents our mapping language vocabulary `http://ontoaccess.org/r3m/` while `ex` is used for the namespace `http://example.com/mapping/` of our example mapping. `ver` represents the namespace of the SEON version control ontology `http://evolizer.org/ontologies/seon/2010/03/versions.owl` (*cf.* Section 4.2). Listing 1.1a) depicts a *TableMap* representing the mapping of a database table to a class in the ontology. A *TableMap* contains the name of the table (line 2) and the class it is mapped to (line 3). The URI pattern (line 4, abbreviated) is used to generate the URIs for instances of this table based on values of table attributes that are specified between double percentage signs (e.g. *%%number%%* where *number* is the name of an unique attribute such as the primary key). A *TableMap* further contains a list of *AttributeMaps* (line 5, abbreviated).

Listing 1.1b) presents an example of an *AttributeMap* that maps a database attribute to a property in the ontology. An *AttributeMap* contains the name of the attribute in the database schema (line 8) and the property it is mapped to (line 9). Additionally, an *AttributeMap* includes information about the datatype of the database attribute (lines 10 and 11) as well as information about (database) constraints defined on that attribute (e.g., a not null constraint; line 12). R3M

---

[3] `http://hibernate.org/`

supports the constraints `r3m:PrimaryKey`, `r3m:ForeignKey`, `r3m:NotNull`, `r3m-`
`:Default`, and `r3m:Check`.

Listing 1.1c) shows a *LinkTableMap* representing the mapping of a link table
to an ontology property. A *LinkTableMap* specifies the name of the link table in
the database (line 15) and the property it is mapped to (line 16). A link table
always contains two foreign key attributes that point to the tables of the N:M
relationship. These attributes are represented as *AttributeMaps* (line 17 and 18)
that provide the names of the attributes, the foreign key references to the tables,
and the direction of the relationship (from subject to object).

4   A Case Study on Bridging Software Analysis Data

To motivate our case study, we present use cases that require interoperability between EVOLIZER and SOFAS. These use cases need a bidirectional data exchange,
i.e., from EVOLIZER to SOFAS and vice versa.

First, EVOLIZER contains data about hundreds of software systems that were
imported over the past years (*cf.* Section 2.1). The SOFAS platform needs to
be able to access this data without the need for re-importing it. This requires
RDF-based **read access** to the EVOLIZER database. Second, EVOLIZER implements importers to import source code and history data from centralized version
control systems, such as CVS and SVN. Lately decentralized version control systems, such as Git or Mercurial, gained popularity. Therefore, respective import
services were developed for the SOFAS platform. The data produced by these
importer services is modeled in RDF, based on the SEON ontologies described
in Section 4.2. This data is also valuable to EVOLIZER because existing tools
could be used to leverage it. This, however, requires RDF-based **write access**
to the EVOLIZER database. Lastly, SOFAS implements an extensible framework
to compute software metrics on the data. Again, this data is modeled in RDF,
but matching relations are available in the EVOLIZER database schema. RDF-
based write access to the RHDB is needed to make the metrics data available to
EVOLIZER.

These use cases indicate that, for making a bridge between EVOLIZER and
SOFAS, a RDB-to-RDF mapper is needed that provides RDF-based read and
write access to relational data. Whereas existing approaches are limited to read-
only queries, we developed ONTOACCESS that provides read and write access. In
the remainder of this section, we present the relational data model of EVOLIZER,
the ontology-based data model of SOFAS, and how the mapping of ONTOACCESS
is used to bridge the conceptual gap between these two data models. We further
discuss challenges we encountered during this case study.

4.1   Data Schema of Software Analysis within Evolizer

The data schema of EVOLIZER consists of several distinct parts covering many
aspects of the Software Engineering domain. For this case study, we focus on
those parts that are concerned with historical aspects and with source code

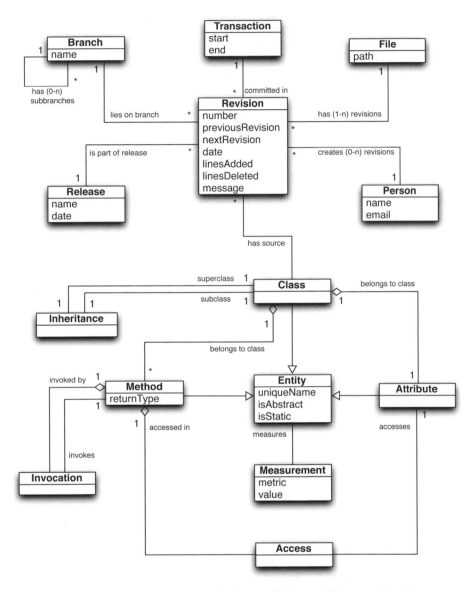

Fig. 2. Evolizer Data Schema for Source Code and Historical Analysis

information. The history model is generic, *i.e.,* it applies to a certain extent to centralized version control systems, as well as to distributed ones. In the following, we describe the most important parts of our schema. An overview of the simplified version of the schema is given in Figure 2; the full schema consists of approximately 40 distinct tables.

One of the core entities is *Revision.* A revision is a particular version of a file; a *Person,* that is to say a software developer, edits a file, and commits the modifications to the version control system. The latter tracks the date of the commit, the reason for the modification (*i.e.,* the commit message provided by the developer), as well as additional information such as the number of lines that were affected. A *Release* constitutes an important milestone in the life-cycle of a software system. It is often identified by a codename and contains a snapshot of the most recent revisions of all the files at the release data. New or experimental features, as well as bug fixes, are often developed on a *Branch.* Once the code is stable, it is merged back into the trunk.

To obtain a meaningful source code model, the system under analysis needs to be in a consistent state. This is generally guaranteed only for a release and therefore, for revisions that are part of a release, we can parse the source code and instantiate a reasonable model accordingly. A revision then corresponds to one top-level *Class* in Java, C#, etc. Classes have a set of members, *i.e.,* they contain *Attributes* and *Methods.* Classes, attributes, and methods are generalized into *Entities.*

Relationships between source code entities, such as *Invocations* between methods, *Accesses* from methods to attributes, and *Inheritance* between classes, are also made explicit by representing them as an association class or link table. While they are hard to distinguish from real entities, this is the only means that the relational model provides when we want to explicitly query for relationships.

Entities can be measured. Such a *Measurement* is specified by a metric, for example 'number of lines of code' for a class or method, or 'number of accesses' for an attribute, and the value that has been measured.

4.2   Ontologies of Software Analysis within SOFAS

To describe the data produced and consumed by SOFAS, we developed a family of Software Evolution ONtologies (SEON). They describe different aspects of software and its evolution, such as version control, issue tracking, static source code structure, change coupling, software design metrics, and so on. SEON is organized as several ontology pyramids. For each of the major subdomains, we have developed higher level ontologies defining their common concepts. For system-specific or language-dependent concepts we developed some concrete low-level ontologies. For instance, there is a high-level version control ontology and several low-level ontologies for concrete version control systems, such as CVS, SVN, and Git, that extend the high-level version control ontology. In this paper, we limit the discussion to the main terms of the source code ontology and the version control ontology. The source code ontology models the static source code structures based on the FAMIX meta model. It is therefore similar to the EVOLIZER data

Table 1. Source Code Ontology Overview

Class: Class
→ declaresMethod : Method
→ declaresField : Field
→ isReturnTypeOf : Method
→ isSubclassOf : Class
→ isSuperclassOf : Class
→ hasName : xsd:string
Class: Field
→ isDeclaredFieldOf : Class
→ isAccessedByMethod : Method
→ hasName : xsd:string

Class: Method
→ accessesField : Field
→ hasParameter : Parameter
→ invokesMethod : Method
→ hasReturnClass : Class
→ isInvokedByMethod : Method
→ isMethodOf : Class
→ hasName : xsd:string
Class: Parameter
→ isParameterOf : Method
→ hasName : xsd:string

Table 2. Version Control Ontology Overview

Class: Version
→ hasID : xsd:string
→ follows : Version
→ precedes : Version
→ hasCreationDate : xsd:date
→ linesAdded : xsd:int
→ linesDeleted : xsd:int
→ hasMessage : xsd:string

Class: ChangeSet
→ hasCommitDate : xsd:date
Class: Branch
→ hasTag : xsd:string
Class: Release
→ hasReleaseDate : xsd:string
→ hasTag : xsd:string

schema described in the previous section. Table 1 provides an overview of the main classes and properties of the SEON source code ontology. The full ontology covers many more concepts such as *interfaces, local variables,* and *exceptions*.

The version control ontology models the structure of version control systems and is based on the data model described in [11]. Table 2 provides an overview of the main classes and properties of the SEON version control ontology.

4.3   OntoAccess as a Bridge to the New Town of Software Analysis

ONTOACCESS bridges the conceptual gap between the RDB-based EVOLIZER and the Semantic Web-enabled SOFAS. It introduces a RDB-to-RDF mapping and provides on-the-fly translation of RDF-based read and write requests to the EVOLIZER RHDB. Table 3 and Table 4 contain an overview of the mapping in a schematic representation. Again, we focus in the presentation of the mapping on the parts of the EVOLIZER RHDB that are relevant to this case study. The mapping uses the following namespace declarations. ver for the SEON version control ontology http://evolizer.org/ontologies/seon/2010/03/versions.owl, top for http://evolizer.org/ontologies/seon/2010/03/top.owl, java for the SEON source code ontology http://evolizer.org /ontologies/seon/2009/ 06/java.owl, and foaf for http://xmlns.com/foaf/0.1/. Table 3 lists the mapping of tables from Figure 2 that represent a domain concept and their attributes. The table consists of four columns. The first names the table as in Figure 2 and the

Table 3. Mapping Overview Part I

table	→ class	attribute	→ property
Revision	→ ver:Version	number	→ ver:hasID
		previousRevision	→ ver:follows
		nextRevision	→ ver:precedes
		date	→ ver:hasCreationDate
		linesAdded	→ ver:linesAdded
		linesDeleted	→ ver:linesDeleted
		message	→ ver:hasMessage
Transaction	→ ver:ChangeSet	start	→ -
		end	→ ver:hasCommitDate
Branch	→ ver:Branch	name	→ ver:hasTag
File	→ top:File	path	→ top:filePath
Release	→ ver:Release	name	→ ver:hasTag
		date	→ ver:hasReleaseDate
Person	→ foaf:Person	name	→ foaf:name
		email	→ foaf:mbox
Entity	→ -	isAbstract	→ java:isAbstract
		isStatic	→ java:isStatic
Class	→ java:Class		
Method	→ java:Method	returnType	→ java:hasReturnType
Attribute	→ java:Field		
Measurement	→ met:SoftwareMetric	metric	→ met:hasName
		value	→ met:hasValue

second the class in the ontology it is mapped to. Column 3 contains the attributes of the respective table and their mapping to properties depicted in Column 4. A dash in the Columns 2 or 4 means that there is no mapping. The table *Entity* is not mapped to a class in the ontology but its attributes are mapped to properties. *Entity* is just a super type of several of the other concepts and only those (sub-)concepts are represented in the ontology (*cf.* Section 4.4).

Table 4 lists the mapping of link tables that represent M:N relationships in RDBs. As RDF provides different means to represent M:N relationships, such helper constructs are not needed and those tables are mapped to ontology properties instead. The table consists of three columns. The first names the link tables that are represented in Figure 2 as connecting lines between two concepts or as explicit concepts themselves. In the first case, the name is composed of the two participating table names separated by an underscore. Column 2 lists the property that each link table is mapped to. Column 3 lists the corresponding inverse property. For instance, the relationship from *Release* to *Revision* is mapped to the property ver:comprises and the inverse relationship from *Revision* to *Release* is mapped to the property ver:appearsIn.

4.4   Discussion

In our case study, we showed how ONTOACCESS has been successfully deployed to *make a bridge to the new town*. It provides a gradual migration path from

Table 4. Mapping Overview Part II

link table	→ property	: inverse property
Release_Revision	→ ver:comprises	: ver:appearsIn
Branch_Revision	→ ver:comprises	: ver:isOn
Transaction_Revision	→ ver:comprises	: ver:commitedIn
File_Revision	→ ver:hasVersion	: ver:belongsTo
Person_Revision	→ -	: ver:committedBy
Class_Revision	→ ver:hasSource	: -
Method_Class	→ java:isDeclaredMethodOf	: java:declaresMethod
Attribute_Class	→ java:isDeclaredFieldOf	: java:declaresField
Measurement_Entity	→ met:isMetricOf	: met:hasMetric
Inheritance	→ java:hasSubClass	: java:hasSuperClass
Invocation	→ java:invokesMethod	: java:isInvokedByMethod
Access	→ java:accessField	: java:isAccessedByMethod

a legacy system such as EVOLIZER, to a new platform, in our example SOFAS. We demonstrated how ONTOACCESS bridges the conceptual gap between the relational data model of EVOLIZER and the RDF-based SOFAS. We further motivated that existing, read-only RDB-to-RDF mapping approaches are unsuitable for this application scenario as they limit RDF-based data access to read-only queries. During this case study, we faced several challenges w.r.t. to the mapping in ONTOACCESS. In the following, we report on two major ones and the solutions we developed to overcome them.

The first challenge is related to the representation of concept inheritance in relational database systems. Inheritance is a central concept in the object-oriented methodology and is therefore commonly used in object-oriented systems, including EVOLIZER. Relational, unlike object-relational or object-oriented databases, do not directly support inheritance. However, there exist three principal strategies to implement inheritance in relational database schemata (*cf.* [14]): *table-per-hierarchy* represents all classes of the inheritance hierarchy in a single table. This table contains columns for the attributes of all classes and a special column, called discriminator, that stores the type (*i.e.,* class) for each instance. *Table-per-concrete-class* represents each class in its own table. Each of those tables contains columns for the attributes of the class and all super-classes up to the root of the inheritance hierarchy. As a result, attributes of a common super-class are duplicated in all of its sub-classes. The third strategy, called *table-per-subclass*, also represents each class in its own table. In contrast to the *table-per-concrete-class* strategy, the attributes of the super-class(es) are not duplicated as columns in the sub-classes. Instead, a shared primary key is used to connect the tables representing classes in the inheritance hierarchy.

EVOLIZER uses different strategies for different inheritance hierarchies, for example the *table-per-hierarchy* strategy to implement inheritance for the *Entity* concept and its subconcepts. For the sake of this case study, we had to add explicit support for mapping inheritance hierarchies to ONTOACCESS. The *table-per-concrete-class* strategy was mappable out-of-the-box since it defines

Listing 1.2. Extended R3M Mapping Examples

```
1  a)   ex:method    a    r3m:TableMap;
2          r3m:hasTableName        "Entity";
3          r3m:mapsToClass         java:Method;
4          r3m:hasDiscriminator    ex:method_type;
5          r3m:uriPattern          "http://.../method_%%id%%";
6          r3m:hasAttribute        ex:method_type, ....
7        ex:method_type    a    r3m:AttributeMap;
8          r3m:hasAttributeName    "ctype";
9          r3m:hasValue            "Method".
10
11 b)   ex:Method    a    r3m:TableMap;
12         r3m:hasTableName        "Method";
13         r3m:mapsToClass         java:Method;
14         r3m:hasParentTable      ex:entity;
15         r3m:uriPattern          "http://.../method_%%id%%";
16         r3m:hasAttribute        ex:method_returnType.
17       ex:entity    a    r3m:TableMap;
18         r3m:hasTableName        "Entity";
19         r3m:hasAttribute        ex:entity_uniqueName, ....
```

one table per class and the tables are independent from each other. Mapping the other two strategies required support for features such as discriminator columns and relating tables in a parent-child relationship. We addressed this limitation by adding explicit mapping constructs to the ONTOACCESS mapping language. First, discriminator columns were added to provide support for the *table-per-hierarchy* strategy. Since support for mapping a subset of the columns in a table already exists, it is possible to provide multiple mappings for tables that represent all classes within an inheritance hierarchy (one mapping for each class). Each mapping only contains the respective subset of the columns and a description of the discriminator column with its name and value. Listing 1.2a) depicts a concrete mapping example that is using a discriminator column. We also added a mapping construct for relating two tables to each other in a parent-child relationship to provide support for the *table-per-subclass* strategy. The mapping of a table can reference another table as its parent table. This enables ONTOACCESS to detect that a concept from the application domain is split among multiple tables in the database schema. As a result, the involved tables can automatically be joined (on the primary key). Listing 1.2b) depicts a concrete mapping example that is using a parent table reference. These two extensions to R3M enable support for mapping the relational representations of concept inheritance with all three strategies.

The second challenge is related to defining the RDB-to-RDF mappings. Mappings in ONTOACCESS are encoded in RDF which makes them well-suited for automatic processing by machines but hinders the accessibility for human users. Manually defining such mappings is a time-consuming and error-prone task, consisting of mostly repetitive steps. Therefore, tool support for defining mapping is indispensable in more complex application scenarios where the number of database tables and columns is of significance. We built a tool [6] to ease the definition of ONTOACCESS mappings. It semi-automatically generates a mapping from a RDB schema in two steps. First, it automatically generates a basic mapping, based on information extracted from the schema catalog of the database system. Terms of the target ontology are also generated in this step, based on table and column names in the database schema. Next, the tool displays a graphical editor for refining the mapping. This step is mainly concerned with replacing the generated terms with actual terms from the target ontology. The tool further provides validation of existing mappings to catch errors from manual editing. The tool is implemented as a plug-in for the ontology editor Protégé[4] to enable quick access to the definition of the target ontology.

5   Related Work

D2R Server [4] is an approach for publishing existing relational databases on the Semantic Web. Based on mappings expressed in the D2RQ [5] mapping language, it enables browsing the relational data as RDF via dereferenceable URIs (*i.e.*, as Linked Data). Further, support for the SPARQL query language is provided. D2R is limited to read-only data access, updating RDF data is not supported.

The Virtuoso Universal Server features RDF Views [9] to expose relational data on the Semantic Web. A declarative Meta Schema Language is used for defining the mapping of SQL data to RDF vocabularies. This enables the use of SPARQL as an alternative query language for the relational data. Likewise, Virtuoso implements a Linked Data interface to these views. RDF Views are limited to read-only queries.

R$_2$O [1] is an extensible and fully declarative language to describe mappings between relational database schemata and ontologies. R$_2$O is aimed at situations where the similarity between the ontology and the database model is low. It has been conceived to be expressive enough to cope with complex mapping cases where one model is richer, more generic/specific, or better structured than the other. This high expressiveness renders R$_2$O mappings read-only.

The W3C has recognized the importance of mapping relational data to the Semantic Web by starting the RDB2RDF Incubator Group[5] (XG) to investigate the need for standardization. The XG recommended [22] that the W3C initiates a working group (WG) to define a vendor-independent RDB-to-RDF mapping language. The RDB2RDF WG[6] started its work on R2RML [7] in late 2009. According to their charter [18], the requirements for updating relational data are out of scope and are therefore not addressed by the WG. It was shown in [15] that adding write support to the R2RML approach is impractical.

---

[4] http://protege.stanford.edu/
[5] http://www.w3.org/2005/Incubator/rdb2rdf/
[6] http://www.w3.org/2001/sw/rdb2rdf/

6   Conclusions

In theory the Semantic Web provides a common framework that greatly facilitates data sharing and reuse across application, enterprise, and community boundaries. In practice its wide adoption is still hampered by the fact that many organizations have locked away their data in relational databases. Business-critical legacy applications rely on these databases to sustain daily operations and newly developed systems often need to run in tandem with their predecessors until the latter can be gradually phased out. Both, the legacy systems, as well as their successors, usually need to operate cooperatively on existing data. This includes reads and updates. A complete paradigm shift in data representation is therefore often extremely difficult and costly to achieve.

In this paper, we presented ONTOACCESS, a RDB-to-RDF mediation platform that enables RDF-based read and write access to relational databases. It greatly facilitates the transition from legacy systems to Semantic Web-enabled applications in practice by providing a semantic layer on top of existing relational databases. Semantic Web query and update requests are translated on-the-fly to SQL for execution in the database system. ONTOACCESS therefore eliminates the need for mirroring and synchronizing relational data with its RDF representation and, in addition, allows one to further exploit the advantages of the well-established database technology, such as query performance, scalability, transaction support, and security.

In our case study, we have described how we successfully deployed ONTO-ACCESS to provide a gradual migration path between two of our own large-scale software systems, namely the legacy application EVOLIZER and its successor, the SOFAS platform. We identified challenges when it comes to mapping inheritance hierarchies with ONTOACCESS and we have extended the latter accordingly to support different inheritance mapping strategies. Further, we established tooling to semi-automate the process of extracting mappings from a relational database schema to an ontology.

In summary, judging from the experiences made in our case study, we are confident that ONTOACCESS is a valuable tool that will foster the acceptance of Semantic Web technology in practice.

References

1. Barrasa, J., Corcho, O., Gómez-Pérez, A.: R2O, an Extensible and Semantically Based Database-to-Ontology Mapping Language. In: Proc. Workshop on Sem. Web and Databases (August 2004)
2. Berners-Lee, T.: Linked Data (2009),
   http://www.w3.org/DesignIssues/LinkedData.html (last visited June 2011)
3. Berners-Lee, T.: Relational Databases on the Semantic Web (2009),
   http://www.w3.org/DesignIssues/RDB-RDF.html (last visited June 2011)
4. Bizer, C., Cyganiak, R.: D2R Server – Publishing Releational Databases on the Semantic Web. In: Proc. Int'l Sem. Web Conf. (November 2006)
5. Bizer, C., Seaborne, A.: D2RQ – Treating Non-RDF Databases as Virtual RDF Graphs. In: Proc. Int'l Sem. Web Conf. (November 2004)

6. Brügger, N.: RDB-RDF Mapping Generation from Relational Database Schemata. Master's thesis, University of Zurich (December 2009)
7. Das, S., Sundara, S., Cyganiak, R.: R2RML: RDB to RDF Mapping Language. W3C Working Draft (October 2010),
   http://www.w3.org/TR/2010/WD-r2rml-20101028/
8. Demeyer, S., Ducasse, S., Nierstrasz, O.: Object Oriented Reengineering Patterns. Morgan Kaufmann Publishers Inc., San Francisco (2002)
9. Erling, O., Mikhailov, I.: RDF Support in the Virtuoso DBMS. In: Proc. of the SABRE Conf. on Social Sem. Web (September 2007)
10. Fielding, R.T.: Architectural Styles and the Design of Network-based Software Architectures. Ph.D. thesis, University of California, Irvine (2000)
11. Fischer, M., Pinzger, M., Gall, H.: Populating a Release History Database from Version Control and Bug Tracking Systems. In: Proc. Int'l Conf. Softw. Maintenance (September 2003)
12. Fürber, C., Hepp, M.: Using SPARQL and SPIN for Data Quality Management on the Semantic Web. In: Abramowicz, W., Tolksdorf, R. (eds.) BIS 2010. LNBIP, vol. 47, pp. 35–46. Springer, Heidelberg (2010)
13. Gall, H.C., Fluri, B., Pinzger, M.: Change Analysis with Evolizer and ChangeDistiller. IEEE Softw. (January/February 2009)
14. Garcia-Molina, H., Ullman, J.D., Widom, J.: Database Systems: The Complete Book. Prentice Hall Press (2008)
15. Garrote, A., Garcia, M.N.M.: RESTful Writable APIs for the Web of Linked Data Using Relational Storage Solutions. In: Proc. WWW 2011 Workshop on Linked Data on the Web (April 2011)
16. Ghezzi, G., Gall, H.C.: Towards Software Analysis as a Service. In: Proc. Int'l ERCIM Workshop on Softw. Evolution and Evolvability (September 2008)
17. Ghezzi, G., Gall, H.C.: SOFAS : A Lightweight Architecture for Software Analysis as a Service. In: Working IEEE/IFIP Conf. on Softw. Architecture (June 2011)
18. Halpin, H., Herman, I.: RDB2RDF Working Group Charter (2009),
    http://www.w3.org/2009/08/rdb2rdf-charter (last visited June 2011)
19. Hert, M.: Relational Databases as Semantic Web Endpoints. In: Proc. European Sem. Web Conf. (June 2009)
20. Hert, M., Reif, G., Gall, H.C.: Updating Relational Data via SPARQL/Update. In: EDBT Workshop Proc. (March 2010)
21. Hert, M., Reif, G., Gall, H.C.: A Comparison of RDB-to-RDF Mapping Languages. In: Proc. Int'l Conf. on Semantic Systems (2011)
22. Malhotra, A.: W3C RDB2RDF Incubator Group Report (January 2009),
    http://www.w3.org/2005/Incubator/rdb2rdf/XGR-rdb2rdf-20090126/
23. Prud'hommeaux, E., Seaborne, A.: SPARQL Query Language for RDF. W3C Recommendation (January 2008),
    http://www.w3.org/TR/2008/REC-rdf-sparql-query-20080115/
24. Seaborne, A., Manjunath, G., Bizer, C., Breslin, J., Das, S., Davis, I., Harris, S., Idehen, K., Corby, O., Kjernsmo, K., Nowack, B.: SPARQL Update – A Language for Updating RDF Graphs. W3C Member Submission (July 2008),
    http://www.w3.org/Submission/2008/SUBM-SPARQL-Update-20080715/

The MetaLex Document Server
Legal Documents as Versioned Linked Data

Rinke Hoekstra[1,2]

[1] Leibniz Center for Law, Faculty of Law, University of Amsterdam
hoekstra@uva.nl
[2] Computer Science Department, VU University Amsterdam
r.j.hoekstra@vu.nl
http://www.rinkehoekstra.nl

Abstract. This paper introduces the MetaLex Document Server (MDS), an ongoing project to improve access to legal sources (regulations, court rulings) by means of a generic legal XML syntax (CEN MetaLex) and Linked Data. The MDS defines a generic conversion mechanism from legacy legal XML syntaxes to CEN MetaLex, RDF and Pajek network files, and discloses content by means of HTTP-based content negotiation, a SPARQL endpoint and a basic search interface. MDS combines a transparent (versioned) and opaque (content-based) naming scheme for URIs of parts of legal texts, allowing for tracking of version information at the URI-level, as well as reverse engineering of versioned metadata from sources that provide only partial information, such as many web-based legal content services. The MDS hosts all 28k national regulations of the Netherlands available since May 2011, comprising some 100M triples.

Keywords: metalex, rdf, legal xml, law, linked open data, open government.

1 Introduction

Where open government data is concerned, the rules and regulations a government imposes on its citizens are arguably close to the top of the list of every open data enthusiast. Law is the oldest form of open government information in existence. For it to be effective, the adage holds that "every citizen is expected to know the law" – 'knowing' in the sense of 'having access to'. Legislation and court rulings grow in importance. Policy makers are increasingly inclined to 'govern by regulation'[1], EU directives form an ever more complex legal framework that shapes national policies and regulations, giving citizens easier access to supra-national legislatures such as the European court. Businesses are subject to highly detailed regulations concerning e.g. financial reporting, safety and security. Gartner estimates that the cost of meeting regulatory compliance needs will

---

[1] A recent example in The Netherlands was the threat of the Minister of the Interior to draft legislation that would force municipalities to accept a budget cut and carry out tasks previously belonging to national government.

L. Aroyo et al. (Eds.): ISWC 2011, Part II, LNCS 7032, pp. 128–143, 2011.
© Springer-Verlag Berlin Heidelberg 2011

pose severe problems for smaller banks by 2013.[2] Compliance affects businesses and government agencies alike: how to ensure the minimally required alignment of internal business processes with (external) regulations?

Regulations are at the heart of modern society, they affect every aspect of our lives, from public safety, to education, health, environment, food, civil disputes, traffic, privacy and democracy itself. It is therefore not surprising that many national governments have been publishing legislation and court rulings on the web for quite some time now. The National Archive's Legislation.gov.uk was at the forefront of the linked open government data wave that hit shore in 2009.[3] It set the standard for what governments should do to provide 5-star access to legal documents.[4].

In the Netherlands, the 'wetten.nl'[5] portal was launched in 2003 with all legislation published since 2002. In the following years, earlier legislation, treaties and other types of regulations were made available through the portal as well. In several respects, the features of the wetten.nl portal are symptomatic for the way in which the Dutch government communicates information to its citizens in the Netherlands: it looks *fancy* and costs a *tonne*, but is *not flexible*. Although current versions of regulations are available in XML, they are stripped of essential information, such as the version date of the document. Wetten.nl presents regulations as books with hyperlinks; the position of an article within the running text of a regulation is the only context provided. Given the highly networked structure of legislation, this traditional restricted presentation is suboptimal: potential alternative ways of serialising one or more regulation texts (e.g. by topic) are discarded. This is not only a potential problem for businesses and citizens trying to understand the norms applying to their case, it is problematic for the civil servants and government organisations that have to apply these norms as well.

This paper describes our efforts to publish the contents of wetten.nl as *5-star open data*: i.e. to extract, aggregate, reconstruct and enrich the datasource underlying the wetten.nl portal using publicly accessible webservices, and publish it both as CEN MetaLex[6], as Linked Data, and in a format suitable for social network analysis. By design, this conversion is independent of the language and XML format in which regulations are published. This conversion is the first large scale effort to transform an existing legacy legislative XML format to MetaLex.

Section 2 describes the requirements and use cases for the information that will be published through the MetaLex Document Server (MDS). Section 3 describes the current situation in the Netherlands in more detail. Section 4 introduces

---

[2] See http://bit.ly/aND1Rj

[3] See http://www.legislation.gov.uk and http://bit.ly/cV2MRu for a discussion of its features.

[4] See http://www.w3.org/DesignIssues/LinkedData.html

[5] Literally 'legislation.nl', see http://wetten.overheid.nl

[6] CEN MetaLex is published as CEN Workshop Agreement, CWA 15710, see ftp://ftp.cen.eu/CEN/Sectors/List/ICT/CWAs/CWA15710-2010-MetaLex2.pdf

MetaLex, and is followed by a description of our methodology in section 5. Section 6 describes the results, followed by a discussion.

2 Use Cases and Requirements

We identify four stakeholders when it concerns the interpretation of legal texts: *citizens, businesses, legal professionals* and *government bodies. Citizens* are expected to 'know the law'. Governments have a duty to make the law known to their citizens. Even though citizens may not be interested or able to understand legal language [6,5], they must at least be offered the opportunity to know their rights and duties. *Businesses* have a vested interest in complying to regulations as the (financial) risks of not complying are high, and governments have the means to check for compliance through audits and obligatory reports. They therefore need to be kept up-to-date with respect to new or changed regulations they are subject to.

Legal *professionals* not only need to be kept up-to-date, but they frequently require access to non-current versions of regulations when dealing with cases that emerged prior to the latest change (retroactive enactment is quite seldom). Even for legal professionals, the texts of regulations are not self explanatory, and they consult a wide variety of additional sources to interpret the law. Examples are the official motivation of the legislator, case law, notes provided by other experts, journals, and reports of parliamentary hearings.

Government bodies enact, enforce, implement and execute regulations. Law is a large interconnected, and therefore interdependent network of norms. Understanding and guiding the effects of new proposed legislation is one of the primary concerns of the legislator. Currently, legislative drafting largely depends on the expertise of civil servants, their access to books and legal search engines. In the Netherlands and Switzerland, no specific editing environment is currently available: Laws are drafted by editing and sending regular Word documents around. Secondly, executive agencies have internal business processes that need to align with all potentially applicable versions of the law. Lastly, government organisations are increasingly required to share information amongst themselves. However, organisations form different and sometimes incompatible speech communities. The term 'income' means something different for determining social benefits as it does for taxation. Legislation (and in particular its structure) forms the ideal 'coat rack' for knowledge management and interchange between government bodies.[7]

2.1 Use Cases

Each of our stakeholders has benefited from the increased transparency offered by web-based search engines. However, their interests and needs go beyond simple search. Businesses are increasingly aware of the importance of streamlining

---

[7] With thanks to Hans Overbeek of ICTU for the metaphor.

their internal operations. The market of business process management suites is expected to grow to \$3.4 billion globally by 2014.[8] This provides opportunities for more fine grained *regulatory compliance* management: business processes that are potentially affected by regulations can be identified by explicitly linking them to applicable norms. Legal professionals working at businesses and government bodies need to *annotate* parts of legal texts with interpretations and guidelines, and share them with their colleagues. *Provenance* information is essential for determining the motivation of a legislation: what parliamentary hearings and led to the current version of an article? Annotation and provenance are a key requirement in the current modernisation of the legislative drafting system of the Swiss Federal Chancellery [10].

Regulations are not *integrated*, different types of 'law' are issued by different government bodies. National government issues legislation, judges produce case law, and municipalities issue local regulations. Regulations are published on different websites, a situation that misrepresents the dependencies between them.

It is hard to *consistently interpret the meaning of concepts*. Regulations contain both 'hyperlink' style references to parts of text, as well as *'imports'* that import the meaning of a term from another regulation. Furthermore, the meaning of a term can be *scoped* to a particular part of the regulation, such as a chapter or article [12]. To complicate matters, regulations contain so-called 'deeming provisions' that, within a specified scope, assign the label of one concept to another concept. For instance, the provision "for the purposes of this chapter, a house boat is deemed to be a house" allows the legislative drafter to use the term 'house' to refer both to houses and house boats. Although legislative drafters are very careful to be specific about their choice of words, not all concepts used are properly introduced.

Not all parts of a regulation are equally *important*. That is, it is often the case that a small number of articles hold for the majority of cases regulated by a law, while the rest deal with more specific cases and exceptions. Furthermore, although related articles are often grouped within a chapter, this grouping does not cross the borders of a single law. Even though articles in distinct laws may be more closely related. The Dutch Immigration and Naturalisation Service (IND) has to deal with highly dynamic legislation. Knowing what parts of a law matter most to them, as well as the dependency between articles and their internal business processes (cf. regulatory compliance) is key in their ability to advice the ministry on the possibilities and difficulties in amending existing immigration laws.

The IND has a hard time dealing with all different *versions* of legislation, caused by dynamic legislation and lengthy procedures. Legislation follows an intricate versioning scheme [3,4,7,11]: enactment, efficacy, publication and repeal dates all interact. This information may even be part of other regulations, e.g. in the Netherlands efficacy of regulations is typically described by Royal Decree. Legislation is frequently modified at the *sentence* level: e.g. a modifying law will

---

[8] Gartner Inc.: "Forecast: Enterprise Software Markets, Worldwide, 2009-2014, 4Q10 Update, December 2010.

replace the second sentence of article X. Finally, references between legal texts can point to a specific version of a regulation, as well as to the 'current' version.

Key in these use cases is the ability to refer to parts of legal texts. It requires persistent *identifiers* for every element of a legal text. These identifiers should be dereferencable to the element they describe, or a description of the element's metadata. It is furthermore a feature if these identifiers are *transparent* and follow a prescribed *naming convention*. This allows third parties to construct valid identifiers without having to first query a name service.

To support *versioning* of legal texts, references, and metadata, requires identifiers that reflect its different versions. The various parts of a text should be versioned *independently*, allowing for transitory regimes. Furthermore, the versioning mechanism should distinguish between a regulation text as it exists at a particular time, and the regulation 'as such'. A likely solution is the adoption of the distinctions made by the IFLA FRBR [15]:[9] the *work* as a "distinguishable intellectual or artistic creation" (e.g. the constitution);the *expression* as the "intellectual or artistic form that a work takes each time it is realised" (e.g. "The Constitution of July 15th, 2008"); the *manifestation* as the "physical embodiment of an expression of a work" (e.g. a PDF version of "The Constitution of July 15th, 2008"); and the *item* as a "single exemplar of a manifestation" (e.g. the PDF version of "The Constitution of July 15th, 2008" residing on my USB stick).

Metadata and annotations should be traceable to the most detailed part of a text, as well as to its version, when needed. The same requirement holds for references between texts, allowing for fine-grained analysis of interdependencies between texts. Current regulation search portals are developed from the perspective of the issuing government body, and are jurisdiction specific. The document server should provide a publishing platform that is independent of language, region and jurisdiction.

3   Wetten.nl and the Basiswettenbestand

Wetten.nl is part of a larger Overheid.nl (government.nl) website that provides access to a wide range of government information, both legal (national regulations, local regulations, permits, and publications, official national publications and disciplinary rulings) and general information such as information about the structure of goverment, addresses of government bodies, and a link to the Dutch open data catalog.[10] Amongst these, the wetten.nl portal is one of the oldest.

Users can perform a full text search through the titles and text of all regulations of the Kingdom of the Netherlands. They can search for a specific article,

---

[9] IFLA: International Federation of Library Associations and Institutions. FRBR: Functional Requirements for Bibliographic Records.
See http://archive.ifla.org/VII/s13/frbr/frbr1.htm for the exact definitions.
[10] See http://data.overheid.nl, a CKAN installation currently containing 40 datasets.

as well as for the version of a text holding at a specified date. Wetten.nl also supports deeplinks, but is not very consistent about it. For instance, both:

`http://wetten.overheid.nl/cgi-bin/deeplink/law1/bwbid=BWBR0005416/article=6/date=2005-01-14`

and

`http://wetten.overheid.nl/BWBR0005416/TitelII698946/HoofdstukII/Artikel16/`
`geldigheidsdatum_14-01-2005`

point to article 6 of the Municipal law, as it was valid on January 14th, 2005.[11] These deeplinks can be considered to be permanent URIs of work level (without date) or expression level (with date) identifiers. Unfortunately, they are not always predictable (cf. the '698946' in the second URI), nor stable, nor part of a government standard.

The string 'BWBR0005416' is the opaque *BWB identifier* (BWB-ID) of the regulation. The Basiswettenbestand (BWB) is the content management system for all Dutch regulations that underlies the Wetten.nl portal. An 'R' following 'BWB' indicates that the document is a regulation, a 'V' indicates a treaty ('verdrag'). The 7-digit number does not carry a specific meaning. The opaqueness of the BWB identifier is unfortunate, but hard to avoid, as the title of a regulation may change over time and cannot be used. An index of all BWB identifiers, with basic attributes such as official and abbreviated titles, enactment and publication dates, retroactivity, etc. is available as a zipped XML dump.[12] Alternatively, a SOAP service allows retrieval of the same information for individual regulations. Unfortunately, the date of the latest change to a regulation is not really the date of the latest modification, but of the latest update of the regulation in the CMS. The two dates often coincide, but not all civil servants work weekends.

The BWB uses its own XML format for storing regulations. BWB XML provides elements for structure as well as annotation elements for capturing version history. It does not separate structural elements (e.g. 'article' or 'chapter'), presentation-type elements (e.g. 'emphasis', 'paragraph') and content-type elements (e.g. 'law', 'treaty'). The text of regulations is contained within meaningless presentation-type elements ('al' for alinea) rather than as separate sentences. The schema does not allow for intermixing with any third-party elements or attributes, ruling out obvious extensions such as RDFa.[13] Finally, the REST web service for obtaining the BWB XML representation of regulations only provides the *latest* version of an entire regulation.[14] The XML document returned is stripped of all version history, and does not even contain the version date of the text itself.

---

[11] Note that 'geldigheidsdatum' is the validity date.
[12] See `http://www.overheid.nl/help/wr/deeplinks`
[13] RDFa: RDF attributes for use in XHTML,
see `http://www.w3.org/TR/rdfa-syntax/`. RDF is the Resource Description Framework, see `http://www.w3.org/standards/techs/rdf`
[14] See `http://bit.ly/kdTniY` and e.g. `http://bit.ly/mQTWwo` for a BWB XML version of the Municipal law.

The BWB-ID forms the basis of the *Juriconnect* standard for referring to parts of regulations.[15] The standard describes a procedure for constructing unique identifiers from the structure of BWB XML documents. BWB XML documents use these identifiers to specify citations between regulations. For instance, the Juriconnect identifier of article 16 of the Municipal law, valid on January 14, 2005 is:

1.0:c:BWBR0005416&artikel=16&g=2005-01-14

Juriconnect does not prescribe a method by which the identifier should be used inside the XML of regulations or referring text: BWB XML elements do not carry Juriconnect identifiers. Neither does the standard specify whether an expression-level reference without validity date points to the latest, or current version, nor does it specify what should be returned for a work-level reference. Furthermore, the standard does not describe a method for dereferencing an identifier to the actual text of (part of) a regulation.

The wetten.nl portal meets most, if not all requirements of the pre open-data day and age. However, more demanding use of the content underlying the portal is not straightforward. The content service is crippled by incomplete information (the version date of retrieved documents, version history), limited functionality (no time travel) and identifiers in a non-standard format. The following section introduces the CEN MetaLex format for representing the text of legal resources, after which section 5 describes our method for republishing the regulations of wetten.nl as MetaLex and Linked Data.

4   CEN MetaLex

CEN MetaLex[16] is a jurisdiction independent XML standard for representing, publishing and interchanging the structure of legal resources. It is the result of a 10 year standardisation project in which multiple European government organisations, publishers and academic partners participated. MetaLex was initially designed by [1] as a syntactic grounding for building elaborate knowledge-based services.[17] At the time, the BWB XML was a proprietary format, and the Dutch government was still in negotiation with legal publishers about freely distributing its self-created content on the web. CEN MetaLex combines the original MetaLex with (primarily) insights from the Italian Norme in Rete project,[18], the Akoma Ntoso legal XML standard of African parliaments[19], the Austrian government and LexDania.[20] Amongst others, adopting CEN MetaLex allows

---

[15] Juriconnect is a consortium of government bodies, legal publishers and academia, see http://www.juriconnect.nl

[16] See http://www.metalex.eu and

[17] See legacy.metalex.eu for more information.

[18] Norme in Rete: laws online portal for the Italian government. The portal itself is no longer available.

[19] See http://www.akomantoso.org/.

[20] LexDania is the XML format behind the Danish ministerial regulations portal, see http://www.ministerialtidende.dk/

the use of generic legislative drafting tools, rather than only jurisdiction (and often vendor) specific solutions.

MetaLex elements are purely *structural*. Syntactic elements (structure) are strictly distinct from the meaning of elements by specifying for each element its name and its *content model* [16]. What this essentially does, is paving the way for a purely semantic description of the types of content of elements in an XML document.

The standard prescribes the *existence* of a *naming convention* for minting URI-based identifiers for all structural elements of a legal document [2]. Names should be guessable from identifying features (attributes, context) of elements, described in the metadata. Names must exist for each of the FRBR levels, and a standard GRDDL[21] transformation for producing an RDF graph of the identifying metadata. MetaLex explicitly encourages the use of *RDFa* attributes on its elements, and provides special metadata-elements for serialising additional RDF triples that cannot be expressed on structural elements themselves. MetaLex includes an *ontology*, which defines the different FRBR levels in the context of legislation, and an event model for legislative modifications.[22]

Elements defined by the MetaLex schema can be *refined* via XML Schema to the jurisdiction specific elements of legacy legal XML formats such as BWB XML, LexDania and CHLexML.[23] These generic elements are: root as the root of every MetaLex document; hcontainer and container for titled and untitled parts of a text; block elements for textual content, and inline for elements that occur in running texts. The htitle block element is used to specify the title of a hcontainer, the cite inline element carries a reference to another element; milestone elements for fixed-position, but content-less inline elements such as page breaks; mcontainer and meta elements for listing additional RDFa metadata inside the body of hcontainer and mcontainer elements. MetaLex is agnostic to non-conflicting third-party XML elements and attributes in block and inline elements, such as HTML markup for rendering tables.

Although most of its predecessors were implemented at enterprise scale, the MetaLex language itself has never been applied to a realistically large corpus. Although the language holds the promise of flexible interchange of legal texts, government institutions are slow movers. This is part of the challenge; does MetaLex live up to its promise as generic schema for legal texts, and does its commitment to Semantic Web standards provide substantial added value to government goals? The following section describes the methodology and vocabularies used for transforming legacy XML to MetaLex, and producing metadata descriptions in RDF. Section 6 discusses the results.

---

[21] GRDDL: Gleaning Resource Descriptions from Dialiects of Languages, see http://www.w3.org/TR/grddl/

[22] See http://www.metalex.eu. The legislation.co.uk portal has adopted the MetaLex event model for representing modifications, but uses the standard FRBR ontology for indicating levels.

[23] CHLexML was designed as an XML standard for the representation of Swiss legal texts, see http://www.svri.ch/CHLexML/CHLexML_Reference_1.0.pdf

5   Conversion and Publication

The transformation of legacy XML to MetaLex and RDF is implemented in the MetaLex converter, an open source Python script available from GitHub.[24] Conversion occurs in four stages: *mapping* legacy elements to MetaLex elements, minting *identifiers* for newly created elements, *producing metadata* for these elements, and *serialising* to appropriate formats. In this section, we briefly discuss how each of these is implemented in the MetaLex converter.

For the transformation of BWB XML files, the converter is sequentially fed with all BWB XML files and identifiers listed in the BWB ID index. Version information, citation titles and other medatada is retrieved through via a custom build scraper of the information pages on the wetten.nl website.[25] The information pages provide more elaborate and reliable information about regulations than can be obtained through the web service, such as the entire version history and types of modification of each law.

5.1   Mapping Legacy Elements to CEN MetaLex

The MetaLex schema is designed to be independent of jurisdiction, which means that it should be possible to map each legacy XML element to a MetaLex element in an unambiguous fashion. For the BWB to MetaLex translation, element mappings were obtained semi-automatically from the BWB DTD. Elements allowed to contain #PCDATA are mapped onto block or inline elements, where inline elements only occur inside the definition of blocks. All hcontainer elements allow a title-element, while container elements are only allowed to contain the block elements identified earlier.

Based on a mapping table, the converter traverses the DOM[26] tree of the source document, and synchronously builds a DOM tree for the target document. There are three special cases for which the converter has to make additional repairs. Sometimes 'obvious' candidates for the MetaLex hcontainer element do not fit the MetaLex schema as the source schema allows them to contain text, e.g. the artikel element in BWB XML. Secondly, footnotes are typically present as block or container-type and occur inside other blocks. The converter moves these to the parent container element of the containing block. On some occasions, inline-type source elements appear directly underneath container-type elements. Their target elements are wrapped inside an extra block element to ensure MetaLex compliance.

Attributes on source elements are passed to the identifier and metadata generators. Target elements receive five standard attributes: name, with the value of the target element name (for MetaLex compliance), class, with the value of

---

[24] See http://github.com/RinkeHoekstra/metalex-converter

[25] See e.g. http://wetten.overheid.nl/BWBR0005416/
geldigheidsdatum_14-01-2005/informatie

[26] DOM: XML Document Object Model,
see http://www.w3.org/TR/REC-DOM-Level-1/

the source element name (for custom CSS rendering), xml:lang, a language tag (if specified on the source element, or one of its parents), id, with an item-level identifier, relative to the xml:base of the document, and about, with an expression-level identifier.

5.2   Minting Identifiers

For every element in the document we create transparent URL-like URIs for the *work*, *expression* and *manifestation* level, and two opaque URIs for the *expression* and *item* level in the FRBR specification.

We use a naming scheme that is based on the URIs used at legislation.gov.uk, with slight adaptations to allow for the Dutch situation.[27] Juriconnect references in the source BWB XML are automatically translated to this naming scheme:

```
{API-URL}/{document-identifier}
   [(/{hcontainer-class}/{index}]*[/{block-class}/{block-id}]*
   [/{authority}][/{extent}][/{lang}][/{version}])|
   (/{opaque-version-hash})]
   [/{repr}]
```

The *API URL* is the first part of all URIs, and the URL at which the URI resolver resides. Examples are `http://legislation.gov.uk` and `http://doc.metalex.eu`, for obvious reasons we use the latter. This part is followed by a *document identifier*, a work-level identifier of the entire legal text. Different countries may have different forms for this identifier, e.g. the Dutch portal uses opaque BWB identifiers, while the UK portal uses {type}/{year}/{number} as document identifier. These two components are followed either by a transparent reflection of the hierarchical structure of the XML document or an opaque hash of the contents of the element.

Hierarchical *work* URIs consist of a path from root node to current element. For hcontainer elements we use its class, i.e. the source elements' name, combined with its official index.[28] For block elements, we use its class combined with a generated index based on the position of the element amongst all children of its parent. The third part of the hierarchical URI consists of an optional indication of the *authority* (issuer) and *extent* (jurisdiction) of the text. Several European member states, such as the UK, have lower governments that can alter or implement specific parts of national regulations. For *expression*-level identification, the work URI is followed by an optional language tag, and the version identifier: the date at which this version became official. Manifestation URIs follow the same conventions as those of 'document URIs' in the legislation.gov.uk portal. Item level identifiers are required by the MetaLex standard, but cannot be generated in any meaningful way. We have therefore chosen to use randomly generated character strings as item identifiers, combined with an empty xml:base.

---

[27] See `http://www.legislation.gov.uk/developer/uris`

[28] Note that a single combination of class and index already provides a locally unique identifier within the legislation, i.e. the relative identifier 'chapter/1/article/1' is identical to 'article/1'. This does not hold for elements below the article level [1].

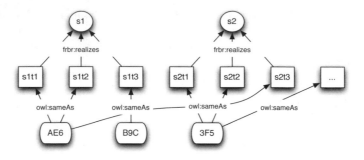

Fig. 1. The benefit of opaque URIs for versioning legal texts

The *opaque version URI* is needed to distinguish different versions of a text. The current webservice does not provide access to all versions of regulations (only to the latest), let alone at a level of granularity lower than entire regulations. We therefore need some way of constructing a version history by regularly checking for new versions, and comparing them to those we looked at before. By including a unique SHA1 hash of the textual content of an XML element in the opaque URI, and simultaneously maintaining a link between the opaque URI and the transparent identifier, different expressions of a work can be automatically distinguished through time. This is needed to work around issues with identifiers based on numbers: the insertion of a new element can change the position (and therefore the identifier) of other elements without a change in the content of the elements. How to find out how the new identifiers correspond to the old ones?

Consider two sentences with work-level URIs *s1* and *s2* (see Figure 1). At time *t1*, these sentences are respectively realised by the transparent expression-level URIs *s1t1* and *s2t1*, and by the opaque version URIs *AE6* and *3F5*. The two identifier-types hold for the same XML element, and are therefore considered to be semantically equivalent, hence the **owl:sameAs** relation. At *t2* the sentences undergo no changes, sentences *s1* and *s2* are realised respectively by *s1t2* and *s2t2*, and again by *AE6* and *3F5*. At *t3*, however, a new text is inserted as sentence before the old version of the first sentence: *s1* is now realised by *s1t3* and *B9C*. Consequently *s2* is now realised by *s2t3* of which the hash is the same as that of *s1t2*: *AE6*.

By this method, globally persistent URIs of every element in a legal text can be consistently generated for both current and future versions of the text. By simultaneously generating an opaque and a transparent expression level URI, identification of these text versions does not have to rely on numbering.

5.3   Producing Metadata

The MetaLex converter produces three types of metadata. First, *legacy* metadata from attributes in the source XML is directly translated to RDF triples with an expression URI as subject, the literal attribute value as object, and an RDF property with the source attribute's name as predicate. Second, metadata describing the *structural* and *identity* relations between elements. This includes typing

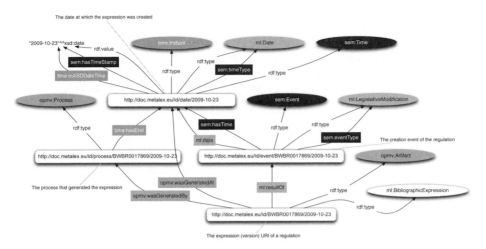

Fig. 2. Event model of the MDS

resources according to the MetaLex ontology, e.g. as ml:BibliographicExpression, creating ml:realizes relations between expressions and works, owl:sameAs relations between opaque and transparent expression URIs, ml:cites relations between citing and cited resources, and ml:partOf relations between expressions and parent elements in the XML. For each expression, we generate additional links to manifestations in RDF, XML and HTML, using rdfs:isDefinedBy, foaf:page and foaf:homePage properties, respectively. The official title, abbreviation and publication date of regulations are respectively represented using the dcterms:title, dcterms:alternative and dct:valid properties.

Events and Processes. Event information plays a central role in determining what version of a regulation was valid when. Processes capture essential provenance information needed for the interpretation of regulations. Traditional methods for assigning validity intervals to parts of regulations use multiple attributes to indicate e.g. enactment, publication, and efficacy dates [4]. Making explicit which events and modifying processes contributed to an expression of a regulation provides for a more flexible and extensible model. Especially since multiple different timestamps for the same element can be grouped together via the opaque URIs described in the preceding section.

The MDS uses the MetaLex ontology for legislative modification events, the Simple Event Model (SEM) of [8] (used in the eCulture domain) and the W3C Time Ontology [9] for an abstract description of events and event types, and the Open Provenance Model Vocabulary (OPMV) of [14] for describing processes and provenance information.[29] As depicted in Figure 2, these vocabularies can be combined in a compatible fashion, allowing for maximal reuse of event and process descriptions by third parties that may not necessarily commit to the MetaLex ontology.

---

[29] See http://www.w3.org/TR/owl-time and http://openprovenance.org

5.4   Serialization

The MetaLex converter supports three formats for serialising a legal text to a manifestation. First of all, it can produce the MetaLex format itself. The converter does not produce HTML since this can be easily obtained from the XML version. MetaLex can be viewed in a browser by linking a CSS stylesheet.[30] Secondly, generated RDF metadata can be serialised as inline RDFa attributes on meta tags. This is a very verbose format similar to N-triples (one element per triple), and it is often preferable to serialise the RDF as separate files using Turtle syntax,[31] unless the use case requires the representation of a legal text to be self contained. If required, the converter can automatically upload RDF to a triple store through either the Sesame API, or SPARQL updates.[32]

During conversion, citations are stored in a separate graph, linking citing resources at the level of articles (rather than at the level of inline elements carrying the reference) to cited resources. This graph can be exported to a '.net' network file, for further analysis in social network software tools such as Pajek and Gephi.[33] The MetaLex converter script optionally generates a network file containing citations of all regulations converted in the same batch.

5.5   Publication

The result of this procedure can be published through the MetaLex Document Server (MDS).[34] The MDS is essentially a Python wrapper for a SPARQL end-point for RDF metadata, and a file-based store for MetaLex documents and network files. It follows the Cool URIs specification,[35] and implements HTTP-based redirects for work- and expression level URIs to corresponding manifestations based on the HTTP accept header. Requests for an HTML mime-type are redirected to a the Marbles[36] HTML rendering of a Symmetric Concise Bounded Description (CBD) of the RDF resource.[37] Similarly, requests for RDF content return the SCBD itself; supported formats are RDF/XML and N3/Turtle. A request for XML will return the MetaLex of an XML snippet corresponding to the requested element. For work level identifiers MDS will only return RDF.

The MDS provides two convenience methods for retrieving manifestations of a regulation. Appending '/latest' to a work URI will redirect to the latest expression present in the triple store. Appending an arbitrary ISO date will return the last expression published before that date if no direct match is available.

---

[30] See http://www.w3.org/Style/CSS/ and
http://doc.metalex.eu/static/css/metalex.css
[31] See http://www.w3.org/TR/turtle/
[32] See http://openrdf.org and http://www.w3.org/TR/sparql11-update/
[33] See http://pajek.imfm.si/doku.php and http://gephi.org respectively.
[34] See http://doc.metalex.eu for the server, and
http://github.com/RinkeHoekstra/metalex-web-converter for the sources.
[35] See http://www.w3.org/TR/cooluris/
[36] See http://marbles.sourceforge.net/
[37] See http://www.w3.org/Submission/CBD/

Table 1. Conversion performance for 300 randomly selected regulations

Substitutions[40]	Number	%	Corrections	Number	%
container	22312	29 %	artikel	2525	72 %
hcontainer	3730	5 %	divisie	519	15 %
htitle	3730	5 %	colspec	289	8 %
block	34325	44 %	illustratie	54	2 %
inline	13527	17 %	*others*	99	3 %
Total	77624		*Total*	3486	
			Total no. of regulations	300	
			Revoked regulations	109	30 %
			Correction %		4 %

Lastly, the MDS offers a simple search interface for finding regulations based on the title and version date.

6   Conclusion and Results

We ran the MetaLex conversion script on all regulations available through the wetten.nl portal in May 2011, resulting in a total of 27.687 versions of regulations being converted, roughly 1 GB in size for BWB XML, and 2.27 GB as MetaLex.[38] The size increase is primarily due to the length and number of identifiers in MetaLex, which aren't present in BWB XML. The generated Turtle files comprise 9.9 GB, and contain 87.9 million triples. At this moment, the MDS runs on a 32GB Dell PowerEdge and a 4Store triple store.[39] New and modified regulations are published almost every other day, which means that the number of regulation versions accumulates with time: currently 28.752 versions and 100M triples (August 2011).

We evaluated the ability of MetaLex to map onto the BWB XML by running the converter on 300 randomly selected BWB identifiers; results are presented in Table 1. The artikel element accounts for 72% of all corrections, and corresponds to 68% of all htitle substitutions (5 % of total). This means that only a very small part of BWB XML does not directly fit onto the MetaLex schema. We have conducted a similar exercise on a single example of a CHLexML document and results were comparable; the main cause for incompatibility is the restriction in MetaLex that hcontainer elements are not allowed to contain block elements.

6.1   Meeting Requirements

Publishing identifiers and metadata of regulations in RDF meets the minimal requirements for facilitating *regulatory compliance* and *annotation*. Third parties can freely and transparently annotate regulations with specialised vocabularies and business rules. Our versioning scheme allows these annotations to be fine-grained and stable through time. For instance, annotating an opaque expression

---

[38] The actual number of regulations available at a single time is typically a bit lower. The conversion was done in several batches, and several modified regulations were published in the meantime.

[39] See http://4store.org

URI ensures that the annotation remains valid until the text of the expression changes, rather than when the official 'version' changes. Versioning and *time travel* is possible through the combination of SEM and the MetaLex ontology on the one hand, and opaque and transparent expression URIs on the other. Adoption of the OPMV vocabulary for expressing *provenance*, allows the construction of elaborate provenance trails, potentially referring even to pre-publication processes in the legislative drafting workflow.

Together with the Dutch Finance Ministry we started a pilot, based on [12], to detect both the definitions and scope of concepts as well as implicitly introduced concepts (nouns and noun phrases) in the domain of inheritance tax. All concepts are linked to both to the Cornetto Wordnet thesaurus in RDF and the relevant elements in law.[41] Although the scope of concepts can be made explicit by using namespaces or suffixes, ensuring *concept consistency* by resolving the scope to a set of elements internal and external to the law is an open issue. Furthermore, the inheritance tax law alone contains 1255 concepts in 72k triples, which will put a further strain on our hardware if concept extraction is let loose on other regulations.

Although we have shown that ingestion of a large corpus of legacy XML is feasible, other regulatory datasets need to be investigated to ensure the genericness of the approach. In particular, the transformation of different types of regulations, such as municipal regulations and EU directives, will contribute to the *integration* of regulations about similar topics. Conceptual annotation of legal sources will certainly improve the integration of these sources across the borders of government organisations.

Social network analysis of reference structures in legal texts allows us to determine certain properties of articles, such as the number of incoming citations (in degree), and the role of an article in connecting other articles (betweenness centrality). We conducted a small experiment at the immigration service (IND) to determine whether these measures corresponded to their intuitions of which articles are most *important* in immigration law: the in degree proved to be an almost perfect match to the most important articles, while betweenness centrality corresponded to articles in regulations that translated legislation to guidelines and procedures for civil servants at IND. Network analysis tools such as Gephi also provide nice visualisations of these reference networks, where closely related articles are grouped together, indicating *themes* in legislation. Gephi can even simulate how citations change through time. Indeed, all this is not new technology, but until now it has been beyond the grasp of civil servants in agencies such as the IND.

The MetaLex Document Server is an important step in opening up Dutch regulations for advanced analysis and semantic annotation. We described a procedure for incrementally rebuilding metadata and version information not made available by publicly accessible regulatory content services. Although the MDS and conversion script has not yet been used for converting other types of regulations, it was designed to be generically applicable to a wide range of legal

---

[41] See http://ckan.net/package/cornetto

XML formats by adopting the CEN MetaLex standard. We have furthermore gathered evidence that MetaLex is indeed able to represent and augment legal resources expressed in legacy XML. Perhaps most importantly, we have made a couple of people at the IND and the Dutch Ministry of Finance rather enthusiastic about the combination of legal information, network analysis, and Semantic Web technology.

References

1. Boer, A., Hoekstra, R., Winkels, R.: METALex: Legislation in XML. In: Bench-Capon, T., et al. (eds.) Jurix 2002: The Fifteenth Annual Conference, pp. 1–10. FAIA, IOS Press (2002)
2. Boer, A.: MetaLex Naming Conventions and the Semantic Web. In: Governatori, G. (ed.) Jurix 2009: The Twenty-Second Annual Conference. IOS Press (December 2009)
3. Boer, A., Winkels, R., van Engers, T., de Maat, E.: A Content Management System based on an Event-based Model of Version Management Information in Legislation. In: Gordon, T. (ed.) Jurix 2004: The Seventeenth Annual Conference, pp. 19–28. IOS Press (2004)
4. Boer, A., Hoekstra, R., Winkels, R., van Engers, T., Breebaart, M.: Time and Versions in METALex XML. In: Proceeding of the Workshop on Legislative XML. Kobaek Strand (2004)
5. Dick, J.P.: Conceptual Retrieval and Case Law. In: Proceedings of the International Conference on Artificial Intelligence & Law (ICAIL), pp. 106–115 (1987)
6. Fillmore, C.J.: The Case for Case. In: Bach, E., Harms, R.T. (eds.) Universals in Linguistic Theory. Holt, Rinehart and Winston (1968)
7. Gangemi, A., Pisanelli, D., Steve, G.: A formal ontology framework to represent norm dynamics. In: Winkels, R., Hoekstra, R. (eds.) Proceedings of the Second International Workshop on Legal Ontologies, LEGONT (2001)
8. van Hage, W., Malaisé, V., Segers, R., Hollink, L.: Design and Use of the Simple Event Model (SEM). Journal of Web Semantics (to appear, 2011)
9. Hobbs, J., Pan, F.: An Ontology of Time for the Semantic Web. Transactions on Asian Language Processing (TALIP): Special issue on Temporal Information Processing 3(1), 66–85 (2004)
10. Hoekstra, R.: Modernisation of the KAV system - A second opinion study on technology and implementation. Tech. rep., University of Amsterdam (2011)
11. Klarman, S., Hoekstra, R., Bron, M.: Versions and Applicability of Concept Definitions in Legal Ontologies. In: Clark, K., Patel-Schneider, P.F. (eds.) Proceedings of OWLED 2008, Washington, DC (metro) (April 2008)
12. de Maat, E., Winkels, R.: Automatic Classification of Sentences in Dutch Laws. In: Jurix 2008: The 21st Annual Conference. IOS Press (December 2008)
13. de Maat, E., Winkels, R., van Engers, T.: Automated Detection of Reference Structures in Law. In: van Engers, T.M. (ed.) Jurix 2006: The Nineteenth Annual Conference. IOS Press (December 2006)
14. Moreau, L., et al.: The Open Provenance Model core specification (v1.1). Future Generation Computer Systems (2010) (in press)
15. Saur, K.: Functional requirements for bibliographic records. IFLA Section on Cataloguing 19 (1998)
16. Vitali, F., Iorio, A., Gubellini, D.: Design patterns for document substructures. In: Extreme Markup 2005 Conference, Montreal (2005)

Leveraging Community-Built Knowledge for Type Coercion in Question Answering

Aditya Kalyanpur, J. William Murdock, James Fan, and Christopher Welty

IBM Research, 19 Skyline Drive, Hawthorne NY 10532
{adityakal,murdockj,fanj,welty}@us.ibm.com

Abstract. Watson, the winner of the Jeopardy! challenge, is a state-of-the-art open-domain Question Answering system that tackles the fundamental issue of answer typing by using a novel type coercion (TyCor) framework, where candidate answers are initially produced without considering type information, and subsequent stages check whether the candidate can be *coerced* into the expected answer type. In this paper, we provide a high-level overview of the TyCor framework and discuss how it is integrated in Watson, focusing on and evaluating three TyCor components that leverage the community built semi-structured and structured knowledge resources -- DBpedia (in conjunction with the YAGO ontology), Wikipedia Categories and Lists. These resources complement each other well in terms of precision and granularity of type information, and through links to Wikipedia, provide coverage for a large set of instances.

Keywords: Question Answering, Type Checking, Ontologies, Linked Data.

1 Introduction

Typing, the task of recognizing whether a given entity is a member of a given class, is a fundamental problem in many AI areas. We focus on the typing problem where both the entity and type are expressed lexically (as strings), which is typically the case in open-domain Question Answering (QA), and present a solution that leverages community-built web knowledge resources.

Many traditional QA systems have relied on a notion of *Predictive Annotation* [10] in which a fixed set of expected answer types are identified through manual analysis of a domain, and a background corpus is automatically annotated with possible mentions of these types before answering questions. These systems then analyze incoming questions for the expected *answer type*, mapping it into the fixed set used to annotate the corpus, and restrict candidate answers retrieved from the corpus to those that match this answer type using semantic search (IR search augmented with search for words tagged with some type).

This approach suffers from several problems. First, restricting the answer types to a fixed and typically small set of concepts makes the QA system brittle and narrow in its applicability and scope. Such a closed-typing approach does not work well when answer types in questions span a broad range of topics, are expressed using a variety of lexical expressions. Second, the QA system performance is highly dependent on the precision and recall of the predictive annotation software used, which acts as a candidate selection filter.

L. Aroyo et al. (Eds.): ISWC 2011, Part II, LNCS 7032, pp. 144–156, 2011.
© Springer-Verlag Berlin Heidelberg 2011

In contrast to this type-and-generate approach, we consider a *generate and type* framework, in which candidate answers are initially produced without use of answer type information, and subsequent stages check whether the candidate answer's type can be *coerced* into the Lexical Answer Type (LAT) of the question. The framework is based loosely on the notion of *Type Coercion* [11] (TyCor). The most notable aspects of the approach are: it does not rely on a fixed type system, however it does not discard one when available and useful; it is a multi-strategy and multi-source approach, gathering and evaluating evidence in a generative way rather than a predictive one; it is not part of candidate generation, rather it is simply another way of analyzing and scoring candidate answers; it is not a hard filter, producing for each candidate answer a probability that it is (or is not) of the right type.

In this paper, we provide a high-level architecture of the TyCor framework and discuss how it is integrated into Watson. We present three TyCor components that leverage community built semi-structured and structured knowledge resources, including DBpedia [2] (in conjunction with YAGO [13]), Wikipedia Categories and Lists.

2 Background: Open Domain Question Answering

Watson is a QA system capable of rivaling expert human performance on answering open-domain questions on the challenging TV quiz show Jeopardy!, whose questions cover a wide range of topics and are expressed using rich, complex natural language expressions. The typing problem for historical Jeopardy! questions is not trivial, as our analysis reveals that nearly any word in the English language can be used as an answer type in Jeopardy! questions, as shown in Table 1.

Given this variability, one of the intuitive problems with predictive annotation is that we cannot reliably predict what types there are going to be and what their instances are. We need to be open and flexible about types, treating them as a property of a question and answer combined. In other words, instead of finding candidates of the right type, we want to find candidates (in some way) and judge whether each one is of the right type by examining it in context with the answer type from the question. Furthermore, we need to accommodate as many sources as possible that reflect the same descriptive diversity as these questions. Furthermore, by not relying on a fixed type system, it is imperative to develop a system that has wide coverage for *rare* types.

Table 1. Sample Jeopardy! Questions showing variability in answer types

Jeopardy! Question	Answer
Invented in the 1500s to speed up the game, this *maneuver* involves 2 pieces of the same color	Castling
The first known airmail service took place in Paris in 1870 by this *conveyance*	Hot-air balloon
When hit by electrons, a phosphor gives off electromagnetic energy in this *form*	Light
A 1968 *scarefest*: The title character made it a family of 3 for the Woodhouses	Rosemary's Baby

2.1 DeepQA

Underlying the Watson system is DeepQA, a massively parallel probabilistic evidence-based architecture designed to answer open domain natural language questions. It consists of the following major stages (more details can be found in [4]):

Question Analysis: The first stage of processing performs a detailed analysis to identify key characteristics of the question (such as focus, lexical answer type, question class, etc.) used by later stages. The focus is the part of the question that refers to the answer, and typically encompasses the string representing the lexical answer type (LAT). The system employs various lexico-syntactic rules for focus and LAT detection, and also uses a statistical machine-learning model to refine the LAT(s). Like all parts of our system, LAT detection includes a confidence, and all type scores are combined with LAT confidence.

Hypothesis (Candidate) Generation: For the candidate generation step, the system issues queries derived from question analysis to search its background information (corpora, data- and knowledge-bases) for relevant content, and uses a variety of candidate generators to produce a list of potential answers.

Hypothesis and Evidence Scoring: Answer scoring is the step in which all candidates, regardless of how they were generated, are evaluated. During this phase, many different algorithms and sources are use to collect and score evidence for each candidate answer. Type information is just one kind of evidence that is used for scoring, other dimensions of evidence include temporal/spatial constraints, n-grams, popularity, source-reliability, skip-bigrams, substitutability, etc.

Candidate Ranking: Finally, machine-learning models are used to weigh the analyzed evidence and rank the answer candidates and produce a confidence. The models generate a confidence that each answer candidate is the correct answer to the given question, and the system answers with the top-ranked candidate. The system can also choose to refrain from answering if it has a low confidence in all of its candidates.

2.2 Type Coercion (TyCor)

The TyCor framework is part of Hypothesis and Evidence scoring, and consists of a suite of answer scoring components that each take a Lexical Answer Type (LAT) and a candidate answer, and return a probability that the candidate's type is the LAT. Each TyCor component uses a source of typing information and performs four steps, each of which is capable of error that impacts its confidence:

Entity Disambiguation and Matching (EDM): The most obvious, and most error-prone, step in using an existing source of typing information is to find the entity in that source that corresponds to the candidate answer. Since the candidate is just a string, this step must account for both polysemy (the same name may refer to many entities) and synonymy (the same entity may have multiple names). Each source may require its own special EDM implementations that exploit properties of the source, for example DBpedia encodes useful naming information in the entity URI. EDM implementations typically try to use some context for the answer, but in purely structured sources this context may be difficult to exploit.

Predicate Disambiguation and Matching (PDM): Similar to EDM, the type in the source that corresponds to the LAT must be found. In some sources this is the same algorithm as EDM, in others, type looking requires special treatment. In a few, especially those using unstructured information as a source, the PDM step just returns the LAT itself. In type-and-generate, this step corresponds to producing a semantic answer type (SAT) from the question. PDM corresponds strongly to notions of word sense disambiguation with respect to a specific source.

Type Retrieval (TR): After EDM, the types of the retrieved entity must be themselves be retrieved. For some TyCors, like those using structured sources, this step exercises the primary function of the source and is simple. In others, like unstructured sources, this may require parsing or other semantic processing of some small snippet of natural language.

Type Alignment: The results of the PDM and TR steps must then be compared to determine the degree of match. In sources containing e.g. a type taxonomy, this includes checking the taxonomy for subsumption, disjointness, etc. For other sources, alignments utilize resources like WordNet for finding synonyms, hypernyms, etc. between the types.

Each of the steps above generates a score reflecting the accuracy of its operation, taking into account the uncertainty of the entity mapping or information retrieval process. The final score produced by each TyCor component is a combination of the four step scores and the confidence in the LAT.

3   Acquiring Community-Built Knowledge for TyCor

We wanted to determine if community-built knowledge resources could be effectively (and cheaply) used to bootstrap a dynamic open-domain typing system, as well as deal with the very long tail of answer types. For this reason, we acquired a broad domain structured knowledge base (DBpedia) and ontology (YAGO), and semi-structured folksonomies with wide topical coverage (Wikipedia Categories and Lists).

3.1   DBpedia and YAGO

The DBpedia knowledge base contains relational information found in the info-boxes of Wikipedia pages. A one-to-one correspondence between all Wikipedia pages and DBpedia entries maps the two resource names (or URIs): e.g., the DBpedia page with URI http://dbpedia.org/resource/IBM corresponds to the Wikipedia page titled "IBM" with relational facts (triples) captured from the infobox.

Additionally, DBpedia has type assertions for many instance objects. The types are assigned from a collection of ontologies, including YAGO, a large taxonomy of more than 100K types. Crucially, the YAGO ontology has mappings to WordNet [7]: every YAGO type corresponding to a WordNet concept has the associated 9-digit WordNet sense id appended to its name/id. Thus the YAGO type "*Plant100017222*" links to the WordNet concept plant (living organism), while the type "*Plant103956922*" corresponds to the concept of an industrial plant or factory.

YAGO types are arranged in a hierarchy, and DBpedia instances are often assigned several low-level types corresponding to Wikipedia categories (e.g. "*CompaniesEstablishedIn1896*"). For these, navigation up the YAGO type tree leads to more general and normalized (via sense-encoding) YAGO WN concepts.

These design points of DBpedia and YAGO enable us to obtain precise type information for many instances, given Wikipedia domain coverage and YAGO-WordNet type/sense coverage.

One downside is that the YAGO ontology does not handle mutually exclusive (disjoint) types – ones that do not share instances. For example, Country and Person types are logically disjoint: no given instance can be both a Country and a Person; on the other hand, Painter and Musician are not disjoint. Type disjointness is useful for QA, to rule out candidate answers with types incompatible with the question LAT. For this reason, we decided to add disjointness relations between YAGO types. Given the size of YAGO ontology (> 100K types), manually asserting such relations between all applicable type pairs is infeasible. Instead, we only specify disjointness between prominent top-level types of the YAGO hierarchy, and use a logical reasoning mechanism to propagate the disjointness to lower subclasses. For example, it follows logically that if the *Person* and *GeoPoliticalEntity* are disjoint, then every subclass of *Person* is disjoint with every subclass of *GeoPoliticalEntity* (e.g. *Musician* is disjoint with *Country*). Our additions to the YAGO Type system comprise approximately 200 explicit disjoint relations, which translate through inference to more than 100K disjoint relations.

3.2 Wikipedia-Based Folksonomies

For the purposes of this paper, we consider Wikipedia Categories and Lists to be folksonomies. Wikipedia categories are true tags that are applied to articles by Wikipedia users without very much centralized control, and new categories can be invented as desired. The categories have some explicit structure, in that category pages can themselves be put into categories.

One may reasonably argue that Wikipedia lists are not true tags, as they are assigned in a more middle-out and sometimes top-down method than bottom up. Lists are generally created and then populated with articles, and frequently one cannot access the lists an article is in from the article. We ignore this difference and treat Wikipedia lists as the same kind of resource as categories.

4 TyCor Algorithms

In this section, we describe the algorithms underlying the three TyCor components that use YAGO (through DBpedia), Wikipedia Categories and Lists respectively as a source of type information.

4.1 Shared EDM Algorithm

The three TyCors described here share one EDM algorithm, that takes as input the candidate answer string and a corresponding context – the question text and (optionally) a text passage containing the candidate – and returns a ranked list of Wikipedia page URIs that match the candidate, with associated match scores. The match scores are computed based on five heuristics:

- **Direct Contextual Match.** In some cases the Wikipedia URI of the candidate is known, and EDM is not performed. For example, if the candidate answer was generated from the title of a Wikipedia page, or if the candidate mention itself is hyper-linked to another Wikipedia page, we store that information in our candidate answer object and use it as the result of our EDM step with a score of 1.0.
- **Title Match.** When there is an exact string match between the candidate string and the title of a Wikipedia page, the URI is returned with a score of 1.0.
- **Redirect Match.** When the candidate string matches the name of a redirect page, the redirect destination URI is returned with a score of 1.0. There are some noisy redirects in Wikipedia, e.g. Eliza Doolittle (character) redirects to Pygmalion (play), but in general we have observed the redirects to be reliable.
- **Disambiguation Match.** When the candidate string matches the title of a Wikipedia disambiguation page all the disambiguation URIs are returned with a score of 1/(the number of disambiguations).
- **Anchor-Link Match.** When a candidate string matches one or more anchor text strings in Wikipedia, all the URIs pointed to by those anchors are returned with a score for each based on the conditional probability of the link pointer given the anchor text (ie how often does the anchor text point to the URI).
- **DBpedia name properties.** DBpedia includes over 100 *name properties*, properties whose objects are some form of name string (firstName, lastName, etc). When a candidate string matches one of these, the triple subject is returned with a score of 1/(number of URIs returned).

The EDM algorithm also contains an optional parameter to rank the results based on the *popularity* of the corresponding Wikipedia page, overriding the confidence set by the heuristics. Popularity is computed using information such as page-rank and IDF of the title string.

4.2 YAGO TyCor

The YAGO TyCor uses the EDM step described above, and transforms the Wikipedia page URLs returned at the end of the step to corresponding DBpedia URIs.

Type Retrieval Using DBpedia. The TR algorithm produces a set of URIs for the Yago types of the candidate entity (the result of the EDM step). DBpedia contains type information for a vast number of Wikipedia entities, represented by the *rdf:type* relation, that come from several ontologies; the largest is YAGO. In many cases, the explicitly assigned type for an entity is a low-level (and highly-specific) type in the YAGO ontology, such as *yago:CompaniesEstablishedIn1898* (typically derived from Wikipedia Categories). We generalize this type by navigating up the hierarchy till we reach a Yago type that has a WordNet sense id associated with it (e.g. *yago:Company108058098*). This generalization helps type alignment (when aligning with types derived from the LAT) and improves the coverage of the TyCor.

PDM in YAGO. The PDM algorithm produces a set of URIs for the Yago types that match the LAT, by matching the LAT to the labels or IDs of Yago types. We then score the matches based on a weighted combination of its WordNet sense rank, and the number of instances of the concept in DBpedia. The latter is an approximation of type popularity, and has performed well in our experiments.

There are two additional features of our PDM algorithm that help improve its precision and recall respectively. The first is the notion of a domain-specific type-mapping file that is optionally input to the algorithm. For example, based on analysis of Jeopardy! question data, we found the LAT *"star"* refers to the sense of star as a *movie star* roughly 75% of the time, with the remaining cases referring to the *astronomical object*.

The second heuristic we use in PDM helps improve its recall. We estimate a statistical relatedness between two types by computing the conditional probability that an instance with type A also has type B, using the metric: NI (A and B) / NI (A), where NI is the number of instances of the concept in DBpedia (including instances of its subtypes). In PDM, if the lexical type matches some YAGO type, we expand it to include related types based on their conditional probabilities exceeding an empirically determined threshold (0.5).

YAGO Type Alignment. The type alignment algorithm produces a single score based on the alignment of the *instance types* from the TR step, and the *LAT types* from the PDM step. The algorithm uses these conditions:

- **Equivalent/Subclass match.** When the instance type and the LAT type are equivalent (synonyms) in the YAGO ontology, or the instance type is a subclass (hyponym) of the LAT type, a score of 1.0 is returned.
- **Disjoint match.** When the instance type and LAT type are found to be disjoint (based on axioms added to YAGO, see Section 3.1) a score of -1.0 is returned.
- **Sibling match.** When the instance type and LAT type share the same parent concept in the YAGO ontology, a score of 0.5 is returned. In this case, we exclude cases in which parent classes' depth < 6, since these high level types (like "Physical Entity") tend to be less useful.
- **Superclass match.** When the instance type is a superclass (hypernym) of the LAT type a score of 0.3 is returned. This may seem counter-intuitive since the candidate answer is supposed to be an instance of the LAT and not vice versa, however, we have seen cases where checking the type alignment in the opposite direction helps, either due to inaccuracies in the EDM or PDM step, or due to source errors, or the question itself asks for the type of a particular named entity.
- **Statistical Relatedness.** When the statistical type relatedness between the instance type and LAT type, computed as described above, exceeds an empirically determined threshold of 0.5, a score of 0.25 is returned.
- **Lowest Common Ancestor (LCA).** When the LCA of the instance type and LAT type is deep in the taxonomy (we use a depth threshold of 6 in the Yago taxonomy), a score of 0.25 is returned. This is based on the intuition that the types are strongly related, even though they may not be a direct subclass or sibling relationship among them, if their LCA is not a very high level class.

The thresholds used in the type matching conditions above and the weights of the respective rules are manually assigned based on an empirical evaluation, conducted by running the algorithm on a large number of test cases. Since the TR phase may produce multiple types per candidate, the maximum type alignment score is returned.

4.3 Wiki-Category and Wiki-List TyCors

The Wiki-Category and Wiki-List TyCors are fundamentally different from YAGO TyCor because the types that they use are natural language strings and not types in a formal ontology. The Wikipedia list pages do not have any explicit structure among them. Wikipedia categories do have some hierarchical structure, but the Wiki-Category TyCor does not use that structure as it is too unreliable.

Both of these TyCors use the same **Entity Disambiguation and Matching** component as YAGO TyCor. They also both use a simple lookup in an RDF store for **Type Retrieval** (we augmented DBpedia with list associations), that returns the category (resp. list) names for the entity.

Wiki-Category and Wiki-List both have a trivial **Predicate Disambiguation and Matching** step that simply returns the LAT itself.

Type Alignment is the most complex portion of these TyCors. It receives as input a natural language LAT and a natural language type (from Type Retrieval) and tries to determine if they are consistent. In both cases, Type Alignment divides this task in to two distinct subtasks:

(1) Is the head word of the LAT consistent with the head word of the category or list names,
(2) Are the modifiers of head word of the LAT consistent with the modifiers of the category or list names.

In both cases, terms are matched using a variety of resources such as WordNet. For example, given a list named "cities in Canada" and a question asking for a "Canadian metropolis" (i.e., with a LAT "metropolis" that has a modifier "Canadian"), Type Alignment will separately attempt to match "city" with "metropolis" and "Canada" with "Canadian." Type Alignment uses a variety of resources to do this matching; for example, in WordNet the primary sense of "city" is synonymous with the primary sense of "metropolis." Wiki-Category and Wiki-List provide separate match scores for the (head word) type match and the modifier match. Those separate scores are used as features by the DeepQA Candidate Ranking mechanism.

5 Experiments

All experiments were done on the March, 2010 version of Wikipedia, and used DBpedia release 3.5. Wikipedia categories were obtained from DBpedia, Wiki lists were scraped from Wikipedia.

5.1 Evaluating EDM on Wikipedia Link Anchors

We evaluated the performance of the EDM component on a Wikipedia link anchor data set. This data set is comprised of 20,000 random pairs of Wikipedia anchor texts and their destination links. Note that the destination of an anchor text is a Wikipedia article whose title (string) may or may not explicitly match the anchor text. For example, the article *Gulf of Thailand* contains the following passage: *"The boundary of the gulf is defined by the line from <u>Cape Bai Bung</u> in southern Vietnam (just south of the mouth of the <u>Mekong</u> river) to the city <u>Kota Baru</u> on the Malayian coast"*. While anchors *Mekong* and *Kota Baru* point to articles whose titles are exactly the same as the anchor text, *Cape Bai Bung*'s link points to the article titled *"Ca Mau Province"*.

We use the anchor texts as inputs to EDM, and the destinations as ground truth for evaluation, similar to [4]. The performance of the EDM is shown in Table 4 with precision, recall (over all candidates returned) and the average number of candidates returned. We tested four versions of EDM, with and without popularity ranking and DBpedia name properties. Note that DBpedia names increase the number of candidates without impacting precision or recall. This is partially a side-effect of the link-anchor based evaluation, which skews the results to prefer alternate names that have been used as link anchor texts, and thus does not really reflect a test of the data the alternate names provide. However, from inspection we have seen the name properties to be extremely noisy, for example finding DBpedia entities named "China" using the name properties returns 1000 results. The TyCor experiments below, therefore, do not use DBpedia name properties in the EDM step.

Table 4. EDM performance on 20,000 Wikipedia anchor texts

Ranking	Names	Precision	Recall	Avg. # of candidates
No	No	74.6%	94.3%	16.97
No	Yes	74.7%	94.3%	13.02
Yes	Yes	75.2%	94.3%	16.97
Yes	No	75.7%	94.3%	13.02

5.2 Evaluating Typing on Ground Truth

Although the TyCor components' impact on end-to-end QA performance can be measured through ablation tests, they do not reflect how well the components do at the task of type checking because wrong answers may also be instances of the LAT. To measure the TyCor components performance on the task of entity-typing alone, we manually annotated the top 10 candidate answer strings produced by Watson from 1,615 Jeopardy! questions, to see whether the candidate answer is of the same type as the lexical answer type. Because some questions contain multiple LATs, the resulting data set contains a total of 25,991 instances. Note that 17,384 (67%) of these are negative, i.e. the candidate answer does not match the LAT.

Table 5. Evaluating TyCor Components on Entity-Type Ground Truth

Tycor Component	Accuracy	Precision	Recall
Yago Tycor	76.9%	64.5%	67.0%
Wikipedia Category Tycor	76.1%	64.1%	62.9%
List Tycor	73.3%	71.9%	31.6%
All Three (Union)	73.5%	58.4%	69.5%

Table 5 shows the performance of the three TyCor components that use community–built knowledge, by counting any candidate with a TyCor score > 0.0 to be a positive judgment; this is not the way the score is used in the end to end Watson system, but gives a sense for how the different components perform. The "All Three" experiment counts any candidate with at least one score from the three TyCors that is > 0.0 to be a positive judgment. The bump in recall shows that they can complement each other, and in the end to end system experiments below, this is validated.

5.3 Impact on End-to-End Question Answering

Table 6 shows the accuracy of the end-to-end DeepQA question answering system with different TyCor configurations.

Table 6. Accuracy on end-to-end question answering with only the TyCors specified in each column

	No TyCor	YAGO TyCor	Wiki-Category TyCor	Wiki-List TyCor	All 3 TyCors
Baseline Accuracy	50.1%	54.4% (+4.3%)	54.7% (+4.6%)	53.8% (+3.7%)	56.5% (+6.5%)
Watson Accuracy	65.6%	68.6% (+3.0%)	67.1% (+1.5%)	67.4% (+1.8%)	69.0% (+3.4%)

The "Baseline Accuracy" shows the performance of a simple baseline DeepQA configuration on a set of 3,508 previously unseen Jeopardy! questions. The baseline configuration includes all of the standard DeepQA candidate generation components, but no answer scoring components other than the TyCors listed in the column headings. The baseline system with no TyCor components relies only on the candidate generation features (e.g., rank and score from a search engine) to rank answers. The subsequent columns show the performance of the system with only that component added (the difference from No Tycor is show in parens), and the final column shows the impact of combining these three TyCors.

The "Watson Accuracy" shows the accuracy of the complete Watson question answering system *except* for the TyCor components. Again, the first column shows the accuracy of the system with no TyCor components, and the next four columns show distinct (*not* cumulative) additions, and the last column shows the accuracy with all three combined. Again, each TyCor alone is better than no TyCor, and effectively combining all three components is better than any one of them.

All the column-wise gains (from Baseline to Watson), and all the differences from "No Tycor" shown in the table are statistically significant (significance assessed for p < .05 using McNemar's test with Yates' correction for continuity). The "All-3 Tycors" improvement is significant over the individuals in all cases except the Watson Yago TyCor.

6 Related Work

QUARTZ [12] is a QA System that uses a statistical mapping from LATs to WordNet for PDM, and collocation counts for the candidate answer with synonyms of the mapped type for Type Retrieval. In [6] the approach has been taken a step further by combining correlation-based typing scores with type information from resources such as Wikipedia, using a machine-learning based scheme to compute type validity. Both [6] and [12] are similar to our TyCor approach in that they defer type-checking decisions to later in the QA pipeline and use a collection of techniques and resources (instead of relying on classical NERs) to check for a type match between the candidate and the expected answer type in the question. However, the fundamental difference with our approach is that the type match information is not used as a filter to throw out candidate answers, instead, the individual TyCor scores are combined with other answer scores using a weighted vector model. Also, our type-coercion is done within a much more elaborate framework that separates out the various steps of EDM, PDM, Type Retrieval and Alignment etc.

A similar approach to our combination of NED and WikiCat is presented in [3]. The traditional type-and-generate approach is used when question analysis can recognize a semantic answer type in the question, and falls back to Wikipedia categories for candidate generation, using it as a hard filter instead of predictive annotation. In our approach we assume any component can fail, and we allow other evidence, from other TyCor components or other answer scoring components, to override the failure of one particular component when there is sufficient evidence.

An approach to overcoming problems of *a-priori* answer type systems in proposed in [9], based on discriminative preference ranking of answers given the question focus and other words from the question. This approach is actually quite similar in spirit to other components of our TyCor suite that we did not discuss here. In other work we have shown that techniques like this provide coverage for infrequent types that may not have been accounted for in some ontology or type system, such as "scarefest" to describe a horror movie, but do not perform nearly as well on the types known by the ontology. Thus we found combining techniques like [9] with those that use structured information provides the best overall performance.

A TyCor component that uses Linked Open Data (LoD), including DBpedia, geo-Names, imdb, and MusicBrainz was presented in [8]. This TyCor component includes a special-purpose framework for scaling LoD type triple datasets combined with *latent semantic indexing* to improve the matching steps (EDM and PDM), and an intermediate ontology designed specifically for the Jeopardy! task. The approach is evaluated on the classification task alone, as our question answering performance was still confidential. The classification performance is considerably lower than shown here, roughly 62% accuracy, though we expect it to improve as more LoD sources are added. In internal experiments, it had no impact on Watson performance, and was not used in the final fielded Watson system.

The idea of using Wikipedia link anchor text as a gold standard was presented in [4], along with a word-sense disambiguation algorithm for EDM using context vectors and Wikipedia categories. Our EDM results (see Table 4) are roughly the same, as the 86-88% accuracy numbers reported in [4] do not include cases where recall fails completely. In our experiments, we found popularity ranking of the results from our heuristics performs just as well as the method in [4], and is significantly faster at run-time.

7 Conclusion

We have presented a novel open-domain type coercion framework for QA that overcomes the brittleness and coverage issues associated with Predictive Annotation techniques. The TyCor framework consists of four key steps involving entity disambiguation and matching (EDM), predicate disambiguation and matching (PDM), type retrieval and type alignment. We have shown how community-built knowledge resources can be effectively integrated into this TyCor framework and provided corresponding algorithms for the four TyCor steps. The algorithms exploit the structure and semantics of the data, and in some cases, benefit from extensions made to existing knowledge to add value (e.g. addition of disjoints to YAGO). Our results show that the TyCors built using Web knowledge resources perform well on the EDM and entity typing tasks (both fundamental issues in NLP and Knowledge Acquisition), as well significantly improving the end-to-end QA performance of the Watson system (which uses machine learning to integrate TyCor) on rich and complex natural language questions taken from Jeopardy!

References

1. Aktolga, E., Allan, J., Smith, D.A.: Passage Reranking for Question Answering Using Syntactic Structures and Answer Types. In: Clough, P., Foley, C., Gurrin, C., Jones, G.J.F., Kraaij, W., Lee, H., Mudoch, V. (eds.) ECIR 2011. LNCS, vol. 6611, pp. 617–628. Springer, Heidelberg (2011)
2. Auer, S., Bizer, C., Kobilarov, G., Lehmann, J., Cyganiak, R., Ives, Z.G.: DBpedia: A Nucleus for a Web of Open Data. In: Aberer, K., Choi, K.-S., Noy, N., Allemang, D., Lee, K.-I., Nixon, L.J.B., Golbeck, J., Mika, P., Maynard, D., Mizoguchi, R., Schreiber, G., Cudré-Mauroux, P. (eds.) ASWC 2007 and ISWC 2007. LNCS, vol. 4825, pp. 722–735. Springer, Heidelberg (2007)
3. Buscaldi, D., Rosso, P.: Mining Knowledge from Wikipedia for the Question Answering task. In: Proceedings of the International Conference on Language Resources and Evaluation (2006)
4. Cucerzan, S.: Large-Scale Named Entity Disambiguation Based on Wikipedia Data. In: Proceedings of EMNLP 2007, Prague, pp. 708–716 (2007)
5. Ferrucci, D., Brown, E., Chu-Carroll, J., Fan, J., Gondek, D., Kalyanpur, A.A., Lally, A., William Murdock, J., Nyberg, E., Prager, J., Schlaefer, N., Welty, C.: Building Watson: An Overview of the DeepQA Project. AI Magazine (2010)
6. Grappy, A., Grau, B.: Answer type validation in question answering systems. In: Proceeding RIAO 2010 Adaptivity, Personalization and Fusion of Heterogeneous Information (2010)
7. Miller, G.A.: WordNet: A Lexical Database for English. Communications of the ACM 38(11), 39–41 (1995)

8. Ni, Y., Zhang, L., Qiu, Z., Wang, C.: Enhancing the Open-Domain Classification of Named Entity Using Linked Open Data. In: Patel-Schneider, P.F., Pan, Y., Hitzler, P., Mika, P., Zhang, L., Pan, J.Z., Horrocks, I., Glimm, B. (eds.) ISWC 2010, Part I. LNCS, vol. 6496, pp. 566–581. Springer, Heidelberg (2010)
9. Pinchak, C., Lin, D., Rafiei, D.: Flexible Answer Typing with Discriminative Preference Ranking. In: Proceedings of EACL 2009, pp. 666–674 (2009)
10. Prager, J.M., Brown, E.W., Coden, A., Radev, R.: Question-Answering by Predictive Annotation. In: Proceedings of SIGIR 2000, Athens, Greece, pp. 184–191 (2000)
11. Pustejovsky, J.: Type Coercion and Lexical Selection. In: Pustejovsky, J. (ed.) Semantics and the Lexicon. Kluwer Academic Publishers, Dordrecht (1993)
12. Schlobach, S., Ahn, D., de Rijke, M., Jijkoun, V.: Data-driven type checking in open domain question answering. J. Applied Logic 5(1), 121–143 (2007)
13. Suchanek, F.M., Kasneci, G., Weikum, G.: Yago: A Core of Semantic Knowledge-Unifying WordNet and Wikipedia. In: Proceedings WWW (2007)

Privacy-Aware and Scalable Content Dissemination in Distributed Social Networks⋆

Pavan Kapanipathi[1,2], Julia Anaya[1], Amit Sheth[2],
Brett Slatkin[3], and Alexandre Passant[1]

[1] Digital Enterprise Research Institute
National University of Ireland, Galway
{julia.anaya,alexandre.passant}@deri.org
[2] Kno.e.sis Center, CSE Department
Wright State University, Dayton, OH - USA
{pavan,amit}@knoesis.org
[3] Google, San Francisco, CA - USA
bslatkin@google.com

Abstract. Centralized social networking websites raise scalability issues — due to the growing number of participants — and policy concerns — such as control, privacy and ownership of users' data. Distributed Social Networks aim to solve those by enabling architectures where people own their data and share it whenever and to whomever they wish. However, the privacy and scalability challenges are still to be tackled. Here, we present a privacy-aware extension to Google's PubSubHubbub protocol, using Semantic Web technologies, solving both the scalability and the privacy issues in Distributed Social Networks. We enhanced the traditional features of PubSubHubbub in order to allow content publishers to decide whom they want to share their information with, using semantic and dynamic group-based definition. We also present the application of this extension to SMOB (our Semantic Microblogging framework). Yet, our proposal is application agnostic, and can be adopted by any system requiring scalable and privacy-aware content broadcasting.

Keywords: Semantic Web, Distributed Social Networks, Social Web, Privacy, FOAF, PubSubHubbub.

1 Introduction

Centralized social networking websites, such as Twitter or Facebook, have raised, on the one hand, scalability issues [14] — due to the growing number of participants — and, on the other hand, policy concerns — such as control, privacy and ownership over the user's published data [2]. Distributed Social Networks, such

⋆ This work is funded by Science Foundation Ireland — Grant No. SFI/08/CE/I1380 (Líon-2) — and by a Google Research Award. We would also like to thank Owen Sacco for his work on PPO and Fabrizio Orlandi for his feedback on SMOB.

L. Aroyo et al. (Eds.): ISWC 2011, Part II, LNCS 7032, pp. 157–172, 2011.
© Springer-Verlag Berlin Heidelberg 2011

as SMOB[1], StatusNet[2], Diaspora[3] or OneSocialWeb[4], aim to solve this issue by enabling architectures where people own their data and share it intentionally. While they use different stack, their goal is to allow users to setup their own "Social Space" — as people can do now by setting up a weblog —. Synchronisation between the different user spaces is performed with tools and protocols ranging from XMPP[5] to SPARQL 1.1 Update and its protocol[6] to OStatus[7] or to Activity Streams[8]. Yet, scalability, and most importantly privacy are still ongoing challenges. New techniques are needed to deal with information overload and to ensure content is directed only to intended recipient. This would enable, for instance, to keep a large list of followers/friends[9], and to limit the distribution of particular content to only a subset of people, on-demand (as opposed to generic policies such as "friends" or "family"). For example, limiting content about project X only to project members, this list being dynamically generated.

To achieve this goal, we have built an extension of Google's PubSubHubbub protocol [3] (aka PuSH, described in the next section), that improves both the scalability and the privacy issues in Distributed Social Networks. We enhanced its original broadcasting feature in order to allow publishers to decide whom they want to share their information with, among the users in their social network. Using our approach, content is delivered on-demand to a list of interested parties, as defined by the publisher, but using dynamic preferences. We do this by combining PuSH (including an RDF ontology to describe its core attributes), SPARQL 1.1 Update and a the Privacy Preference Ontology [17] — a lightweight vocabulary for modeling user privacy on the Social Web. Therefore, our approach aims at combining "the best of both worlds", re-using efficient and pragmatic Web 2.0 approaches (PuSH and RSS) with outcomes from the Semantic Web community (lightweight vocabularies and SPARQL). In the rest of this paper we first discuss some background information used in our work (Section 2). We then describe our motivation for, and how we extended the PuSH protocol (Section 3). Further we detail an implementation use-case (Section 4) and conclude with the related work (Section 5).

2   Background

2.1   Distributed Social Networks and PubSubHubbub

Centralized Social Networks (CSN) such as Facebook, Myspace and Twitter suffer drawbacks such as those mentioned in Section 1. For instance, the growing

---

[1] http://smob.me
[2] http://status.net
[3] http://joindiaspora.com
[4] http://onesocialweb.org
[5] http://xmpp.org
[6] http://www.w3.org/TR/sparql11-http-rdf-update/
[7] http://ostatus.org
[8] http://activitystrea.ms/
[9] I.e. people allowed to see your content and information.

number of Twitter users has been a continuous concern for the performance of
the service[10]. Issues related to the sharing of personal information with third
party websites by Facebook[11] or Twitter retweet issues [11] have defeated the
privacy mechanisms of these services. These lead to new approaches to engineer
Online Social Networks (OSN), termed as Distributed Social Networks (DSN)
[10]. While CSNs store users' data in their own servers and owns user's data
as per Terms of Service, DSNs distribute the data across users, emphasizing on
data portability and interoperability. In addition, they promote ownership of
users' data, as data resides either on a trusted server or on a local computer.

Implementing DSN requires various layers, including one to transmit data
between users' platform. A common way to do this is Google's PubSubHubbub
(PuSH), a decentralized publish-subscribe protocol which extends Atom/RSS to
enable real-time streams. It allows one to get near-instant notifications of the
content (s)he is subscribed to, as PuSH immediately "pushes" new data from
publisher to subscriber(s) where RSS readers must periodically "pull" new data.
The PuSH ecosystem consist of a few *hubs*, many *publishers*, and a large number
of *subscribers*. Hubs enable (1) publishers to offload the task of broadcasting new
data to subscribers; and (2) subscribers to avoid constantly polling for new data,
as the hub pushes the data updates to the subscribers. In addition, the protocol
handles the communication process between publishers and subscribers:

1. A subscriber pulls a feed (Atom/RSS) from a publisher (a "topic" in the
   PuSH terminology). In its header, the feed refers to a hub where the subscriber must register to get future notifications about publisher's updates;
2. The subscriber registers to the feed at the hub's URL. This process is automatically done by the subscriber the first time the feed is accessed;
3. The publisher notifies the hub whenever new content is published — also,
   the hub can check for updated directly from the publisher by pulling its feed;
4. Finally, when new content is generated by the publisher, the hub sents updates to all its subscribers for this feed.

PuSH is a scalable protocol, and Google provides a public hub that people can
use to broadcast their content[12]. This public hub delivers for approximately
40 million unique active feeds, with 117 million subscriptions. In two years,
approximately 5.5 billion unique feeds have been delivered, fetching 200 to 400
feeds and delivering 400 to 600 of them per second. Its largest subscribers get
between 20 and 120 updates per second from the hub.

2.2   Semantics in Distributed Social Networks

Within DSN, individuals mapped to each other with their social relationships
form what is generally termed as a "social graph", that became popular with

---

[10] http://mashable.com/2010/06/11/twitter-engineering-fail/
[11] http://www.adweek.com/news/advertising-branding/
     facebook-facing-more-privacy-issues-126296
[12] http://pubsubhubbub.appspot.com/

OSNs such as Facebook. OSNs and other Social Web services take advantage of the relationships between individuals to provide better and more personalized online experience. In [8], Brad Fitzpatrick, founder of the LiveJournal blogging community[13], discussed his views on building a *decentralized social graph* and the aggregation of individual's friends across sites.

Lightweight semantics can play an important role in social graphs and DSN, allowing to share content between users whether or not they are on the same system. FOAF [5] — Friend of a Friend — is generally used to represent information about individuals (name, e-mail, interests, etc.) and their social relations in a machine readable format. Generally combined with FOAF, SIOC [4] — Semantically-Interlinked Online Communities — is a lightweight vocabulary used (in combination with several of its modules) to represent social data and user activities (blog posts, wiki edits, etc.) in RDF. To a larger extent, vocabularies such as the Open Graph Protocol[14], or schema.org[15] could be considered to model the objects being manipulated and shared by users (movies, photos, etc.) — especially as they may have a larger uptake than the previous ones at Web-scale.

2.3   WebID

To enable users privacy and secure communications in a DSN, an authentication protocol is required. WebID [19] is an decentralized authentication protocol that allows users to manage their own identities and data privacy. It uses X.509 certificates and SSL certificate exchange mechanisms to provide an encrypted communication channel and ensures that users are who they claim, represented by a WebID URI — generally identifying a `foaf:Person`. Hence, FOAF relation may be enhanced with trust descriptions so as to create a reputation network. Moreover, this trust network can be backed by the use of cryptographic keys and digital signatures, so as to form a secure *Web of Trust*[16].

It can also be used for authorization purposes in conjunction with other vocabularies and ontologies such as, Privacy Preference Ontology (PPO) [17] to provide a fine grained access control. In a nutshell, the protocol works as follows:

1. A client sends its X509 certificate (including his WebID URI) to a server;
2. The server extracts the public key and the URI entries from certificate;
3. The server dereferences the URI and extracts a public FOAF profile;
4. The server attempts to verify the public key information. If the public key in the certificate is part of the public keys associated with the URI, the server assumes that the client uses this public key to verify their ownership of the WebID URI;
5. The client is authenticated, authorization mechanism can be applied.

---

[13] http://livejournal.com
[14] http://ogp.me
[15] http://schema.org
[16] http://xmlns.com/wot/0.1/

2.4   PPO - The Privacy Preference Ontology

By itself, WebID does not determine what kind of access an authenticated user has on a resources. Yet, it can be combined with authorization mechanisms to provide such access control. The Privacy Preference Ontology [17] (PPO) is a lightweight vocabulary built on Web Access Control ontology [13] to provide fine-grained restrictions to access RDF data. It consists of a main class `PrivacyPreference` with properties defining (1) the resource to be restricted; (2) the conditions to create the privacy preferences; (3) the access privileges and; (4) the attribute patterns that must be satisfied by the requester — also known as access space. Access Spaces are SPARQL queries, checking FOAF profiles of the requesters to grant access (or not) to some data, so that FOAF plays a central role in the authorization process.

For instance, in a scenario when Alice requests to access to Bob's information (*e.g.* a microblog post), Bob's privacy preference for the corresponding resource are checked. If the access spaces for this preference matches Alice's description (from her FOAF profile), she will gain access to the requested data. A resource can have multiple access spaces, and access is granted if one's profile matches at least one of the access spaces.

3   Extending PubSubHubbub for Privacy-Aware Content Dissemination

3.1   Motivations for Extending PuSH

PuSH provides a distributed architecture and hence more scalability compared to a centralized architecture, but it still does not implement any privacy policies. In CSN such as Twitter, minimal privacy settings are provided to users. Users can either make their account *public* (by default, everyone can view their content) or *protected* (only approved followers can view their content). Yet, the lack of fine-grained privacy policies caused several incidents, such as people being fired because some content reached undesired people in their network[17].

Using PuSH in OSNs brings similar patterns where a publisher can either broadcast his data to all the subscribers or not. Although it would be possible to enable finer-grained access control in PuSH by creating one feed per subscriber; this is considered to be difficult at significant scale. Therefore, we extend PubSubHubbub to feature user-controlled data dissemination. This allows one user to dynamically create groups of people who will receive a private post that remains hidden to other users.

3.2   PuSH Extension

The Publisher and the Hub are extended with respect to their counter parts in the original PubSubHubbub protocol, while the Subscriber functionality is kept

---

[17] http://www.huffingtonpost.com/2010/07/15/
fired-over-twitter-tweets_n_645884.html

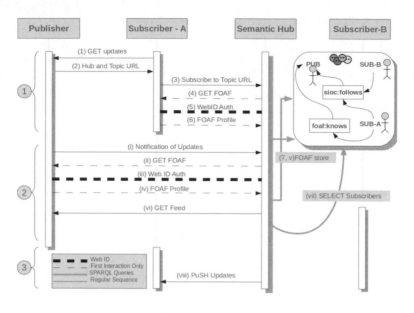

Fig. 1. Sequence of Interactions

intact. Semantic Web technologies such as RDF, SPARQL and tools such as Triple-Stores are the primary modifications we brought. Following the original design principles of PuSH, the Hub manages most of the complexity of user content dissemination. Therefore, it is solely responsible for pushing feeds to the subscribers explicitly targeted by the publisher. We term this a "Semantic Hub" since it uses Semantic Web technologies and tools to perform this dynamic and private dissemination feature. This is detailed in Sections 3.3 and 3.5.

Fig. 1 illustrates the sequence of interactions between the main participants of the protocol. The ecosystem comprises of the Publisher, the Subscriber and the Semantic Hub. The sequence is divided into three parts (1) *Subscription process* by Subscriber-A (Sub-A) to the Publisher's feeds; (2) *Updates notifications* by the Publisher (Pub) to the Semantic Hub (SemHub); (3) *Updates pushes* to the Subscribers by the Semantic Hub.

The ***Subscription Process*** is independent of the other two whereas the *Updates notifications* and *Updates pushes* happens in sequence. The communication steps in a subscription process begins with Sub-A requesting Pub for its feeds (S-1)[18]. Pub answers with a feed that includes the Topic URL and SemHub URL (S-2). Sub-A then requests the Semantic Hub to subscribe to Pub's feeds Topic URL (S-3). The first communication between Sub-A or any Subscriber with the Semantic Hub leads to the access of Sub-A/Subscriber's FOAF profile by the Semantic Hub (S-4 to S-7). This is further explained in Section 3.3. The interactions that take place only in the first communication are illustrated by dashed lines in the Fig. 1.

---

[18] S-X refers to Step X in Fig. 1.

In the **Updates Notification** the flow starts with an *item* generated by Pub. Once a new *item* is created, Pub embeds its privacy preference for the *item* in the feed. We detail the generation of privacy preferences and how they are embed in the feed in Section 3.4. The preference is a set of SPARQL queries (also known as *access space*) and represents a subset of the semantic social graph hosted in SemHub (Section 3.3). Once the privacy preferences are embed, Pub notifies an update to SemHub (S-i). Similar to Sub-A's first interaction with SemHub, Pub must also provide its FOAF profile to the Hub in order to enable privacy-aware dissemination (S-ii to S-v). This interaction happens only once between a Publisher and a Semantic Hub and is illustrated in Fig. 1 using dotted lines. As soon as the Semantic Hub is notified with the update, SemHub fetches the feed from the Pub (S-vi). Each *access space* for an *item* is executed on the semantic social graph (S-vii). Only the matched Subscribers are eligible to receive the updated *item* from the Publisher.

Updates pushes sequence (Fig. 1) represents the privacy-aware dissemination of the updates only to Sub-A because of the privacy settings of the Pub (S-viii). On the other hand, Sub-B does not receive the updates even though it is subscribed to Pub's topic.

3.3   Distributed Social Graph

FOAF profiles play a crucial role in our architecture. They provide a means for authenticating (WebID) to a platform where users can dynamically create groups for a privacy-aware microblogging. The latter requires a Semantic Social Graph (SSG) where SPARQL queries represent a subset of the SSG — which in-turn forms the dynamic group of people. Generation of the SSG in our protocol consists of collecting and linking the FOAF profiles of people who communicate via the Semantic Hub.

Although collecting the FOAF profiles can be done with secure connections and authorizations, the linking of FOAF profiles in terms of PubSubHubbub protocol required a vocabulary. Since SIOC does not consider communication protocols to represent users' social activities, it is not enough for linking the FOAF profiles using PuSH protocols for further usage. Hence, we created a lightweight vocabulary for PubSubHubbub on top of SIOC [12]. The description of the vocabulary and its usage is explained in the use case (Section 4).

The Semantic Hub uses a triplestore with a SPARQL endpoint to store the RDF data such as the FOAF profiles. The detailed process of collecting, storing and linking the FOAF profiles to enable a Semantic Social Graph is as follows As illustrated in Fig. 1, the Semantic Hub gathers user's profiles during the registration for publishing/subscribing in the following sequence.

1. The user sends a requests for publishing/subscribing at the Semantic Hub.
2. Before acknowledging for publishing/subscription, the Semantic Hub requests the user's FOAF profile.
3. The user authenticates to the Semantic Hub using WebID, further providing a secure connection to the user's personal information stored in FOAF

format. As it can be inferred from the sequence, the Semantic Hub has its own WebId URI and certificate for the users to authenticate.

4. The Semantic Hub stores the FOAF profiles with added necessary information about subscriber and publisher in the RDF store. The generation of necessary information in case of SMOB is presented in Section 4.

3.4   Generating Privacy Preference

Creating SSG helps to dynamically extract groups of subscribers from the publisher's social network who are eligible to receive the publisher's post. To do so, the publisher must create preferences to restrict which users can access the data. These preferences are defined using PPO based on (i) SPARQL queries defining *access spaces* to represent a subset of the SSG that can access the data (e.g. people interested in "Semantic Web" among my friends) and (ii) conditions to match *items* that must be delivered with their corresponding *access space*.

Our implementation provides a user-friendly interface to formulate SPARQL queries where no knowledge about SPARQL or Semantic Web is required (Section 4). Formulating *access spaces* for each *item* to be published is not practical. The privacy settings for *items* can have conditions to categorize an item and assign *access space* for the corresponding category. For example, the privacy preference in Fig. 2 restricts any document tagged with *Semantic Web* (categorizing by tags) from the publisher to only those users who share an interest in *Semantic Web*. Since the Privacy Preferences are represented in RDF, a triple store is necessary at the publisher to store and retrieve the privacy settings.

```
<http://example.org/privacy/3> a ppo:PrivacyPreference;
    ppo:appliesToResource
        <http://xmlns.com/foaf/0.1/Document>;
    ppo:hasCondition [
      ppo:hasProperty tag:Tag;
      ppo:resourceAsObject
        dbpedia:Semantic_Web
    ];
  ppo:assignAccess acl:Read;
  ppo:hasAccessSpace [
    ppo:hasAccessQuery "SELECT ?user WHERE {
      ?user foaf:topic_interest dbpedia:Semantic_Web }"
  ] .
```

Fig. 2. Example SMOB Privacy Preference Filtering

Each *access space* is a subset of the SSG in the Semantic Hub, which in turn is the list of subscribers who are eligible to receive the post. Our implementation of privacy settings for a SMOB user embeds a default *access space* for the microposts that do not have any predefined privacy settings. If there are more *access spaces* then all are embedded into the RSS. Embedding *access spaces* into RSS will include another element `<privacy>` for each item/post in the RSS/Atom feed. An example of RSS *item* including the privacy preferences is shown in Fig. 3. The `<privacy>` element comprises of each *access space* as a child element `<accessspace>`. Also, it allows us to simply reuse and extend RSS / Atom to pass the policies from the publisher to the Hub. The Semantic Hub then uses the embedded *access spaces* to filter the list of subscribers to only those who can receive the *item*.

```
<item>
  <title >Only  Friends</title >
  <description>
    Send  this  to  only  people  I  know  and
    interested  in  Semantic  Web
  </description>
  <link >http://example.org/rss</link >
  <guid >123123123123</guid>
  <pubDate>March  06  2001</pubDate>
  <privacy >
    <accessspace>
      SELECT  ?user  WHERE  {
      foaf:me  foaf:knows  ?user  .
      ?user  foaf:topic_interest  dbpedia:Semantic_Web  .  }
    </accessspace>
      . . .
  </privacy>
</item>
```

Fig. 3. Access space embedded in an RSS feed

3.5   Semantic Dissemination of Content

The final step of the protocol is to disseminate the publisher's content based on the privacy settings for the content. Once the post and its privacy settings are updated in the feed by the publisher, the publisher notifies the Semantic Hub for updates. The Semantic Hub pulls the updated feed and parses it to fetch the updates and the corresponding *access spaces*.

Every updated item in the feed has its own set of *access spaces*. The process for multicasting each item is as follows:

1. Each *access space* for the item is queried on the SSG at the Semantic Hub's RDF store. The result is a subset of the SSG that matches the *access space* executed.
2. Union of all the subsets of the SSG retrieved by executing the *access spaces*, comprises of the subscribers who are eligible for receiving the *item*.

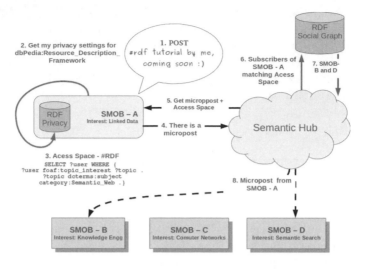

Fig. 4. Sequence of Interactions SMOB

3. As illustrated in the Fig. 3 the RSS/Atom *items* in the feed also comprises of its own *access spaces* in the `<privacy>` element. However, the *item* with the *access spaces*, if broadcasted will let the subscribers be aware of the privacy settings of the publishers. Therefore, to maintain the privacy of the publishers, the `<privacy>` element is deleted from each *item* at the Semantic Hub.

4. The modified *item* is then sent to only the callback URLs of the filtered subscribers list from step 2.

4 Implementation and Use Case in SMOB

SMOB - `http://smob.me` is a open and distributed semantic microblogging application combining Semantic Web standards and Linked Data principles with State-of-the-art social networking protocols. In the microblogging context, it is common to use the *follower* and *followee* terms, where the *follower* is a user who follows another user's timeline (i.e. her/his microblog posts) and the *followee* is the user generating content and being followed. A user can be both a follower and a followee, and can have multiple followers and followees. Combining the PuSH terminology with this one, we have: a follower is a PuSH *Subcriber*, a followee is a PuSH *Publisher*, a PuSH feed/topic is the *User's Timeline* and each micropost is a PuSH *item* in the feed.

SMOB used the PuSH implementation (using Google's public hub[19]) to broadcast feeds. But privacy was still a concern. We now present the step by step implementation of our protocol in SMOB, enabling both privacy-awareness and scalability when distributing microblog posts.

There is an screencast of the full process available at `http://smob.me/video`

---

[19] `http://pubsubhubbub.appspot.com/`

4.1   SMOB Hub-User Initial Interaction

The Semantic Hub creates a SSG of the users communicating via the Hub using
the same process than the one explained in Section 3.3. We had to distinguish
between the user's social connections on FOAF and the ones in SMOB. For
example, in Fig. 4 the access query just mentions all users interested in Semantic
Web, whereas we need only (in this subset) those who are followers of SMOB-A.
Hence, we modeled the social activities over the Semantic Hub.

We created a lightweight vocabulary for PuSH and SMOB. Fig. 5 depicts our
vocabulary, where classes and properties are mapped to the terminology used
in core specification of PubSubHubbub [3]. The full specification is available at
`http://vocab.deri.ie/push`. As per specification, the PuSH Topic represents
the feed of the Publisher. In SMOB, there is a one to one relationship between
the Publisher and his Feed. We are thus using `sioc:UserAccount` to each user
communicating via the SemanticHub. Instances of `sioc:UserAccount` are linked
to PuSH Topics with appropriate properties. We introduced:

- `push:has_hub` to express the relationship between the PuSH Topic and the
  Semantic Hub. Semantic Hub is considered to be a FOAF Agent.
- `push:has_callback` to express the relationship between Subscriber and its
  callback URL which is of type `rdfs:Resource`.

Added   relations   stored   with   followee's   FOAF   are
(1) newly created unique `sioc:UserAccount`; (2) relation to the PuSH Topic
(Followee's Timeline) she/he is creating; and (3) PuSH Topic related to the
Semantic Hub. The follower also has similar properties, i.e. (1) newly created
unique `sioc:UserAccount`; (2) relation to the PuSH Topic (Follower's timeline)
he/she is subscribing to; and (3) Callback URL of the Subscriber. When a follower wants to unfollow a followee, the follower's relation with the PuSH Topic
(Followee's timeline) is simply removed.

4.2   SMOB Followee - Publishing

Since SMOB is a semantic microblogging service, it already includes an Arc2
triple store during installation. Therefore the requirement for the Publishers to
include a triple store in our protocol was fulfilled. The same RDF store is used
to store and retrieve the privacy preferences of the corresponding user.

Further, we frequently refer to Fig. 4 to explain the use case. SMOB-A is the
Followee, SMOB-B SMOB-C and SMOB-D are followers of SMOB-A.

A SMOB user (SMOB-A) has to generate privacy preferences (Section 3.4) for
controlling his content distribution. In SMOB, the categorization of microposts
is done using "semantic hashtags" mentioned in the micropost (i.e. tags linked to
resources from the LOD cloud using models such as MOAT[20] or CommonTag[21]).
Hence, we map these tags to the privacy preferences to enable the privacy in content distribution. As shown in the Fig. 6, we have build a simple user-interface

---

[20] `http://moat-project.org`
[21] `http://commontag.org`

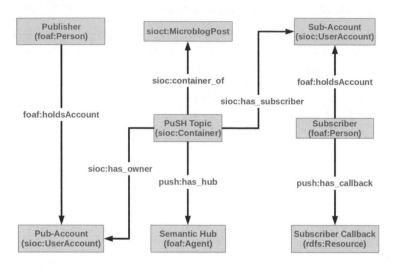

Fig. 5. Vocabulary to represent PuSH information in RDF

to let users create such a setting without prior knowledge of SPARQL or the Semantic Web. Both the hashtag and the interest are looked-up in DBPedia[22] and Sindice[23] and the concepts are suggested to the user. Once the user selects the intended concepts, the resulting Privacy Preference is stored in the local triple store. For example, Fig. 6 shows the interface to create the privacy preferences where the microposts tagged with #rdf, should be sent only to those followers who are interested in Semantic Web. Also, we offer the ability to restrict based on the relation that people share together, using the RELATIONSHIP vocabulary. The main advantage compared to pre-defined groups is that the settings are automatically updated based on the user's attribute, in almost real-time since the Semantic Hub stores users' profiles[24] Once again, these two use-cases are just an example, and the privacy settings could be further enhanced.

When SMOB-A is creating a new micropost and tags it with the #rdf hashtag (Step 1 of Fig. 4) , the SMOB interface suggests links to the resources with that label from DBPedia and Sindice in a similar way as for the privacy settings. Once the micropost is posted, SMOB-A queries the local triple store for *access spaces* matching the URI representing the hashtag #rdf (Step 2, 3 of Fig. 4). The *access space* is embeded into the RSS feed of SMOB-A.

The SMOB Followee then notifies the Semantic Hub about the update (Step 4 in Fig. 4).

---

[22] http://dbpedia.org/

[23] http://sindice.com/

[24] As said earlier, so far, we assume that the information provided in FOAF profile is correct, or at least should be trusted - which is legally the case with WebID as it requires a certificate which implies some real trust settings.

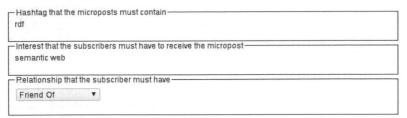

Privacy settings

Hashtag that the microposts must contain
rdf

Interest that the subscribers must have to receive the micropost
semantic web

Relationship that the subscriber must have
Friend Of ▼

Generating Privacy preferences ...

Privacy Preferences generated

```
http://mysite.org/preference/rdf a ppo:PrivacyPreference;
    ppo:appliesToResource
      http://rdfs.org/sioc/ns#MicroblogPost;
    ppo:hasCondition [
      ppo:hasProperty tag:Tag;
      ppo:resourceAsObject
        http://dbpedia.org/resource/Resource\_Description\_Framework
    ];
    ppo:assignAccess acl:Read;
    ppo:hasAccessSpace [
      ppo:hasAccessQuery "SELECT ?user WHERE {
?user foaf:topic_interest ?topic .
?topic dcterms:subject category:Semantic_Web .}"
    ] .
```

Fig. 6. Privacy Settings interface in SMOB

4.3   SMOB Semantic Hub - Distribution

Querying data to and from the Semantic Hub triple store is performed using the Python SPARQLWrapper[25]. The wrapper is entirely HTTP-based, thus gives the flexibility to deploy the RDF store in any environment and access it remotely. To maintain privacy of the profiles, the SPARQL endpoint is accessible only by the Semantic Hub.

After SMOB-A notifies the Semantic Hub about the update, the Semantic Hub fetches the feed and, for the newly created microposts extracts the *access space*. Before the access query is executed, more conditions are added based on the corresponding relations added during FOAF profiles storage (Section 4.1). In this use-case conditions are added to retrieve the callback URLs of the followers, who holds a sioc account subscribed to SMOB-A's topic. The *access space* is executed against the SSG of SMOB-A in conjunction with the above conditions (Step 5 in Fig. 4) and the list of users (SMOB-B and SMOB-D) matching the SPARQL query are returned (Step 6 in Fig. 4). SMOB-B and SMOB-D matched because their interests are in the category of "Semantic Web" where as SMOB-C has interest "Computer Networks" which does not fall into the category restricted by SMOB-A.

---

[25] http://sparql-wrapper.sourceforge.net/

5   Related Work

Related work can be considered from two different perspectives (1) Distributed Social Networks and Semantics, and (2) Privacy in Social Networks. This combination should lead to a distributed and privacy aware Social Web, as envisioned in the final report of the W3C Social Web Incubator Group "Standards-based, Open and Privacy-aware Social Web" [1].

Research on privacy in OSNs are prominently about misuse of private information and therefore providing more security [6] [9]. Very few have done research where broadcasting of user content is taken into consideration. Work by B. Meeder et al. [11] analyses the privacy breach for protected tweets by retweets. Our work more or less falls in the category of awareness in distribution of content. Next, gathering information about users from online social networks to form social graphs [22] [16] has been really useful to leverage the powers of social graphs. Work by Matthew Rowe shows the creation of Social Graphs by exporting information from online social networks into semantic form using FOAF vocabulary [16]. Further to this, few of the areas where the power of such Semantic Social Graphs is leveraged are in finding friends in federated social networks [7] and to disambiguate identity of users online [15].

Efforts such as P3P or more recently policy systems such as Protune also enable rule-based access to content on the Web. However, they are not necessary suited to Web-based environment. Cuckoo [21] is a decentralized P2P socio-aware microblogging system. It focuses on providing a reliable and scalable system to increasing the performance with respect to traditional centralized microblogging systems. Yet, they rely on a more complex architecture which is not only HTTP based and makes it difficult to deploy in practical Web-based environments. FETHR [18] is another open and distributed microblogging system, that emphasizes about the scalability, privacy and security. While using a publish-subscribe approach, information is sent from the provider to all its subscriber (one of the reason while PuSH was build). More recently, [20] is at a close proximity to our work but on mobile platforms. They provide a semantic social network for mobile platforms, but do not tackle the privacy aspect. In addition, larger open-source projects such as StatusNet or diaspora also enable distributed Social Networks using similar stacks (notably including PuSH and the Salmon protocol[26] for pinging-back comments), but neither directly focus on on-demand and dynamic privacy settings.

6   Conclusion

In this paper, we described how we extended Google's PubSubHubbub protocol to cope with privacy-by-design in Distributed Social Network. We showed how we defined a way to broadcast content only to a subset of one's social graph, without having to hardcode user-group or specify static policies. Hence, our system can cope with dynamic groups and organizations without putting any burden on the

---

[26] http://www.salmon-protocol.org/

user. Moreover, we have shown how we implemented this into SMOB, a Semantic and Distributed microblogging platform. Overall, our approach also shows how to combine pragmatic Social Web protocols and formats (PuSH and RSS/ Atom) with Semantic Web standards and vocabularies (SPARQL Update and PPO).

In the future, we will focus on enabling our architecture for mobile devices. Here, challenges will be to send information from and to devices that can be off-line from time to time, but still need to be notified. Also, we are considering to apply this architecture to Sensor Data, in order to deal with the management of sensitive information such as geolocation.

References

1. A Standards-based Open and Privacy-aware Social web. Technical report. W3c incubator group report December 6th 2010, World Wide Web Consortium (2010), http://www.w3.org/2005/Incubator/socialweb/XGR-socialweb-20101206/
2. Boyd, D., Hargittai, E.: Facebook privacy settings: Who cares? First Monday 15(8) (2010)
3. Fitzpatrick, B., Slatkin, B., Atkins, M.: PubSubHubbub Core 0.3 – Working Draft, http://pubsubhubbub.googlecode.com/svn/trunk/ pubsubhubbub-core-0.3.html
4. Breslin, J., Bojars, U., Passant, A., Fernàndez, S., Decker, S.: SIOC: Content Exchange and Semantic Interoperability Between Social Networks. In: W3C Workshop on the Future of Social Networking (January 2009)
5. Brickley, D., Miller, L.: FOAF Vocabulary Specification 0.98. Namespace Document August 9, 2010 - Marco Polo Edition (2010), http://xmlns.com/foaf/spec/
6. Cutillo, L., Molva, R., Strufe, T.: Safebook: A privacy-preserving online social network leveraging on real-life trust. IEEE Communications Magazine 47(12), 94–101 (2009)
7. Dhekane, R., Vibber, B.: Talash: Friend Findin in Federated Social Networks. In: Proceedings of the Fourth Workshop on Linked Data on the Web (LDOW 2011) at WWW 2011 (2011)
8. Fitzpatrick, B., Recordon, D.: Thoughts on the Social Graph (August 2007), http://bradfitz.com/social-graph-problem
9. Krishnamurthy, B., Wills, C.E.: Characterizing privacy in online social networks. In: Proceedings of the First Workshop on Online Social Networks - WOSP 2008 (2008)
10. Yeung, C.m.A., Liccardi, I., Lu, K., Seneviratne, O., Berners-Lee, T.: Decentralization: The future of online social networking. In: W3C Workshop on the Future of Social Networking Position Papers (2009)
11. Meeder, B., Tam, J., Kelley, P.G., Cranor, L.F.: Rt@iwantprivacy: Widespread violation of privacy settings in the twitter social network (2010)
12. Passant, A., Breslin, J., Decker, S.: Rethinking Microblogging: Open, Distributed, Semantic. In: Benatallah, B., Casati, F., Kappel, G., Rossi, G. (eds.) ICWE 2010. LNCS, vol. 6189, pp. 263–277. Springer, Heidelberg (2010)
13. Prud'hommeaux, E.: W3C ACL System, http://www.w3.org/2001/04/20-ACLs.html
14. Rothschild, J.: High Performance at Massive Scale Lessons learned at Facebook

15. Rowe, M.: Applying Semantic Social Graphs to Disambiguate Identity References. In: Aroyo, L., Traverso, P., Ciravegna, F., Cimiano, P., Heath, T., Hyvönen, E., Mizoguchi, R., Oren, E., Sabou, M., Simperl, E. (eds.) ESWC 2009. LNCS, vol. 5554, pp. 461–475. Springer, Heidelberg (2009)
16. Rowe, M.: Interlinking distributed social graphs. In: Linked Data on the Web Workshop, WWW 2009 (April 2009)
17. Sacco, O., Passant, A.: A Privacy Preference Ontology (PPO) for Linked Data. In: Proceedings of the Linked Data on the Web Workshop, LDOW 2011 (2011)
18. Sandler, D.R., Wallach, D.S.: Birds of a fethr: Open, decentralized micropublishing. In: IPTPS 2009: Proc. of the 8th International Workshop on Peer-to-Peer Systems, Boston, MA, USA (April 2009)
19. Story, H., Harbulot, B., Jacobi, I., Jones, M.: FOAF+SSL: RESTful Authentication for the Social Web. In: European Semantic Web Conference, Workshop: SPOT 2009 (2009)
20. Tramp, S., Frischmuth, P., Arndt, N., Ermilov, T., Auer, S.: Weaving a Distributed, Semantic Social Network for Mobile Users. In: Antoniou, G., Grobelnik, M., Simperl, E., Parsia, B., Plexousakis, D., De Leenheer, P., Pan, J. (eds.) ESWC 2011, Part I. LNCS, vol. 6643, pp. 200–214. Springer, Heidelberg (2011)
21. Xu, T., Chen, Y., Fu, X., Hui, P.: Twittering by cuckoo: decentralized and socio-aware online microblogging services. In: Proceedings of the ACM SIGCOMM 2010 Conference on SIGCOMM 2010, pp. 473–474. ACM, New York (2010)
22. Ye, S., Lang, J., Wu, F.: Crawling online social graphs. In: 2010 12th International Asia-Pacific Web Conference (APWEB), pp. 236–242 (April 2010)

BookSampo—Lessons Learned in Creating a Semantic Portal for Fiction Literature

Eetu Mäkelä[1], Kaisa Hypén[2], and Eero Hyvönen[1]

[1] Semantic Computing Research Group (SeCo),
Aalto University and University of Helsinki, Finland
first.last@aalto.fi
http://www.seco.tkk.fi/
[2] Turku City Library, Turku, Finland
first.last@turku.fi

Abstract. BookSampo is a semantic portal in use, covering metadata about practically all Finnish fiction literature of Finnish public libraries on a work level. The system introduces a variety of semantic web novelties deployed into practise: The underlying data model is based on the emerging functional, content-centered metadata indexing paradigm using RDF. Linked Data principles are used for mapping the metadata with tens of interlinked ontologies in the national FinnONTO ontology infrastructure. The contents are also linked with the large Linked Data repository of related cultural heritage content of CultureSampo. BookSampo is actually based on using CultureSampo as a semantic web service, demonstrating the idea of re-using semantic content from multiple perspectives without the need for modifications. Most of the content has been transformed automatically from existing databases, with the help of ontologies derived from thesauri in use in Finland, but in addition tens of volunteered librarians have participated in a Web 2.0 fashion in annotating and correcting the metadata, especially regarding older litarature. For this purpose, semantic web editing tools and public ONKI ontology services were created and used. The paper focuses on lessons learned in the process of creating the semantic web basis of BookSampo.

1 Introduction

With the advent of the Internet, the role of libraries as primary sources of factual knowledge has diminished, particularly among younger people. This is reflected in many of the analyses published in a 2011 study of library use in Finland [13]. Even taken as a whole, 83% of the respondents of the study said they rather sought factual knowledge from the Internet than from the other tallied channels of public libraries, television, magazines or friends. Even for deeper learning on a subject, 40% favored the Internet, as opposed to 38% who still preferred the library. At the same time, the role of the library as a provider of fiction literature has remained constant. While 34% of the respondents said they still benefited from factual information stored in libraries, the corresponding percentage for fiction was 45%.

L. Aroyo et al. (Eds.): ISWC 2011, Part II, LNCS 7032, pp. 173–188, 2011.
© Springer-Verlag Berlin Heidelberg 2011

These results encourage libraries to improve their services related to fiction literature, such as the search and recommendation facilities available for such content. However, the special nature of fiction necessitates a move from the old library indexing traditions, i.e. mainly classifying books by genre and by cataloguing their physical location, to describing their content. This is a very recent development in the centuries long timescale of libraries,

In Finland, content keywords for fiction have been systematically entered only since 1997, using the Finnish fiction content thesaurus Kaunokki[1], developed since 1993. The reason for this is twofold. First, there is a long running tradition in libraries of favoring assigning a singular classification to each book. This is simply due to the relatively recent advent of library computer systems, starting in Finland in the 1980s. Before, when book information was stored on physical index cards, any added keywords past the necessary single classification necessitated also adding another physical card to the indexing cabinets. For fiction, this has always been somewhat problematic, as it appeared quite impossible to arrive at a single universal best classification system, even though various attempts at such have been proposed from as early as the 1930s [11].

Even after the single classification issue was resolved, fiction content description was still considered almost impossible due to the interpretational ambiguity of texts. It was also feared that content description with only a few words would abridge the connotations of a work, and could actually do more harm than good to literature and its multidimensionality. Experiments in indexing consistency from 1999 however found that there was much uniformity in how individual readers paid attention to certain fictional works, and that most fictional content could be adequately described by about 10 to 15 indexing terms [11].

On the other hand, the study also concluded that customers descriptions of the pertinent points of, and questions about fiction literature tended to combine details related to for example the author, contents and publication history of a given work. Based on this, the author of the study compiled a wide or ideal model for describing fiction, which in his mind should include not only the factual publication data on the book, but also descriptions of the content, information on any intertextual references contained therein and data about the author, as well as information about the reception of the book by readers at different times, interpretations by researchers and other connections that help position the publication in its cultural historical context.

In 1999, this model was considered an ideal, impossible to implement in reality. However, times change, and when the Finnish public libraries in the summer of 2008 started a joint venture to experiment with new ways of describing fiction, the model was chosen as a concrete goal to strive for. At this point, based on knowledge gained from prior collaboration, the libraries approached the Semantic Computing Research Group at the Aalto University and University of Helsinki, and the BookSampo project started as part of the national FinnONTO initiative[2]. Because the model that was strived for placed much emphasis on the

---

[1] http://kaunokki.kirjastot.fi/

[2] http://www.seco.tkk.fi/projects/finnonto/

interconnections between heterogeneous information items, it seemed a good fit for semantic web technologies. For example, the cultural historical context of fiction literature is not restricted to other literature, but spans the whole of culture and history from the time of their writing onwards. Books are nowadays also often sources for movies etc., further demanding a broadening of horizons.

The research group had also much prior experience in publishing cultural heritage content on the semantic web, having created the MuseumFinland portal [6] in 2004 and the CultureSampo portal [5,8] in 2009. The BookSampo project could make use of much of the infrastructure created for CultureSampo in converting legacy data and thesauri to RDF and ontologies respectively, in editing and annotating data semantically, and in providing intelligent end-user services based on that data.

A beta-version of the end-user portal developed is currently available at http://www.kirjasampo.fi/. Already, the portal contains information on virtually all fiction literature published in Finland since the mid 19th century, a total of some 70 000 works of fiction, 25 000 authors and 2 000 publishers.

In the following, lessons learned during the BookSampo project will be presented. First discussed are the many insights gained in modelling fiction, both from an abstract viewpoint as well as how it relates to the semantic web. After that, experiences gained while transforming Kaunokki into an ontology are given. Then the paper presents a technical description of various parts of the system, focusing on the challenges faced and benefits gained from applying semantic web technologies. Finally, the paper discusses the further potential and reception of the developed system in library circles.

2   The BookSampo Data Model

From the start, the BookSampo project was consciously geared towards an ambitious, disruptive experiment as opposed to an incremental improvement. Thus, early on it was decided that the MARC format[3] still currently used in most libraries in Finland would not be used in the project on account of its restrictions, nor would the system be built on top of the current library indexing systems. Instead, the data model of BookSampo was to be based purely on RDF and Semantic Web principles, with both the indexing machinery as well as the public end user interface operating directly on that data to best fulfil its potential.

One of the benefits of the RDF data model is its flexibility. This proved an asset in the project. Because of the experimental nature of the project, there have been multiple times when the model has needed amendment and modification. In addition to simple addition of fields or object types, the schema has undergone two larger alterations during the project.

First, the way the biographical information of the authors was encoded was changed from events to attributes. Initially, details about, among others, the times and places of authors' births, deaths and studies were saved in BookSampo

---

[3] http://www.loc.gov/marc/

as events, in the spirit of the cultural heritage interchange model of CIDOC-CRM [2] and the BIO-schema of biographical information [1].

User research, as well as interviewing library indexers revealed, however, that events as primary content objects are not easily understood by those indexing them or by end-users on a cognitive level. Bringing events to the fore, the approach fractured and distributed the metadata of the original primary objects. For example, people wanted much more to see information on authors' birth and death dates and places as simply attribute-object values of the author, instead of as events where the author was involved in.

Description thus adopted back the more traditional model, where data about times and places of occurrences are directly saved as author attributes. In the case of studies, this did lead to some loss of data specificity, as the original information related for example the dates and places to each individual degree attained. This information could not be maintained in a flat attribute value description centered on the author. However, the indexers deemed the simplicity to outweigh the costs in this situation.

An even larger change however was made to the book schema. It has been a conscious policy that BookSampo should only concentrate on the description and data concerning the contents of the work itself, irrespective of editions. But right from the start, details about translators, publication years, publishers and publishing series crept in. The guidelines at the time were to save only the details of the first Finnish edition.

For a very long time, the model of a single object worked well, until it was decided that the project should also extend to include Swedish literature[4], as well as maintain distinctions between different translations. It then became necessary to reconsider how the different conceptual levels of a work could be separated from each other. Advice was sought from the FRBRoo Model [10], which identifies four conceptual levels, among which the different properties of a work can be divided:

1. Work. The abstract contents of the work—the platonic idea of the work (primary creator, keywords).
2. Expression. The concrete contents of the work — original/translated text, stage script and film script (author, translator and director).
3. Manifestation. The concrete work/product—book, compilation book, entire concept of a stage performance and film DVD (publisher, issuer and ISBN).
4. Item. The physical copy—single book/compilation/performance/DVD.

The idea in the model is that a single abstract conceptual work may be written down in different texts, which may then be published in multiple editions, each comprised of a multitude of physical copies. Each type of item down the scale inherits the properties given on the levels above it. Translated into actual indexing work, this means that for example the content of a work need be described

---

[4] Finland is a bilingual country, the official languages being Finnish and Swedish. This is why a web service maintained by the public libraries must be available in both languages.

only once, with each different language edition merely referring to the resource describing the qualities of the abstract works printed therein.

After what had been learnt from the biography schema, it was not deemed desirable to replace a simple model with the complexity of four entire levels. Also, more generally, experience had proven that the BookSampo indexing model focusing on the contents of the work was already quite a leap to librarians, who were thus far mostly familiar with single level MARC indexing of mostly manifestation level information.

Since data in BookSampo never reaches the level of a single item, it was easy to simply drop the item level. On the other hand, the work level had to be kept separate, so translations in different languages could refer to the same content keywords. It was decided, however, to combine the expression and manifestation levels, since, one translation has on the average one publisher and physical form, and repetitive descriptions would hence not be needed on a large scale.

As a result, works are described at two levels in BookSampo: as an abstract work, which refers to the contents of the work, which is the same in all translations and versions and as a physical work, which describes features inherent to each translation or version.

The following fields are used to describe an abstract work:

- work titles in all the languages in which the physical work is stored
- author(s)
- type (novel, novella, poem, etc.)
- genre (fantasy literature, one-day novel, detective novel, etc.)
- subjects and themes addressed in the work
- actors (in general, children, doctors and middle-aged people)
- main characters (proper names, e.g. Adam Dalgliesh)
- time of events (in general, summer, middle ages or a specific dating)
- place of events (in general, libraries, countryside or a real-world place)
- textual summary or presentation
- language of the original work
- work adaptations (movies, operas, plays, etc.)
- related works (librarian recommendations)
- information related to the work on the web (critiques, descriptions, etc.)
- awards

The following fields are related to the description of a physical work:

- title and subtitle
- original title
- language of the work
- publisher
- date published
- number of pages
- translator
- illustrator
- editor or other contributor
- additional information on publication history (edition, illustrations, etc.)
- serial information

In addition, a book cover image can be included with a physical work, and it is also possible to describe the cover with keywords, provide a brief description of it and list the cover designer. It is also possible to associate review objects to the abstract works, describing for example contemporary reader reactions.

The physical works are linked with the abstract work and their data are presented in the context of the abstract work. This way it is possible to demonstrate, for example, that Hopealaiva and Nostromo are both Finnish translations of Nostromo, a Tale of the Seaboard by Joseph Conrad.

While it can be argued that not using the whole FRBR model diminishes the interoperability of the content in BookSampo with regard to other FRBR collections, it turns out that also others have independently come to a similar simplification of the model, particularly in systems where distributed editing and understandability of content is important, as opposed to for example systems doing automatic conversion of MARC records to FRBR. For example, the Open library[5] recognizes work and edition levels, with the latter also combining expression and manifestation. Exactly the same situation is present also in the LibraryThing portal[6], only naming the entities as "work" and "book". On the other hand, even systems that claim to support separate expression level items on their data model level, such as The Australian Music Centre[7], and the OCLC WorldCat system[8], do not allow these to be shown or searched for independently of their linked work or manifestation entities in their actual user interfaces, thus further corroborating that at least from an end-user perspective, the distinction between an expression and a manifestation is not very important.

In any case, it has already been established by others that separation of expressions from even the original combined MARC fields is possible by mostly automated processing along with manual correction [3,9], so should a need for further separation arise, one could just repeat a similar split procedure as just presented for the BookSampo data.

In BookSampo, the experience of moving from the solution of one conceptual level to that of two was mainly simple and painless. A minor problem was, however, short stories and their relationship with short story collections. Originally, two objects here were turned into four, and their internal relationships required precise inspection. Finally, it was decided to choose a model where each short story had an abstract work level, which was "embodied" as a "part of a physical work". This "part of a physical work" was then included in a physical work, which in turn was the "embodiment" of the short story collection as an abstract work. This set of relationships is depicted in a more visual form in figure 1.

This way both the individual short story and the short story collection overall may separately have content keywords. Whereas most of the data at the manifestation level belongs to the physical work of the short story collection, the data of an individual short story at the expression level, e.g. details of the translator, the name in the collection or page numbers, belongs to the part of the physical work. This same structure is also applied to other similar part-object instances, for example single poems in poem collections.

---

[5] http://www.openlibrary.org/
[6] http://www.librarything.com/
[7] http://www.australianmusiccentre.com.au/about/websitedevelopment
[8] http://frbr.oclc.org/pages/

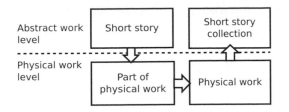

Fig. 1. Relationship between short story and short story collection in the BookSampo data model

As a benefit of the flexibility of the RDF data model, all the transformations described here could be done by quite simple transformation scripts that operated only on the data, without having to change editor or database code.

There is one area where the RDF data model caused problems. For the most part, the model is simple. Each resource describes an independently existing thing, such as a book, award, author or a place, that has relationships with other things. Yet even this was sometimes hard for people who were used to each annotation being completely separate. For example, at one point it was discovered that two different URIs for the 1970s had crept into the database. Upon closer inspection, it was discovered that one of them was actually the URI for 1960s, only someone had changed the label when they were trying to correct a particular book's annotation as happening in the 1970s instead of the 1960s.

However, a much greater problem was the confusion arising from cases where a particular resource actually didn't note an independently existing, understandable thing. This has to do with cases where a relation has metadata of its own, such as when one wants to record the year a book has been given an award or the serial number of a book in a series. In RDF, these situations are usually resolved by creating the link through an anonymous node where this information can be recorded. For example, a book can be related to the resource "Part 7 in the Yellow Library", which is in turn annotated as a part of the Yellow Library series with a part number of 7.

In BookSampo, this caused problems because these auxiliary resources appeared to the user exactly like resources describing something independently extant, yet their function was different—i.e. it doesn't really make sense to think that "Part 7 of the Yellow Library series" exists in any sense separate from the book that holds that place in the series. In BookSampo, there was no way around using these auxiliary resources for certain things, but their use certainly did muddy the primary concept of the graph model. Luckily, in practice the effects of this could be somewhat mitigated by training and annotation instructions. However, further developments in generalized visualization, browsing and editor environments should do well to provide support for special handling of such auxiliary resources, so that such inconsistencies arising from technical limitations can be hidden behind better user interfaces.

3   Ontology Creation and Participation in the FinnONTO Infrastructure

To maximally leverage the possibilities of semantic web technologies in the project, the Finnish and Swedish thesauri for fiction indexing (Kaunokki and Bella) were converted into a bilingual ontology KAUNO [12]. This was done in order to allow the use of inferencing in the recommendation and search functionalities of the end-user portal.

This ontology was also linked with other available ontologies and vocabularies used in CultureSampo, to improve the semantic linking of the content to their cultural historical context. Here, the project leveraged and relied on the work done in the wider FinnONTO project [4], which aims to make uptake of the semantic web as cost-effective as possible in Finland by creating a national infrastructure for it. At the core of the FinnONTO model is the creation of a national ontology infrastructure termed KOKO, which aims to join and link together under an upper ontology as many domain specific ontologies as possible.

The primary core of KOKO is currently comprised of 14 domain ontologies joined under the the Finnish national upper ontology YSO[9]. Among these are for example the museum domain ontology MAO, the applied arts ontology TAO, the music ontology MUSO, the photography domain ontology VALO and the ontology for literature research KITO. Thus, by linking the fiction literature ontology KAUNO to KOKO, links would immediately be created bridging the material to all this other cultural heritage.

Thus far, with the exception of MAO and TAO, these ontologies are each joined only through common links to YSO. This has been possible because almost all common concepts of the domain specific are also in YSO, and domain concepts appear mostly only in that domain. This was the approach taken also with regard to KAUNO. To create the links, automatic equivalency statements between the concepts of the KAUNO and YSO ontologies were generated by tools created in the FinnONTO project. After this, the combined ontology file was loaded into the Protégé ontology editor[10]. All automatically created links were checked by hand, as well as new links created.

The experience of the librarians who ontologized Kaunokki was that it brought in a very welcome additional structuring to the vocabulary. For example, the term "american dream", in the thesaurus only contained information that it belonged to the theme facet. In the ontology however, it had to find a place in the ontology's class hierarchy: a lifestyle, which in turn is a social phenomena, which at the root level is an enduring concept (as opposed to a perduring or abstract concept). This forced additional work ensures that no keyword floats around in isolation, but is always surrounded by co-ordinate, broader and narrower concepts that help define it and relate it to other phenomena. This also beneficially forces the vocabulary keeper to narrow down and elucidate their definition of the keyword, which in turn helps in ensuring uniform use of keywords by indexers.

---

[9] http://www.yso.fi/onki3/en/overview/yso
[10] http://protege.stanford.edu/

The linking to YSO was also deemed extremely beneficial, as before, even all general keywords were maintained in each vocabulary separately. Now, their management could be centralized, while having the work done still be usable as part of systems utilizing the domain ontologies.

4   System Description

In this section, the paper presents the technical solutions created for the Book-Sampo System, focusing on the challenges faced and benefits gained from applying semantic web technologies. First, the data import and integration functionality used to both bootstrap as well as update the system is discussed. Then presented is the primary editing environment created for the dozens of volunteers distributed in Finnish libraries who do content enrichment for the project. Finally, the functionality used to built the end-user portal search and browsing functionality is discussed.

4.1   Data Import and Integration

Contrary to existing library systems, the project was not to focus on the characteristics of individual physical editions of books, but equally to the content as well as the context of the different conceptual works themselves. However, it still made sense to bootstrap the database from existing collections, where possible.

The BookSampo project needed first to do data importing, conversion and integration. Data on books was to be sourced primarily from the HelMet cataloguing system[11] used in the Helsinki metropolitan area libraries, which stored its information in the ubiquitous MARC format. Also, from very early on, the vision of the project included storing information not only of the books, but also of the authors of those books. Data on these were to be sourced from three different databases maintained at various county libraries across Finland. Thus, at the very beginning, the project already had at least two quite different content types, and multiple data sources.

The data converters were created following principles established in the CultureSampo project [8], which focus on first converting a data set into RDF as is in an isolated namespace, and then using common tools to map them to each other, as well as to external ontologies and place and actor registries on a best-effort basis. Doing the mapping at the end on the RDF data model and ontology language level, such as using owl:sameAs mappings, brings multiple benefits. First, it ensures that no primary information is lost in the conversion, what often happens when trying to map text-based keywords in the original data to a curated ontology immediately in place. Second, it also allows any automatic mappings to be later reversed if found problematic, as well as iteratively adding more mappings as resources permit. This can be either done manually, or when centrally developed automatic mapping tools improve.

---

[11] http://www.helmet.fi/search~S9/

In the case of BookSampo, the format of the original author records quite closely matched the end schema sought also in the BookSampo system. However, the book records to be imported were in the edition-centric MARC format. Here, each edition in the source data was simply converted into an abstract work in the BookSampo schema. A large number of volunteers in libraries then poured through the data, manually removing and joining duplicate works that had resulted from multiple editions of a single work in the source.

The conversion of the records from MARC to linked RDF already bought an instant benefit to the project: Before, the fiction content descriptions had been stored in the HelMet library system only as text fields containing the Finnish language versions of the keywords. Now, when they had been converted into URI references in the bilingual ontology, they could instantly be searched using either language. Also, because YSO was available also in English, much of the content could additionally now also be searched in that language. In addition, the use of the CultureSampo authority databases allowed the automatic unification of different forms of author names found in the system, while the place registries of CultureSampo instantly added geo-coordinate information to the place keywords for later use in creating map-based user interfaces to the data.

Recently, the BookSampo project also bought rights to descriptions of newly released books from BTJ Finland Ltd, a company that provides these descriptions to Finnish library systems for a price. These descriptions are fetched from the BTJ servers each night in the MarcXML format used also for HelMet, automatically converted to RDF using the CultureSampo tools, and added to the RDF project with tags indicating they should be verified. The librarians then regularly go through all such tagged resource in the editing environment, removing the "unverified" tags as they go along.

4.2   Collaborative Semantic Web Editing Environment

As BookSampo didn't have a data store or editor environment of its own ready, the project decided to adopt the SAHA RDF-based metadata editor [7] developed by the FinnONTO project as its primary editing environment.

SAHA is a general-purpose, adaptable editing environment for RDF data, capable of utilizing external ontology services. It centers on projects, which contain the RDF data of a single endeavour. The main screen of SAHA provides a listing of the class hierarchy defined in the project, from which new instances can be created. Alternatively, existing instances of a particular type can be listed for editing, or a particular instance sought through text search facilities. Navigation from resource to resource is also supported in order to examine the context of a particular resource, with pop-up preview presentations allowing for even quicker inspection of the resources linked.

The editing view of the SAHA editor is depicted in figure 2. Each property configured for the class the resource is an instance of is presented as an editor field, taking in either literal values or object references. For object references, SAHA utilizes semantic autocompletion. When the user tries to find a concept, SAHA uses at the same time web services to fetch concepts from connected external ONKI ontology repositories [15], as well as the local project. Results are shown in one autocompletion result list regardless of origin, and their properties can also be inspected using pop-up preview presentations. In the example depicted in figure 2 for example, this is extremely useful when the user must choose which of the many Luxors returned from both local and external sources is the correct annotation for this book.

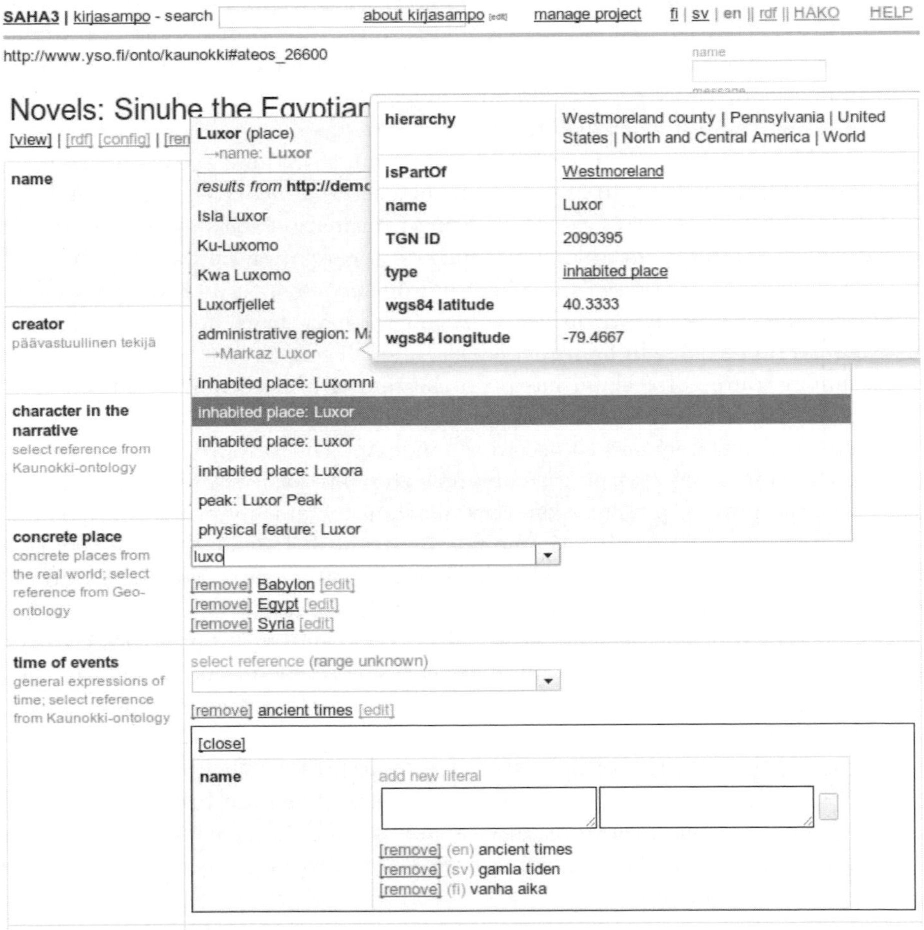

Fig. 2. The SAHA metadata editor, showing both semantic autocompletion as well as a pop-up preview presentation of one of the autocompletion results

For the purposes of the BookSampo project, the SAHA editor was improved with an inline editor feature. The idea is simple: a resource referenced through an object property can be edited inline in a small version of the editor inside the existing editor. Specifically, this functionality was developed to ease the use of the necessary auxiliary resources discussed before. However, there seemed no reason to restrict the functionality to those alone, so this possibility is now available for all linked object resources. In figure 2, this is shown for the property 'time of events" whose value "ancient times" has been opened inline for editing.

From the library indexers point of view, a major source of excitement in the RDF data model and the SAHA editor has been their support for collaborative simultaneous editing of a rich, semantically linked network. Libraries in Finland have shared MARC records between each other for a long time, but these go from a silo to another, and always as whole records focused on individual book editions. In SAHA by contrast, newly added authors or publishers for example, along with all their detailed information are immediately available and usable for all the dozens of voluntary BookSampo indexers across Finland. Once entered, publisher information need also not be repeated again for all new books, which adds an incentive to provide richer detail about also these secondary sources. Similarly, adding a detail to any node in the graph immediately adds value also to all items linked to that node, benefiting information seekers everywhere. In the words of the indexers, this has been both a revelation and a revolution. To further foster co-operation in the SAHA editor between peer indexers, a project-wide chat facility is shown on the top right of each page, facilitating instant discussions (not visible in figure 2 because of occlusion by the pop-up preview).

A similar source of acclaim has been the semantic autocompletion function-ality of SAHA. While previously keywords had to be looked up in two different applications separately and copied to the indexing software by hand, or entered from memory leading to typing errors, now they are easily available in a joined list, with the pop-up presentation view allowing for quickly evaluating possible keywords in place. Also valued is the possibility in SAHA for creating new key-words inside the project if no existing keyword suffices. Previously, this would have gone completely against the thought benefits of having a controlled search vocabulary in the first place. However, in SAHA and with ontologies creating e.g. new concepts or locations is not detrimental, provided that the indexer then uses the inline editing functionality of SAHA to link the newly created resource to the existing ontology hierarchies.

All in all, the majority of indexers taking part in BookSampo indexing have found the SAHA editor to be both intuitive, as well as even inspiring. In many cases however, this happened only after an initial confusion caused by the system having both a foreign data model as well as employing a new functional and content indexing paradigm.

4.3   End-User Portal

The complexity of the BookSampo data model entailed problems for the search service side of CultureSampo, which was to be used in the project as the un-derlying content service. The user interface of BookSampo is being built on top of the Drupal[12] portal system. However, the intention of the project has always been that all primary information be kept and served by CultureSampo, with the

---

[12] http://drupal.org/

Drupal layer only adding commenting and tagging functionality, forums, blogs etc. on the client side.

The problem then became that the CultureSampo search interfaces at the time could only efficiently deal with text queries targeting a single RDF resource, while the BookSampo model required much more complex queries. Particularly, the model required keyword queries to span multiple resources, and combine with relation constraints in the style of [14]. This is because in BookSampo, besides splitting each book into two objects, the model already contained many objects related to the book that needed to be parsed in order produce a search result as well as to render a complete visualization of the book in a user interface. For example, the intent of the BookSampo search interface was that for example the plain text query "Waltari Doctor Inscriptions" would match the abstract work "Sinuhe the Egyptian", because Waltari is the book's author, because it has a main character who is a doctor and because one of its editions has a cover that contains hieroglyphs, which are a type of inscription.

However, inside the data model, this is quite a complex pattern, as visualized in figure 3. First, each resource with a label matching any of the keywords must be found. This results in a result set with (among others) Mika Waltari the author and the keywords doctor and inscriptions. Then, all resources relating to these or their subconcepts must be added to the result set. This results in (among others) the abstract work Sinuhe the Egyptian (whose author is Mika Waltari), the fictional character Sinuhe (who is a doctor), and a particular cover for Sinuhe the Egyptian, which has the keyword hieroglyphs.

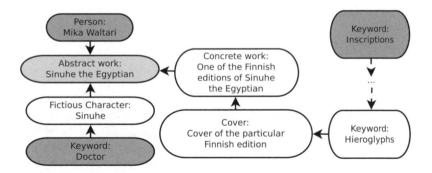

Fig. 3. Mapping to a final search result after text matching in BookSampo. Dark grey resources are those returned from label text matching, while the light grey resource is the final search result.

Finally, all resources that are not already abstract works must be mapped to any that they refer to, and finally an intersection taken between all abstract works found to reveal the final result. Here, the character Sinuhe relates directly to the abstract work, but the cover is still two steps away. One must follow the link from the cover to the particular physical edition that it is the cover for, and from there finally to the abstract work.

To resolve this, the search functionality of CultureSampo was split into multiple stages, each taking in SPARQL queries. First, multiple "select" queries are run, one for each incoming keyword, acting on a dedicated CultureSampo index. Using this index, it is easy to efficiently return any resource that is related in any way to a literal or another resource with a label matching a particular text query. In addition, this index also performs subclass inference. Thus, from this stage, in the case of the example queries one would already get Sinuhe the Egyptian the book, Sinuhe and the cover, in addition to the more direct resource hits.

Then, a "mapping" query is run separately for each select query result set. In BookSampo, these map for example any returned covers, reviews, physical works, fictional characters and so on to the abstract works to which they refer. After this, the system automatically takes an intersection of the mapped results returned from each select query. A further "filter" query is also run. In Book-Sampo, this makes sure that only abstract books and authors ever make it to the final result set returned.

After the result set is finally obtained, it is paged and returned. This can still be manipulated by a "grouping" query. This can be used to ensure that for example a set amount of both authors and books matching a particular query are returned. To make sure all information to be shown in the search listing for each matched resource is included (such as cover images, years of first publication, etc.), the system still runs any given "describe" queries for each returned resource, before finally returning answers.

Because of the efficient indexes of CultureSampo as well as caching of e.g. the mapping query results, the average processing time for even these complex queries is still 100-400 milliseconds on a modern desktop server.

5   Discussion

Libraries have centuries of history in describing books as physical objects, particularly as pertains their physical location in the library. This leads to a large amount of institutional friction in applying new forms of indexing. For example, while libraries have talked of functional indexing (FRBR) from the early 1990s, actual systems have started to appear only concurrently with BookSampo.

Yet, before publishing the end-user portal, the benefits of using semantic web technologies in BookSampo have remained in part elusive to the library professionals. Particularly, there has been a noted scepticism with regard to the added value of ontologies versus the added cost of their maintenance. However, after the end-user portal was published, the search and recommendation functionalities afforded by the CultureSampo engine and the network model of information have been lauded as revolutionary, fulfilling the ideal model of fiction. For example, for a query of "Crime and Punishment", the system not only returns a single work, but actually places it in its literary historical context, also listing all authors that say they have been influenced or touched by the work, all other works that are compared to Crime and Punishment in their reviews, all kindred works and so on. Similarly, each work on its own page is automatically linked to other relevant works and the work's context by recommendation algorithms.

As far as the books and authors in BookSampo are concerned, they are also automatically integrated into the CultureSampo system with some 550,000 cultural objects in it. This makes it possible for the user of CultureSampo to approach the entire Finnish culture from a thematic starting point instead of starting with data type or a data producing organisation. For example, on can retrieve instantly data of museum objects, photographs, paintings, contemporary newspaper articles as well as literature dealing with, for example, agriculture in Finland in the nineteenth century. This way it is also possible, for example, to demonstrate the influences between different arts.

The present wish is to combine the BookSampo database semi-automatically with the Elonet database[13] of the Finnish National Audiovisual Archive, which would offer enormous amounts of additional data to both those in BookSampo who are interested in the film versions of books and those in Elonet who would like to know more about the source work of their favourite film.

Since the contents of BookSampo adhere to the principles of linked open data, they also automatically combine in a wider context with all other such material. For example, further information on both authors and books could be sourced from DBPedia. This way, BookSampo gradually approaches the entire context space of literature described in the ideal model for fiction, where "linking carries on ad infinitum".

There has also already been an example where the linked data of BookSampo could be used in a context outside the original environment it was designed for. On 23 May 2011, the major Finnish newspaper Helsingin Sanomat organized an open data hacking event, which utilized the BookSampo database through an inferring SPARQL endpoint. The analyses and visualization of the materials revealed, for example, that international detective stories have become longer since the beginning of the 1980s—from an average of 200 pages to 370 pages— but Finnish detective stories did not become longer until the 2000s. Other results combined BookSampo data with external grant data, showing for example what types of topics most likely receive grant funding or awards. Even new interactive applications were created, allowing users to discover which years were statistically similar from a publishing viewpoint, or locating all the places associated with Finnish fiction on a map.

Acknowledgements. Thanks to Erkki Lounasvuori, Matti Sarmela, Jussi Kurki, Joeli Takala, Joonas Laitio, and many others for co-operation. This research is part of the National Finnish Ontology Project (FinnONTO) 2003–2012, funded by the National Technology Agency (Tekes) and a consortium of 38 companies and public organizations. The BookSampo project itself is funded by the Finnish Ministry of Education and Culture.

---

[13] http://www.elonet.fi/

References

1. Davis, I., Galbraith, D.: Bio: A vocabulary for biographical information, http://vocab.org/bio/0.1/.html
2. Doerr, M.: The CIDOC CRM – an ontological approach to semantic interoperability of metadata. AI Magazine 24(3), 75–92 (2003)
3. Hickey, T.B., O'Neill, E.T., Toves, J.: Experiments with the IFLA functional requirements for bibliographic records (FRBR). D-Lib Magazine 8(9) (September 2002)
4. Hyvönen, E.: Developing and using a national cross-domain semantic web infrastructure. In: Sheu, P., Yu, H., Ramamoorthy, C.V., Joshi, A.K., Zadeh, L.A. (eds.) Semantic Computing, IEEE Wiley - IEEE Press (May 2010)
5. Hyvönen, E., Mäkelä, E., Kauppinen, T., Alm, O., Kurki, J., Ruotsalo, T., Seppälä, K., Takala, J., Puputti, K., Kuittinen, H., Viljanen, K., Tuominen, J., Palonen, T., Frosterus, M., Sinkkilä, R., Paakkarinen, P., Laitio, J., Nyberg, K.: CultureSampo – Finnish culture on the semantic web 2.0. Thematic perspectives for the end-user. In: Proceedings, Museums and the Web 2009, Indianapolis, USA, April 15-18 (2009)
6. Hyvönen, E., Mäkelä, E., Salminen, M., Valo, A., Viljanen, K., Saarela, S., Junnila, M., Kettula, S.: MuseumFinland—Finnish museums on the semantic web. Web Semantics: Science, Services and Agents on the World Wide Web 3(2-3), 224–241 (2005)
7. Kurki, J., Hyvönen, E.: Collaborative metadata editor integrated with ontology services and faceted portals. In: Workshop on Ontology Repositories and Editors for the Semantic Web (ORES 2010), the Extended Semantic Web Conference ESWC 2010, Heraklion, Greece. CEUR Workshop Proceedings (June 2010)
8. Mäkelä, E., Ruotsalo, T., Hyvönen, E.: How to deal with massively heterogeneous cultural heritage data – lessons learned in culturesampo. Semantic Web – Interoperability, Usability, Applicability (2011) (accepted for publication)
9. Nelson, J., Cleary, A.: FRBRizing an e-library: Migrating from dublin core to FRBR and MODS. code{4}lib (12) (December 2010)
10. Riva, P., Doerr, M., Zumer, M.: FRBRoo: enabling a common view of information from memory institutions. In: World Library and Information Congress: 74th IFLA General Confrence and Council (August 2008)
11. Saarti, J.: Aspects of Fictional Literature Content Description: Consistency of the Abstracts and Subject Indexing of Novels by Public Library Professionals and Client (in Finnish). Ph.D. thesis, University of Oulu (November 1999)
12. Saarti, J., Hypen, K.: From thesaurus to ontology: the development of the kaunokki Finnish fiction thesaurus. The Indexer 28, 50–58(9) (2010)
13. Serola, S., Vakkari, P.: Yleinen kirjasto kuntalaisten toimissa; Tutkimus kirjastojen hyödyistä kuntalaisten arkielämässä. Finnish Ministry of Education and Culture (May 2011)
14. Tran, T., Cimiano, P., Rudolph, S., Studer, R.: Ontology-Based Interpretation of Keywords for Semantic Search. In: Aberer, K., Choi, K.-S., Noy, N., Allemang, D., Lee, K.-I., Nixon, L.J.B., Golbeck, J., Mika, P., Maynard, D., Mizoguchi, R., Schreiber, G., Cudré-Mauroux, P. (eds.) ASWC 2007 and ISWC 2007. LNCS, vol. 4825, pp. 523–536. Springer, Heidelberg (2007)
15. Viljanen, K., Tuominen, J., Hyvönen, E.: Ontology Libraries for Production Use: The Finnish Ontology Library Service ONKI. In: Aroyo, L., Traverso, P., Ciravegna, F., Cimiano, P., Heath, T., Hyvönen, E., Mizoguchi, R., Oren, E., Sabou, M., Simperl, E. (eds.) ESWC 2009. LNCS, vol. 5554, pp. 781–795. Springer, Heidelberg (2009)

SCMS – Semantifying Content Management Systems

Axel-Cyrille Ngonga Ngomo[1], Norman Heino[1],
Klaus Lyko[1], René Speck[1], and Martin Kaltenböck[2]

[1] University of Leipzig
AKSW Group
Johannisgasse 26, 04103 Leipzig
[2] Semantic Web Company
Lerchenfeldergürtel 43
A-1160 Vienna

Abstract. The migration to the Semantic Web requires from CMS that they integrate human- and machine-readable data to support their seamless integration into the Semantic Web. Yet, there is still a blatant need for frameworks that can be easily integrated into CMS and allow to transform their content into machine-readable knowledge with high accuracy. In this paper, we describe the SCMS (Semantic Content Management Systems) framework, whose main goals are the extraction of knowledge from unstructured data in any CMS and the integration of the extracted knowledge into the same CMS. Our framework integrates a highly accurate knowledge extraction pipeline. In addition, it relies on the RDF and HTTP standards for communication and can thus be integrated in virtually any CMS. We present how our framework is being used in the energy sector. We also evaluate our approach and show that our framework outperforms even commercial software by reaching up to 96% F-score.

1 Introduction

Content Management Systems (CMS) encompass most of the information available on the document-oriented Web (also referred to as Human Web). Therewith, they constitute the interface between humans and the data on the Web. Consequently, one of the main tasks of CMS has always been to make their content as easily processable for humans as possible. Still, with the migration from the document-oriented to the Semantic Web, there is an increasing need to insert machine-readable data into the content of CMS so as to enable the seamless integration of their content into the Semantic Web. Given the sheer volume of data available on the document-oriented Web, the insertion of machine-readable data must be carried out (semi-) automatically. The frameworks developed for the purpose of automatic knowledge extraction must therefore be *accurate* (i. e., display high F-scores) so as to ensure that humans need to curate a minimal amount of the knowledge extracted automatically. This criterion is central for the use of automatic knowledge extraction, as approaches with a low recall lead

L. Aroyo et al. (Eds.): ISWC 2011, Part II, LNCS 7032, pp. 189–204, 2011.
© Springer-Verlag Berlin Heidelberg 2011

to humans having to find the false negatives[1] by hand, while a low precision forces the same humans to have to continually check the output of the knowledge extraction framework. A further criterion that determines the usability of a knowledge extraction framework is its *flexibility*, i. e., how easy it is to integrate this framework in CMS. This criterion is of high importance as the current CMS landscape consists of hundreds of very heterogeneous frameworks implemented in dozens of different languages[2].

In this paper, we describe the SCMS framework[3]. The main goal of our framework is to allow the extraction of structured data (i. e., RDF) out of the unstructured content of CMS, the linking of this content with the Web of Data and the integration of this wealth of knowledge back into the CMS. SCMS relies exclusively on RDF messages and simple Web protocols for its integration into existing CMS and the processing of their content. Thus, it is *highly flexible* and can be used with virtually any CMS. In addition, the underlying approach implements a *highly accurate* knowledge extraction pipeline that can be configured easily for the user's purposes. This pipeline allows to merge and improve the results of state-of-the-art tools for information extraction, to manually post-process the results at will and to integrate the extracted knowledge into CMS, for example as RDFa. The main contributions of this paper are the following:

1. We present the architecture of our approach and show that it can be integrated easily in virtually any CMS, provided it offers sufficient hooks into the life-cycle of its managed content items.
2. We give an overview of the vocabularies we use to represent the knowledge extracted from CMS.
3. We present how our approach is being used in a use case centered around renewable energy.
4. We evaluate our approach against a state-of-the-art commercial system for knowledge extraction in two practical use cases and show that we outperform the commercial system with respect to F-score while reaching up to 96% F-score on the extraction of locations.

The rest of this paper is structured as follows: We start by giving an overview of related work from the NLP and the Semantic Web community in Section 2. Thereafter, we present the SCMS framework (Section 3) and its main components (Section 4) as well as the vocabularies they use. Subsequently, we epitomize the renewable energy use case within which our framework is being deployed in Section 5. Section 6 then presents the results of an evaluation of our framework in two use cases against an enterprise commercial system (CS) whose name cannot be revealed for legal reasons. Finally, we give an overview of our future work and conclude.

---

[1] i. e., The entities and relations that were not found by the software
[2] A list of CMS on the market can be found at
    http://en.wikipedia.org/wiki/List_of_content_management_systems
[3] http://www.scms.eu

2   Related Work

Information Extraction is the backbone of knowledge extraction and is one of the core tasks of NLP. Three main categories of NLP tools play a central role during the extraction of knowledge from text: Keyphrase Extraction (KE), Named Entity Recognition (NER) and relation extraction (RE). The automatic detection of keyphrases (i. e., multi-word units or text fragments that capture the essence of a document) has been an important task of NLP for decades. Still, due to the very ambiguous definition of what an appropriate keyphrase is, current approaches to the extraction of keyphrases still display low F-scores [16]. According to [15], the majority of the approaches to KE implement combinations of statistical, rule-based or heuristic methods [11,21] on mostly document [17], keyphrase [28] or term cohesion features [23].

NER aims to discover instances of predefined classes of entities (e. g., persons, locations, organizations or products) in text. Most NER tools implement one of three main categories of approaches: dictionary-based [29,3], rule-based [6,26] and machine-learning approaches [18]. Nowadays, the methods of choice are borrowed from supervised machine learning when training examples are available [32,7,10]. Yet, due to scarcity of large domain-specific training corpora, semi-supervised [24,18] and unsupervised machine learning approaches [19,9] have also been used for extracting named entities from text.

The extraction of relations from unstructured data builds upon work for NER and KE to determine the entities between which relations might exist. Some early work on pattern extraction relied on supervised machine learning [12]. Yet, such approaches demanded large amount of training data. The subsequent generation of approaches to RE aimed at bootstrapping patterns based on a small number of input patterns and instances [5,2]. Newer approaches aim to either collect redundancy information from the whole Web [22] or Wikipedia [30,31] in an unsupervised manner or to use linguistic analysis [13,20] to harvest generic patterns for relations.

In addition to the work done by the NLP community, several tools and frameworks have been developed explicitly for extracting RDF and RDFa out of NL [1]. For example, the Firefox extension Piggy Bank [14] allows to extract RDF from web pages by using screen scrapers. The RDF extracted from these webpages is then stored locally in a Sesame store. The data being stored locally allows the user to merge the data extracted from different websites to perform semantic operations. More recently, the Drupal extension OpenPublish[4] was released. The aim of this extension is to support content publishers with the automatic annotation of their data. For this purpose, OpenPublish utilizes the services provided by OpenCalais[5] to annotate the content of news entries. Epiphany [1] implements a service that annotates web pages automatically with entities found in the Linked Data Cloud. Apache Stanbol[6] implements similar functionality on

---

[4] http://www.openpublish.com
[5] http://www.opencalais.org
[6] http://incubator.apache.org/stanbol

a larger scale by providing synchronous RESTful interfaces that allow Content Management Systems to extract annotations from text.

The main drawback of current frameworks is that they either focus on one particular task (e. g., finding named entities in text) or make use of NLP algorithms without improving upon them. Consequently, they have the same limitations as the NLP approaches discussed above. To the best of our knowledge, our framework is the first framework designed explicitly for the purposes of the Semantic Web that combines flexibility with accuracy. The flexibility of the SCMS has been shown by its deployment on Drupal[7], Typo3[8] and conX[9]. In addition, our framework is able to extract RDF from NL with an accuracy superior to that of commercial systems as shown by our evaluation. Our framework also provides a machine-learning module that allows to tailor it to new domains and classes of named entities. Moreover, SCMS provides dedicated interfaces for interacting (e. g., editing, querying, merging) with the triples extracted, making it usable in a large number of domains and use cases.

3   The SCMS Framework

An overview of the architecture behind SCMS is given in Figure 1. The framework consists of two layers: an *orchestration and curation* layer and an *extraction and storage* layer. The CMS that is to be extended with semantic capabilities resides upon our framework and must be extended minimally via a *CMS wrapper*. This extension implements the in- and output behavior of the CMS and communicates exclusively with the first layer of our framework, thus making the components of the extraction and storage layer of our framework swappable without any drawback for the users.

The overall goal of the first layer of the SCMS framework is to coordinate the access to the data. It consists of two tools: the orchestration service and the data wiki OntoWiki. The *orchestration service* is the input gate of SCMS. It receives the data that is to be annotated as a RDF message that abides by the vocabulary presented in Section 4.2 and returns the results of the framework to the endpoint specified in the RDF message it receives. *OntoWiki* provides functionality for the manual curation of the results of the knowledge extraction process and manages the data flow to the *triple store* Virtuoso[10], the first component of the *extraction and storage layer*. In addition to a triple store, the second layer contains the Federated knOwledge eXtraction Framework FOX[11], that uses machine learning to combine and improve upon the results of NLP tools as well as converts these results into RDF by using the vocabularies displayed in Section 4.3. Virtuoso also contains a crawler that allows to retrieve supplementary knowledge from the Web and link it to the information already available in the CMS by integrating it

---

[7] http://drupal.org
[8] http://typo3.org
[9] http://conx.at
[10] http://virtuoso.openlinksw.com
[11] http://fox.aksw.org

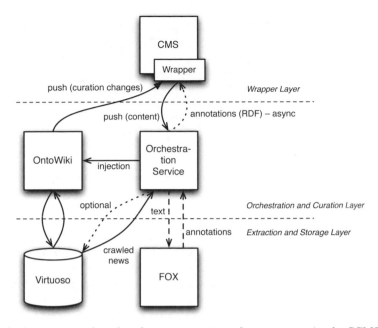

Fig. 1. Architecture and paths of communication of components in the SCMS content semantification system

into the CMS. In the following, we present the central components of the SCMS stack in more detail.

4   Tools and Vocabularies

In this section we describe the main components of the SCMS stack and how they fit together. As running example, we use a hypothetical content item contained in a Drupal CMS. This node (in Drupal terminology) that consists of two parts:

- The `title` *"Prometeus"* and
- a `body` that contains the sentence *"The company Prometeus is an energy provider located in the capital of Hungary, i. e., Budapest."*.

Only the body to the content item is to be annotated by the SCMS stack. Note that for reasons of brevity, we will only show the results of the extraction of named entities. Yet, SCMS can also extract keywords, keyphrases and relations.

4.1   Wrapper

A CMS wrapper (short wrapper) is a component that is tightly integrated into a CMS (see Figure 2) and whose role is to ensure the communication between the

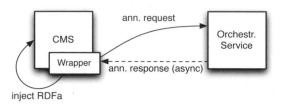

Fig. 2. Architecture of communication between wrapper, CMS and orchestration service

CMS and the orchestration module of our framework. In this respect, a wrapper has to fulfill three main tasks:

1. *Request generation*: Wrappers usually register for change events to the CMS editing system. Whenever a document has been edited, they generate an annotation request that abides by the vocabulary depicted in Figure 3. This request is then sent to the orchestration service.
2. *Response receipt*: Once the annotation has been carried out, the annotation results are sent back to the wrapper. The second of the wrapper's main tasks is consequently to react to those annotation responses and to store the annotations to the document appropriately (e. g., in a triple store). Since the annotation results are sent back asynchronously (i. e., in a separate request), the wrapper must provide a callback URL for this purpose.
3. *Data processing*: Once the data have been received and stored, wrappers usually integrate the annotations into the content items that were processed by the CMS. The integration of annotations is most commonly carried out by "injecting" the annotations as RDFa into the document's HTML rendering. The data injection is mostly realized by registering to document viewing events in the respective CMS and writing the RDFa from the wrapper's local triple store into the content items that are being viewed.

An example of a wrapper request for our example is shown in Listing 1. The `content:encoded` of the Drupal node `http://example.com/drupal/node/10` is to be annotated by FOX. In addition, the whole node is to be stored in the triple store for the purpose of manual processing. Note that the wrapper can choose not to send portions of the content item that are not to be stored in the triple store, e. g., private data. In addition, note that the description of a document is not limited to certain properties or to a certain number thereof, which ensures the high level of flexibility of the SCMS stack. Moreover, the RDF data extracted by SCMS can be easily merged with any structured information provided natively by the CMS (i.e., metadata such as author information). Consequently, SCMS enables CMS that already provide metadata as RDF to answer complex questions that combine data and metadata, e.g., `Which authors wrote documents that are related to Budapest?`

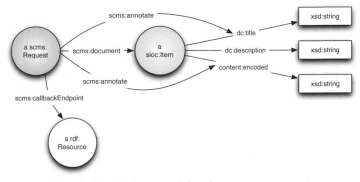

Fig. 3. Vocabulary used by the wrapper requests

```
1  @prefix content: <http://purl.org/rss/1.0/modules/content/> .
2  @prefix dc: <http://purl.org/dc/elements/1.1/> .
3  @prefix sioc: <http://rdfs.org/sioc/ns#> .
4  @base <http://ns.aksw.org/scms/> .
5
6  <http://example.com/wrapperRequest/1> a <Request> ;
7      <document> <http://example.com/drupal/node/10> ;
8      <callbackEndpoint> <http://example.com/wrapper> ;
9      <annotate> content:encoded .
10
11 <http://example.com/drupal/node/10> a sioc:Item ;
12     dc:title "Prometeus" ;
13     content:encoded "The company Prometeus is an energy provider located
                in the capital of Hungary, i.e., Budapest." .
```

Listing 1. Example annotation request as sent by the Drupal wrapper

4.2   Orchestration Service

The main tasks of the orchestration service are to capture state information and to distribute the data across SCMS' layers. The first of the tasks is due to the FOX framework having been designed to be stateless. The orchestration service captures state information by splitting up each document-based annotation requests by a wrapper into several property-based annotation requests that are sent to FOX. In our example, the orchestration service detects that solely the content:encoded property is to be annotated. Then, it reads the content of that property from the wrapper request and generates the annotation request *"The company Prometeus is an energy provider located in the capital of Hungary, i. e., Budapest."* for FOX. Note that while this property-based annotation request consists exclusively of text or HTML and does not contain any RDF, the response returned by FOX is a RDF document serialized in Turtle or RDF/XML.

The annotation results returned by FOX are combined by the orchestration service into the annotation response. Therewith, the relation between the input document and the annotations extracted by FOX is re-established. When all annotations for a particular request have been received and combined, the annotation response is sent back to the wrapper via the provided callback URL. In addition, the results sent back to the wrapper are stored in OntoWiki to facilitate the curation of annotations extracted automatically. The annotation response generated by the orchestration service for our example is shown in Listing 2. It relies upon the output sent by FOX. The exact meaning of the predicates used by FOX and forwarded by the orchestration service are explained in Section 4.3

```
1   @prefix scmsann:    <http://ns.aksw.org/scms/annotations/> .
2   @prefix ctag:       <http://commontag.org/ns#> .
3   @prefix xsd:        <http://www.w3.org/2001/XMLSchema#> .
4   @prefix rdf:        <http://www.w3.org/1999/02/22-rdf-syntax-ns#> .
5   @prefix ann:        <http://www.w3.org/2000/10/annotation-ns#> .
6   @prefix scms:       <http://ns.aksw.org/scms/> .
7
8   []      a           ann:Annotation , scmsann:LOCATION ;
9           scms:annotates <http://example.com/drupal/node/10> ;
10          scms:property <http://purl.org/rss/1.0/modules/content/encoded> ;
11          scms:beginIndex "70"^^xsd:int ;
12          scms:endIndex "77"^^xsd:int ;
13          scms:means <http://dbpedia.org/resource/Hungary> ;
14          scms:source <http://ns.aksw.org/scms/tools/FOX> ;
15          ann:body "Hungary"^^xsd:string .
16
17  []      a           ann:Annotation , scmsann:ORGANIZATION ;
18          scms:annotates <http://example.com/drupal/node/10> ;
19          scms:property <http://purl.org/rss/1.0/modules/content/encoded> ;
20          scms:beginIndex "12"^^xsd:int ;
21          scms:endIndex "21"^^xsd:int ;
22          scms:means <http://scms.eu/Prometeus> ;
23          scms:source <http://ns.aksw.org/scms/tools/FOX> ;
24          ann:body "Prometeus"^^xsd:string .
25
26  []      a           ann:Annotation , scmsann:LOCATION ;
27          scms:annotates <http://example.com/drupal/node/10> ;
28          scms:property <http://purl.org/rss/1.0/modules/content/encoded> ;
29          scms:beginIndex "85"^^xsd:int ;
30          scms:endIndex "93"^^xsd:int ;
31          scms:means <http://dbpedia.org/resource/Budapest> ;
32          scms:source <http://ns.aksw.org/scms/tools/FOX> ;
33          ann:body "Budapest"^^xsd:string .
```

Listing 2. Example annotation response as sent by the orchestration service

4.3   FOX

The FOX framework is a stateless and extensible framework that encompasses all the NLP functionality necessary to extract knowledge from the content of CMS. Its architecture consists of three layers as shown in Figure 4.

FOX takes text or HTML as input. This data is sent to the *controller layer*, which implements the functionality necessary to clean the data, i.e., remove HTML and XML tags as well as further noise. Once the data has been cleaned,

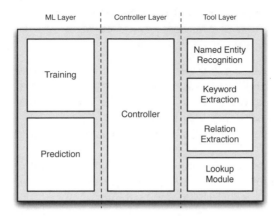

Fig. 4. Architecture of the FOX framework

the controller layer begins with the orchestration of the tools in the *tool layer*. Each of the tools is assigned a thread from a thread pool, so as to maximize usage of multi-core CPUs. Every thread runs its tool and generates an event once it has completed its computation. In the event that a tool does not complete after a set time, the corresponding thread is terminated. So far, FOX integrates tools for KE, NER and RE. The KE is realized by PoolParty[12] for extracting keywords from a controlled vocabulary, KEA[13] and the Yahoo Term Extraction service[14] for statistical extraction and several other tools. In addition, FOX integrates the Stanford Named Entity Recognizer[15] [10], the Illinois Named Entity Tagger[16] [25] and commercial software for NER. The RE is carried out by using the CARE platform[17].

The results from the tool layer are forwarded to the *prediction module* of the *machine-learning layer*. The role of the prediction module is to generate FOX's output based on the output the tools in FOX's backend. For this purpose, it implements several ensemble learning techniques [8] with which it can combine the output of several tools. Currently, the prediction module carries out this combination by using a feed-forward neural network. The neural network inserted in FOX was trained by using 117 news articles. It reached 89.21% F-Score in an evaluation based on a ten-fold-cross-validation on NER, therewith outperforming even commercial systems[18].

Once the neural network has combined the output of the tool and generated a better prediction of the named entities, the output of FOX is generated by

---

[12] http://poolparty.biz

[13] http://www.nzdl.org/Kea/

[14] http://developer.yahoo.com/search/content/V1/termExtraction.html

[15] http://nlp.stanford.edu/software/CRF-NER.shtml

[16] http://cogcomp.cs.illinois.edu/page/software_view/4

[17] http://www.digitaltrowel.com/Technology/

[18] More details on the evaluation are provided at http://fox.aksw.org

using the vocabularies shown in Figure 5. These vocabularies extend the two broadly used vocabularies Annotea[19] and Autotag [20]. In particular, we added the constructs explicated in the following:

- scms:beginIndex denotes the index in a literal value string at which a particular annotation or keyphrase begins;
- scms:endIndex stands for the index in a literal value string at which a particular annotation or keyphrase ends;
- scms:means marks the URI assigned to a named entity identified for an annotation;
- scms:source denotes the provenance of the annotation, i.e., the URI of the tool which computed the annotation or even the system ID of the person who curated or created the annotation and
- scmsann is the namespace for the annotation classes, i.e, location, person, organization and miscellaneous.

Given that the overhead due to the merging of the results via the neural network is of only a few milliseconds and thank to the multi-core architecture of current servers, FOX is almost as time-efficient as state-of-the-art tools. Still, as our evaluation shows, these few milliseconds overhead can lead to an increase of more than 13% F-Score (see Section 6). The output of FOX for our example is shown in Listing 3. This is the output that is forwarded to the orchestration service, which adds provenance information to the RDF before sending an answer to the callback URI provided by the wrapper. By these means, we ensure that the wrapper can write the RDFa in the write segment of the item content.

4.4   OntoWiki

OntoWiki is a semantic data wiki [4] that was designed to facilitate the browsing and editing RDF knowledge bases. Its browsing features range from arbitrary concept hierarchies to facet-based search and query building interfaces. Semantic content can be created and edited by using the RDFauthor system which has been integrated in OntoWiki [27].

OntoWiki plays two key roles within the SCMS stack. First, it serves as entry point for the triple store. This allows for the triple store to be exchanged without any drawback for the user, leading to an easy customization of our stack. In addition, OntoWiki plays the role of an annotation consolidation and curation tool and is consequently the center of the curation pipeline. To ensure that OntoWiki is always up-to-date, the orchestration service sends its annotation responses to both OntoWiki and the wrapper's callback URI. Thus, OntoWiki is also aware of the wrapper (i.e., its callback URI) and can send the results of any manual curation process back to wrapper. Note that manually curated annotations are saved with a different (if manually created) or supplementary (if manually curated) value in their scms:source property. This gives consuming

---

[19] http://www.w3.org/2000/10/annotation-ns#
[20] http://commontag.org/ns#

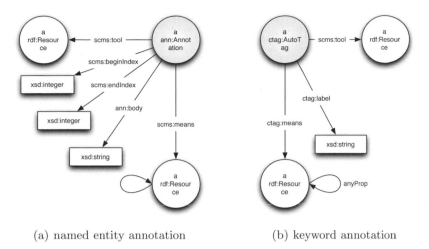

(a) named entity annotation   (b) keyword annotation

Fig. 5. Vocabularies used by FOX for representing named entities (a) and keywords (b)

tools (e. g., wrappers) a chance to assign higher trust values to those annotations. In addition, if a new extraction run is performed on the same document, manually created and curated annotations can be kept for further use. Note that the crawler in Virtuoso can be used to fetch even more data pertaining to the annotations computed by FOX. This data can be sent directly to FOX and inserted in Virtuoso so as to extend the knowledge base for the CMS.

5 Use Case

The SCMS framework is being deployed in the renewable energy sector. The renewable energy and energy efficiency sector requires a large amount of up-to-date and high-quality information and data so as to develop and push the area of clean energy systems worldwide. This information, data and knowledge about clean energy technologies, developments, projects and laws per country worldwide helps policy and decision makers, project developers and financing agencies to make better decisions on investments as well as clean energy projects to set up. The REEEP – the Renewable Energy and Energy Efficiency Partnership[21] is a non-governmental organization that provides the aforementioned information to the respective target groups around the globe. For this purpose, REEEP has developed the `reegle.info` Information Gateway on Renewable Energy and Energy Efficiency [22] that offers country profiles on clean energy, an Actors Catalog that contains the relevant stakeholders in the field per country. Furthermore, it supplies energy statistics and potentials as well as news on clean energy.

---

[21] http://www.reeep.org
[22] http://www.reegle.info

```
1   @prefix scmsann:   <http://ns.aksw.org/scms/annotations/> .
2   @prefix ctag:      <http://commontag.org/ns#> .
3   @prefix xsd:       <http://www.w3.org/2001/XMLSchema#> .
4   @prefix rdf:       <http://www.w3.org/1999/02/22-rdf-syntax-ns#> .
5   @prefix ann:       <http://www.w3.org/2000/10/annotation-ns#> .
6   @prefix scms:      <http://ns.aksw.org/scms/> .
7
8   []      a          ann:Annotation , scmsann:LOCATION ;
9           scms:beginIndex "70"^^xsd:int ;
10          scms:endIndex "77"^^xsd:int ;
11          scms:means <http://dbpedia.org/resource/Hungary> ;
12          scms:source <http://ns.aksw.org/scms/tools/FOX> ;
13          ann:body "Hungary"^^xsd:string .
14
15  []      a          ann:Annotation , scmsann:ORGANIZATION ;
16          scms:beginIndex "12"^^xsd:int ;
17          scms:endIndex "21"^^xsd:int ;
18          scms:means <http://scms.eu/Prometeus> ;
19          scms:source <http://ns.aksw.org/scms/tools/FOX> ;
20          ann:body "Prometeus"^^xsd:string .
21
22  []      a          ann:Annotation , scmsann:LOCATION ;
23          scms:beginIndex "85"^^xsd:int ;
24          scms:endIndex "93"^^xsd:int ;
25          scms:means <http://dbpedia.org/resource/Budapest> ;
26          scms:source <http://ns.aksw.org/scms/tools/FOX> ;
27          ann:body "Budapest"^^xsd:string .
```

Listing 3. Annotations as returned by FOX in Turtle format

The motivation behind applying SCMS to the REEEP data was to facilitate the integration of this data in semantic applications to support efficient decision making. To achieve this goal, we aimed to expand the `reegle.info` information gateway by adding RDFa to the unstructured information available on the website and by making the same triples available via a SPARQL endpoint. For our current prototype, we implemented a CMS wrapper for the Drupal CMS and imported the actors catalog of reegle within in (see Figure 6). This data was then processed by the SCMS stack as follows: All actors and country descriptions were sent to the orchestration service, which forwarded them to FOX. The RDF data extracted by FOX were sent back to the Drupal Wrapper and written via OntoWiki into Virtuoso. The Drupal wrapper then used the keyphrases to extend the set of tags assigned to the corresponding profile in the CMS. The named entities were integrated in the page by using the positional information returned by FOX. By these means, we made the REEEP data accessible for humans (via the Web page) but also for machines (via OntoWiki's integrated SPARQL endpoint and via the RDFa written in the Web pages).

Our approach also makes the automated integration of novel knowledge sources in REEEP possible. To achieve this goal, several selected sources (web sources, blogs and news feeds) are currently being crawled and then analyzed by FOX to extract structured information out of the masses of unstructured text from the Internet.

Fig. 6. Screenshots of SCMS-enhanced Drupal

6   Evaluation

The usability of our approach depends heavily on the quality of the knowledge returned via automated means. Consequently, we evaluated the quality of the RDFa injected into the REEEP data by measuring the precision and recall of SCMS and compared it with that of a state-of-the-art commercial system (CS) whose name cannot be revealed for legal reasons. We chose CS because it outperformed freely available NER tools such as the Stanford Named Entity Recognizer[23] [10] and the Illinois Named Entity Tagger[24] [25] in a prior evaluation on a newspaper corpus. Within that evaluation, FOX reached 89.21% F-score and was 14% better than CS w.r.t. F-score[25]. As it can happen that only segments of multi-word units are recognized as being named entities, we followed a token-wise evaluation of the SCMS system. Thus, if our system recognized *United Kingdom of Great Britain* as a `LOCATION` when presented with *United Kingdom of Great Britain and Northern Ireland*, it was scored with 5 true positives and 3 false negatives.

Our evaluation was carried out with two different data sets. In our first evaluation, we measured the performance of both systems on country profiles crawled from the Web, i.e., on information that is to be added automatically to the REEEP knowledge bases. For this purpose, we selected 9 country descriptions randomly and annotated 34 sentences manually. These sentences contained 119 named entities tokens, of which 104 were locations and 15 organizations. In our

---
[23] http://nlp.stanford.edu/software/CRF-NER.shtml
[24] http://cogcomp.cs.illinois.edu/page/software_view/4
[25] More details at http://fox.aksw.org

second evaluation, we aimed at measuring how well SCMS performs on the data that can be found currently in the REEEP catalogue. For this purpose, we annotated 23 actors profiles which consisted of 68 sentences manually. The resulting reference data contained 20 location, 78 organization and 11 person tokens. Note that both data sets are of very different nature as the first contains a large number of organizations and a relatively small number of locations while the second consists mainly of locations.

The results of our evaluation are shown in Table 1. CS follows a very conservative strategy, which leads to it having very high precision scores of up to 100% in some experiments. Yet, its conservative strategy leads to a recall which is mostly significantly inferior to that of SCMS. The only category within which CS outperforms SCMS is the detection of persons in the actors profile data. This is due to it detecting 6 out of the 11 person tokens in the data set, while SCMS only detects 5. In all other cases, SCMS outperforms CS by up to 13% F-score (detection of organizations in the country profiles data set). Overall, SCMS outperform CS by 7% F-score on country profiles and almost 8% F-score on actors.

Table 1. Evaluation results on country and actors profiles. The superior F-score for each category is in bold font.

Entity Type	Measure	Country Profiles		Actors Profiles	
		FOX	CS	FOX	CS
Location	Precision	98%	100%	83.33%	100%
	Recall	94.23%	78.85%	90%	70%
	F-Score	**96.08%**	88.17%	**86.54%**	82.35%
Organization	Precision	73.33%	100%	57.14%	90.91%
	Recall	68.75%	40%	69.23%	47.44%
	F-Score	**70.97%**	57.14%	**62.72%**	62.35%
Person	Precision	–	–	100%	100%
	Recall	–	–	45.45%	54.55%
	F-Score	–	–	62.5%	**70.59%**
Overall	Precision	93.97%	100%	85.16%	98.2%
	Recall	91.60%	74.79%	70.64%	52.29%
	F-Score	**92.77%**	85.58%	**77.22%**	68.24%

7   Conclusion

In this paper, we presented the SCMS framework for extracting structured data from CMS content. We presented the architecture of our approach and explained how each of its components works. In addition, we explained the vocabularies utilized by the components of our framework. We presented one use case for the SCMS system, i. e., how SCMS is used in the renewable energy sector.

The SCMS stack abides by the criteria of accuracy and flexibility. The flexibility of our approach is ensured by the combination of RDF messages that can be easily extended and of standard Web communication protocols. The accuracy of SCMS was demonstrated in an evaluation on actor and country profiles, within which SCMS outperformed even commercial software. Our approach can be extended by adding support for negative statements, i. e., statements that are not correct but can be found in different knowledge sources across the data landscape analyzed by our framework. In addition, the feedback generated by users will be integrated in the training of the framework to make it even more accurate over time.

References

1. Adrian, B., Hees, J., Herman, I., Sintek, M., Dengel, A.: Epiphany: Adaptable rDFa Generation Linking the Web of Documents to the Web of Data. In: Cimiano, P., Pinto, H.S. (eds.) EKAW 2010. LNCS, vol. 6317, pp. 178–192. Springer, Heidelberg (2010)
2. Agichtein, E., Gravano, L.: Snowball: Extracting relations from large plain-text collections. In: ACM DL, pp. 85–94 (2000)
3. Amsler, R.: Research towards the development of a lexical knowledge base for natural language processing. SIGIR Forum 23, 1–2 (1989)
4. Auer, S., Dietzold, S., Riechert, T.: OntoWiki – A Tool for Social, Semantic Collaboration. In: Cruz, I., Decker, S., Allemang, D., Preist, C., Schwabe, D., Mika, P., Uschold, M., Aroyo, L.M. (eds.) ISWC 2006. LNCS, vol. 4273, pp. 736–749. Springer, Heidelberg (2006)
5. Brin, S.: Extracting Patterns and Relations from the World Wide Web. In: Atzeni, P., Mendelzon, A.O., Mecca, G. (eds.) WebDB 1998. LNCS, vol. 1590, pp. 172–183. Springer, Heidelberg (1999)
6. Coates-Stephens, S.: The analysis and acquisition of proper names for the understanding of free text. Computers and the Humanities 26, 441–456 (1992) 10.1007/BF00136985
7. Curran, J.R., Clark, S.: Language independent ner using a maximum entropy tagger. In: HLT-NAACL, pp. 164–167 (2003)
8. Dietterich, T.G.: Ensemble Methods in Machine Learning. In: Kittler, J., Roli, F. (eds.) MCS 2000. LNCS, vol. 1857, pp. 1–15. Springer, Heidelberg (2000)
9. Etzioni, O., Cafarella, M., Downey, D., Popescu, A.-M., Shaked, T., Soderland, S., Weld, D.S., Yates, A.: Unsupervised named-entity extraction from the web: an experimental study. Artif. Intell. 165, 91–134 (2005)
10. Finkel, J., Grenager, T., Manning, C.: Incorporating non-local information into information extraction systems by gibbs sampling. In: ACL, pp. 363–370 (2005)
11. Frank, E., Paynter, G.W., Witten, I.H., Gutwin, C., Nevill-Manning, C.G.: Domain-specific keyphrase extraction. In: Proceedings of the Sixteenth International Joint Conference on Artificial Intelligence, IJCAI 1999, pp. 668–673. Morgan Kaufmann Publishers Inc., San Francisco (1999)
12. Grishman, R., Yangarber, R.: Nyu: Description of the Proteus/Pet system as used for MUC-7 ST. In: MUC-7. Morgan Kaufmann (1998)
13. Harabagiu, S., Bejan, C.A., Morarescu, P.: Shallow semantics for relation extraction. In: IJCAI, pp. 1061–1066 (2005)
14. Huynh, D., Mazzocchi, S., Karger, D.R.: Piggy Bank: Experience the Semantic Web Inside Your Web Browser. In: Gil, Y., Motta, E., Benjamins, V.R., Musen, M.A. (eds.) ISWC 2005. LNCS, vol. 3729, pp. 413–430. Springer, Heidelberg (2005)

15. Kim, S.N., Kan, M.-Y.: Re-examining automatic keyphrase extraction approaches in scientific articles. In: MWE 2009, pp. 9–16 (2009)
16. Kim, S.N., Medelyan, O., Kan, M.-Y., Baldwin, T.: Semeval-2010 task 5: Automatic keyphrase extraction from scientific articles. In: SemEval 2010, pp. 21–26. Association for Computational Linguistics, Stroudsburg (2010)
17. Matsuo, Y., Ishizuka, M.: Keyword Extraction From A Single Document Using Word Co-Occurrence Statistical Information. International Journal on Artificial Intelligence Tools 13(1), 157–169 (2004)
18. Nadeau, D.: Semi-Supervised Named Entity Recognition: Learning to Recognize 100 Entity Types with Little Supervision. PhD thesis, University of Ottawa (2007)
19. Nadeau, D., Turney, P., Matwin, S.: Unsupervised Named-Entity Recognition: Generating Gazetteers and Resolving Ambiguity. In: Lamontagne, L., Marchand, M. (eds.) Canadian AI 2006. LNCS (LNAI), vol. 4013, pp. 266–277. Springer, Heidelberg (2006)
20. Nguyen, D.P.T., Matsuo, Y., Ishizuka, M.: Relation extraction from wikipedia using subtree mining. In: AAAI, pp. 1414–1420 (2007)
21. Nguyen, T.D., Kan, M.-Y.: Keyphrase Extraction in Scientific Publications. In: Goh, D.H.-L., Cao, T.H., Sølvberg, I.T., Rasmussen, E. (eds.) ICADL 2007. LNCS, vol. 4822, pp. 317–326. Springer, Heidelberg (2007)
22. Pantel, P., Pennacchiotti, M.: Espresso: Leveraging generic patterns for automatically harvesting semantic relations. In: ACL, pp. 113–120 (2006)
23. Park, Y., Byrd, R.J., Boguraev, B.K.: Automatic glossary extraction: beyond terminology identification. In: COLING, pp. 1–7 (2002)
24. Pasca, M., Lin, D., Bigham, J., Lifchits, A., Jain, A.: Organizing and searching the world wide web of facts - step one: the one-million fact extraction challenge. In: Proceedings of the 21st National Conference on Artificial Intelligence, vol. 2, pp. 1400–1405. AAAI Press (2006)
25. Ratinov, L., Roth, D.: Design challenges and misconceptions in named entity recognition. In: CONLL, pp. 147–155 (2009)
26. Thielen, C.: An approach to proper name tagging for german. In: Proceedings of the EACL 1995 SIGDAT Workshop (1995)
27. Tramp, S., Heino, N., Auer, S., Frischmuth, P.: RDFauthor: Employing RDFa for Collaborative Knowledge Engineering. In: Cimiano, P., Pinto, H.S. (eds.) EKAW 2010. LNCS, vol. 6317, pp. 90–104. Springer, Heidelberg (2010)
28. Turney, P.D.: Coherent keyphrase extraction via web mining. In: IJCAI, San Francisco, CA, USA, pp. 434–439 (2003)
29. Walker, D., Amsler, R.: The use of machine-readable dictionaries in sublanguage analysis. Analysing Language in Restricted Domains (1986)
30. Wang, G., Yu, Y., Zhu, H.: PORE: Positive-Only Relation Extraction from Wikipedia Text. In: Aberer, K., Choi, K.-S., Noy, N., Allemang, D., Lee, K.-I., Nixon, L.J.B., Golbeck, J., Mika, P., Maynard, D., Mizoguchi, R., Schreiber, G., Cudré-Mauroux, P. (eds.) ASWC 2007 and ISWC 2007. LNCS, vol. 4825, pp. 580–594. Springer, Heidelberg (2007)
31. Yan, Y., Okazaki, N., Matsuo, Y., Yang, Z., Ishizuka, M.: Unsupervised relation extraction by mining wikipedia texts using information from the web. In: ACL 2009, pp. 1021–1029 (2009)
32. Zhou, G., Su, J.: Named entity recognition using an hmm-based chunk tagger. In: Proceedings of the 40th Annual Meeting on Association for Computational Linguistics, ACL 2002, pp. 473–480. Association for Computational Linguistics, Morristown (2002)

Zhishi.me - Weaving Chinese Linking Open Data

Xing Niu[1], Xinruo Sun[1], Haofen Wang[1], Shu Rong[1], Guilin Qi[2], and Yong Yu[1]

[1] APEX Data & Knowledge Management Lab, Shanghai Jiao Tong University
{xingniu,xrsun,whfcarter,rongshu,yyu}@apex.sjtu.edu.cn
[2] Southeast University
gqi@seu.edu.cn

Abstract. Linking Open Data (LOD) has become one of the most important community efforts to publish high-quality interconnected semantic data. Such data has been widely used in many applications to provide intelligent services like entity search, personalized recommendation and so on. While DBpedia, one of the LOD core data sources, contains resources described in multilingual versions and semantic data in English is proliferating, there is very few work on publishing Chinese semantic data. In this paper, we present Zhishi.me, the first effort to publish large scale Chinese semantic data and link them together as a Chinese LOD (CLOD). More precisely, we identify important structural features in three largest Chinese encyclopedia sites (i.e., Baidu Baike, Hudong Baike, and Chinese Wikipedia) for extraction and propose several data-level mapping strategies for automatic link discovery. As a result, the CLOD has more than 5 million distinct entities and we simply link CLOD with the existing LOD based on the multilingual characteristic of Wikipedia. Finally, we also introduce three Web access entries namely SPARQL endpoint, lookup interface and detailed data view, which conform to the principles of publishing data sources to LOD.

1 Introduction

With the development of Semantic Web, a growing amount of open structured (RDF) data has been published on the Web. Linked Data[3] initiates the effort to connect the distributed data across the Web and there have been over 200 datasets within Linking Open Data (LOD) community project[1]. But LOD contains very sparse Chinese knowledge at the present time. To our knowledge, only Zhao[20] published some Chinese medicine knowledge as Linked Data. However, all data is represented in English thus Chinese language users can hardly use it directly. Some multilingual datasets exist in LOD. Freebase[2], a collection of structured data, contains a certain number of lemmas with Chinese labels. DBpedia only extracts labels and short abstract in Chinese[4]. UWN[11] is another effort in constructing multilingual lexical knowledge base, which maps Chinese

---

[1] http://www.w3.org/wiki/SweoIG/TaskForces/
CommunityProjects/LinkingOpenData
[2] http://www.freebase.com/

L. Aroyo et al. (Eds.): ISWC 2011, Part II, LNCS 7032, pp. 205–220, 2011.
© Springer-Verlag Berlin Heidelberg 2011

words to corresponding vocabulary entries in WordNet. In other words, all of these datasets attach English descriptions to Chinese lemmas. Lack of real-world useful Chinese semantic data cramps the development of semantic applications as well as the Semantic Web itself in the Chinese language community.

When building Chinese Linking Open Data, we have to face some challenging problems that exist in building LOD. In our work, we focus on dealing with two of these challenging problems, that is, managing the heterogeneity of knowledge in different data sources and efficiently discovering `<owl:sameAs>` relations between millions of instances.

It is worthwhile to note that many knowledge bases are not original data providers but extracted from textual articles of other independent data sources. DBpedia[1], which structures Wikipedia knowledge, is a representative one. Fortunately, Chinese textual knowledge is abundant. Chinese Web-based collaborative encyclopedias together contain even more articles than the largest English one: Wikipedia. Besides Wikipedia Chinese version, two Web-based local encyclopedias, Baidu Baike[3] and Hudong Baike[4] has about 3.2 million and 2.8 million[5] articles respectively.

While there is plenty of Chinese textual knowledge, there is very few Chinese semantic data extracted from these sources of knowledge. We are making efforts to build the first Chinese LOD. A fusion of three largest Chinese encyclopedias is our initial achievement which has practical significances: we wish it can help in attracting more efforts to publish Chinese semantic data linked to Zhishi.me. The potential applications of this effort include Chinese natural language processing (entity recognition, word disambiguation, relation extraction), Chinese semantic search, etc.

The work we present in this paper includes:

- We take the three largest Chinese encyclopedias mentioned above as our original data and extract structured information from them. In total, about 6.6 million lemmas as well as corresponding detailed descriptions such as abstracts, infoboxes, page links, categories, etc. are parsed and presented as RDF triples. Procedure of structured data extraction will be introduced in Section 2.
- Since these three encyclopedias are operated independently and have overlaps, we integrate them as a whole by constructing `<owl:sameAs>` relations between every two encyclopedias. Some parallel instance-level matching techniques are employed to achieve this goal. Detailed methods' descriptions can be found in Section 3.1.
- In order to make connections with existing linked data and build a bridge between the English knowledge base and the Chinese knowledge base, we also use `<owl:sameAs>` to link resources in CLOD to the ones in DBpedia, a central hub in LOD. We will discuss it in Section 3.2.

---

[3] http://baike.baidu.com/
[4] http://www.hudong.com/
[5] Statistics collected in March, 2011.

– The first Chinese Linking Open Data around the world has been published as RDF triples on the Web via Zhishi.me. Access mechanisms are presented in Section 4. Finally, we will make some conclusion and outline future work in Section 5.

2   Semantic Data Extraction

We do not generate knowledge data from scratch, but structure existing Web-based encyclopedia information. In this section, we will first introduce the strategies we used to extract semantic data in the form of RDF triples and then give a general view of our knowledge base.

2.1   Extraction Approach

We have three original data sources: Wikipedia Chinese version, Baidu Baike, and Hudong Baike. They provide different ways to edit articles and publish them. Thus there is no one-plan-fits-all extraction strategies.

Wikipedia provides database backup dumps[6], which embed all wiki articles in the form of wikitext source and meta data in XML. The techniques for extracting information from Wikipedia dumps are rather mature. DBpedia Extraction Framework[4] is the most typical efforts. We employ a similar extraction algorithm to reveal structured content from infobox templates as well as their instances.

Wikipedia uses the wikitext language, a lightweight markup language, while both Baidu Baike and Hudong Baike provide the WYSIWYG (what you see is what you get) HTML editors. So all information should be extracted from HTML file archives. Article pages come from these three encyclopedias are alike with minor differences in layout as shown in Figure 1. Currently, we extract 12 types of article content: abstracts, aliases, categories, disambiguation, external links, images, infobox properties, internal links (pagelinks), labels, redirects, related pages and resource ids. They will be explained in detail as follows:

Abstracts. All of these three encyclopedias have separate abstract or summary sections and they are used as values of `zhishi:abstract` property.

Aliases. In Wikipedia, editors can customize the title of an internal link. For example, `[[People's Republic of China|China]]` will produce a link to "People's Republic of China" while the displayed anchor is "China". We call the displayed anchors as the aliases of the virtual article and represent them using `zhishi:alias`. Users cannot rename internal links in Hudong Baike and Baidu Baike, so aliases are not included in these two sources.

Categories. Categories describe the subjects of a given article, `dcterms:subject` is used to present them for the corresponding resources in Zhishi.me. Categories have hyponymy relations between themselves which are represented using `skos:broader`.

---

[6] Dumps of zhWiki: `http://dumps.wikimedia.org/zhwiki/`

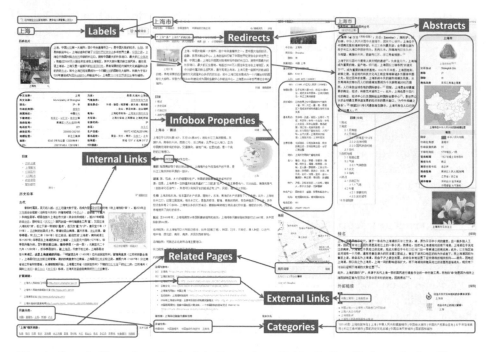

Fig. 1. Sample Encyclopedia Article Pages from Baidu Baike, Hudong Baike, Chinese Wikipedia Respectively

Disambiguation. We say a single term is ambiguous when it refers to more than one topic.

- Hudong Baike and Wikipedia (mostly) use disambiguation pages to resolve the conflicts. However, disambiguation pages are written by natural language, it is not convenient to identify every meaning of homonyms accurately. Thus we only consider a topic, that linked from one disambiguation page and which label shares the same primary term, as a valid meaning. For example, "Jupiter (mythology)" and "Jupiter Island" share the same primary term: "Jupiter".
- Baidu Baike puts all meanings as well as their descriptions of a homonym in a single page. We put every meaning in a pair of square brackets and add it to the primary term as the final Zhishi.me resource name, such as "Term[meaning-1]".

All disambiguation links are described using `zhishi:pageDisambiguates`.

External Links. Encyclopedia articles may include links to web pages outside the original website, these links are represented using `zhishi:externalLink`.

Images. All thumbnails' information includes image URLs and their labels are extracted and represented using `zhishi:thumbnail`.

Infobox Properties. An infobox is a table and presents some featured properties of the given article. These property names are assigned IRIs as the form `http://zhishi.me/[sourceName]/property/[propertyName]`.

Internal Links. An internal link is a hyperlink that is a reference in an encyclopedia page to another encyclopedia page in the same website. We represent these links using predicate `zhishi:internalLink`.

Labels. We use Hudong Baike and Wikipedia article's titles as labels for the corresponding Zhishi.me resources directly. Most Baidu Baike articles adopt this rule, but as explained above, meanings of homonyms are renamed. Labels are represented by `rdfs:label`.

Redirects. All of these three encyclopedias use redirects to solve the synonymous problem. Since Wikipedia is a global encyclopedia while the other two are encyclopedias from Mainland China, Wikipedia contains redirects between Simplified and Traditional Chinese articles. There are rules of word-to-word correspondence between Simplified and Traditional Chinese, so we just convert all Traditional Chinese into Simplified Chinese by these rules and omit redirects of this kind. Redirect relations are described by `zhishi:pageRedirects` to connect two Zhishi.me resources.

Related pages. "Related pages" sections in Baidu Baike and Hudong Baike articles are similar to "see also" section of an article in Wikipedia but they always have fixed positions and belong to fixed HTML classes. Predicate `zhishi:relatedPage` is used to represent this kind of relation between two related resources.

Resource IDs. Resource IDs for Wikipedia articles and most Baidu Baike articles are just the page IDs. Due to the reason that every Zhishi.me resource of homonyms in Baidu Baike is newly generated, they are assigned to special values (negative integers). Articles from Hudong Baike have no page IDs, so we assign them to private numbers. `zhishi:resourceID` is used here.

2.2   Extraction Results and Discussions

Encyclopedia articles from Baidu Baike and Hudong Baike are crawled in March, 2011 and Wikipedia Dump version is 20110412. From the original encyclopedia articles (approximately 15 GB compressed raw data), we totally extracted 124,573,857 RDF triples. Table 1 shows statistics on every extracted content in detail.

Baidu Baike accounts for the most resources, the most categories, as well as the most resources that have categories. It indicates that this data source has a wide coverage of Chinese subjects. The other two data sources, Hudong Baike and Chinese Wikipedia, are superior in relative number of infobox and abstract information respectively. In Table 1, the decimals follows the absolute number is the fractions that divided by the total resources number in each data source.

Comparing other types of content extraction results, each data source has it's own advantage. Baidu Baike has 0.80 images per resource, which means 0.80 picture resources are used on average in one article. Analogously, Hudong

Table 1. Overall Statistics on Extraction Results

Items	Baidu Baike		Hudong Baike		Chinese Wikipedia	
Resources	**3,234,950**		2,765,833		559,402	
∼ that have abstracts	393,094	12.2%	469,009	17.0%	324,627	**58.0%**
∼ that have categories	2,396,570	**74.1%**	912,627	33.0%	314,354	56.2%
∼ that have infoboxes	56,762	1.8%	197,224	**7.1%**	24,398	4.4%
Categories	**516,309**		38,446		93,191	
Properties	**13,226**		474		2,304	
		per res.		per res.		per res.
Article Categories	6,774,442	**2.09**	2,067,349	0.75	796,679	1.42
External Links	2,529,364	0.78	827,145	0.30	573,066	**1.02**
Images	2,593,856	**0.80**	1,765,592	0.64	221,171	0.40
Infobox Properties	477,957	0.14	1,908,368	**0.69**	120,509	0.22
Internal Links	15,462,699	4.78	19,141,664	6.92	9,359,108	**16.73**
Related Pages	2,397,416	0.74	17,986,888	**6.50**	—	—
Aliases	—		—		**362,495**	
Disambiguation Links	28,937		13,733		**40,015**	
Redirects	97,680		37,040		**190,714**	

Table 2. Most Used Properties in Each Data Source

Baidu Baike		Hudong Baike		Chinese Wikipedia	
Chinese Name	37,445	Chinese Name	152,447	Full Name	3,659
Nationality	22,709	Sex	74,374	Population	3,500
Date of Birth	21,833	Occupation	71,647	Area	3,272
Birthplace	19,086	Nationality	70,260	Website	3,061
Occupation	18,596	Era	61,610	Language	2,875
Foreign Name	16,824	Date of Birth	57,850	Height	2,710
Alma Mater	10,709	Home Town	52,518	Kana	2,577
Representative Works	9,751	English Name	52,504	Hiragana	2,203
Nationality[1]	9,621	Kingdom[2]	41,126	Director	2,116
Achievements	7,791	Scientific Name	41,039	Romanization	2,100
Kingdom[2]	7,749	Achievements	40,751	Prefectures	2,099
Category	7,732	Category	40,709	Japanese Name	2,096
Alias	7,725	Family[2]	39,858	Starring	2,015
Family[2]	7,715	Phylum	39,637	Scenarist	1,949
Scientific Name	7,355	Class[2]	39,412	Address	1,949

[1] An ethnic group.
[2] A category in the biological taxonomy.

Baike has relative more infobox properties and related pages than others. While Wikipedia Chinese version articles contain more external and internal links.

Alias, disambiguation links and redirects constitute a valuable thesaurus that can help people to search out most relevant knowledge. Wikipedia Chinese version performs better in this aspect, for the overwhelming superiority it achieves in number of these attributes, even if it has a narrower resource coverage.

Infobox information is the most worthy knowledge in encyclopedias, so we carry out further discussions on this issue here. In Table 2, we list some most frequently used infobox properties. The original properties are written in Chinese but we translated them into English.

The types of resources that are more likely to use infobox can be easily inferred from these frequently used properties. Hudong Baike, which has abundant infobox information, has a large quantity of persons and organisms described in minute detail. Similarly, most listed Baidu Baike properties manifest different facets of somebodies or living things. Chinese Wikipedia also describes lots of people, but in a little different perspective. In addition, featured properties for geographical regions (population, area, etc.) and films (director, starring, etc.) can be seen.

All data sources have their own characteristics, nevertheless they represent subjects in a similar manner, which makes it possible to integrate these attributes. We will give specific examples in Section 4.

Resources that have infobox information are much less than ones with categories. Unfortunately, the quality of these categories are not very high due to the reason that encyclopedia editors usually choose category names casually and many of them are not used frequently. Thus we adopt some Chinese words segmentation techniques to refine these categories, and then choose some common categories to map them to YAGO categories[18] manually.

Top 5 categories in each data source are listed in Table 3. Total number of instances of these categories accounts for over one third resources that have category information. Also notice the top categoires have many overlaps in the three data sources. This suggests the integration of these knowledge bases that we will discuss in the next section is based on good sense.

Table 3. Top 5 Categories with the Number of Their Instances in Each Data Source

Baidu Baike		Hudong Baike		Chinese Wikipedia	
Persons	376,509	Persons	93,258	Persons	50,250
Works	266,687	Works	81,609	Places	28,432
Places	109,044	Words and Expressions	70,101	Organisms	15,317
Words and Expressions	69,814	Places	40,664	Organizations	12,285
Organisms	55,831	Pharmaceuticals	22,723	Works	8,572
Subtotal	877,885	Subtotal	308,355	Subtotal	114,856
Account for	36.6%	Account for	33.8%	Account for	36.5%

3   Data-Level Mapping among Different Datasets

Baidu Baike, Hudong Baike and Wikipedia have their own adherents. Most of
the time, users edit a certain article by their personal knowledge and that lead
to heterogeneous descriptions. Mapping these articles with various description
styles can help to integrate these separated data sources as a whole. At the
same time, we try to bridge the gap between our Chinese knowledge base and
English one (the Linking Open Data). Descriptions in different languages of a
same subject can supplement each other. We will introduce the methods we use
to achieve these goals in next two sub-sections.

3.1   Mappings within CLOD

Finding mappings between datasets in Linking Open Data is usually done in
two levels. One is the practice of schema-level ontology matching, as Jain *et
al.*[9] and Raimond *et al.*[14] did. The other one aims at matching instances and
we mainly focus on this kind of mapping discovery. Not all existing instance
matching algorithms are suitable for finding <owl:sameAs> links between large-
scale and heterogeneous encyclopedias. For example, KnoFuss[13] need instance
data represented as consistent OWL ontologies, however, our extracted semantic
data does not meet this requirement. Raimond *et al.*[17] proposed an interlinking
algorithm which took into account both the similarities of web resources and of
their neighbors but had been proved to be operative in a really small test set.

Silk[19] is a well-known link discovery framework, which indexes resources
before detailed comparisons are performed. Pre-matching by indexes can dramatically reduce the time complexity on large datasets, thus we also match our
encyclopedia articles based on this principle.

Simply indexing resources by their labels has some potential problems. One
is that the same labels may not represent the same subject: different subjects
having the same label is quite common. The other one is opposite: same subject
may have different labels in some cases. These two possible situation would affect
the precision and recall in matching respectively.

We will introduce how we deal with this problem in practice by proposing
three reasonable but not complex strategies to generate the index.

Using Original Labels. The first index generation strategy is just using original labels. This strategy normally has a high precision except it comes with the
problem of homonyms. Fortunately, we extract different meanings of homonyms
as different resources, which has been introduced in Section 2.1. In other words,
it is impossible to find two resources that have different meanings with the same
label if all homonyms are recognized. This fact ensures the correctness of this
strategy.

There is no denying that the performance of this method depends on the
quality of existing encyclopedia articles: whether the titles are ambiguous.

Punctuation Cleaning. When it comes to the second problem: discovering mappings between resources with different labels, one of the most efficient methods we used is punctuation cleaning. Figure 2 shows some examples of same entity having different labels due to the different usage of Chinese punctuation marks. These cases can be handled by the punctuation cleaning method.

(1) 肖申克的救赎 = 《肖申克的救赎》
(2) 海尔波普彗星 = 海尔·波普彗星 = 海尔-波普彗星
(3) 奋进号航天飞机 = "奋进号" 航天飞机

Fig. 2. Examples of Same Entity Has Different Labels

1. Some Chinese encyclopedias encourage editors to use guillemets (《》) to indicate the title of a book, film or album etc. However, guillemets are not imperative to be part of titles. Example (1) in Figure 2 illustrates a same film with/without guillemets.
2. In Chinese, we often insert an interpunct (·) between two personal name components. In some certain cases, people may insert a hyphen instead or just adjoin these components. Example (2) in Figure 2 shows three different labels to indicate a comet named after two persons.
3. According to the usage of Chinese punctuation marks, it is a good practice to quote a cited name by double styling quotation marks (""). However, it is not a mandatory requirement. Example (3) in the figure indicates a space shuttle with its name *Endeavour* quoted and not quoted.

Punctuation marks may have special meanings when they constitute a high proportion of the whole label string. So we calculate the similarity between two labels using Levenshtein distance[10] and attach penalty if strings are too short.

Extending Synonyms. The third strategy we use in index generation also deals with the problem of linking resources with different labels. This one is making use of high quality synonym relations obtained from redirects information (A redirects to B means A and B are synonyms). We can treat redirects relations as approximate `<owl:sameAs>` relations temporarily and thereupon find more links based on the transitive properties of `<owl:sameAs>`.

Usually, the title of a redirected page is the standard name. So we just link two resources with standard names to avoid redundancy. Resources with aliases can still connect to other data source via `pageRedirects`.

Since our dataset is very large, it still has a great time and space complexity even we adopt the pre-matching by index method. We utilize distributed MapReduce[5] framework to accomplish this work. All resources are sorted by their index term in a map procedure, and naturally, similar resources will gather together and wait for detailed comparisons. In practice, totally approximately 24 million index terms are generated from our data sources. This distributed algorithm makes it easier to discovering links within more datasets because pairwise comparisons between every two datasets are avoided.

We say two resources are integrated if they are linked by <owl:sameAs> property. Thus two data sources have intersections when they provide descriptions for mutual things. The number of links found by our mapping strategies is reflected in Figure 3. It confirms the nature of heterogeneity in these three data sources. Original 6.6 million resources are merged into nearly 5 million distinct resources, while only a small proportion (168,481, 3.4%) of them are shared by all.

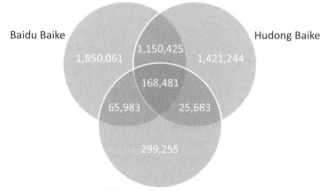

Fig. 3. Intersections of Our Three Data Sources Shown in Venn Diagram

3.2 Linking CLOD with LOD

DBpedia, a structured Wikipedia, is now one of the central data sources in Linking Open Data. Mapping resources between Zhishi.me and DBpedia is a cross-lingual instance matching problem, which is remaining to be solved. Ngai et al.[12] tried to use a bilingual corpus to align WordNet[7] and HowNet[8]. However, their experimental results showed that the mapping accuracy, 66.3%, is still not satisfiable.

Fu et al.[7] emphasized that translations played a critical role in discovering cross-lingual ontology mapping. Therefore, we make use of the high-quality Chinese-English mapping table at hand: Wikipedia interlanguage links. Wikipedia interlanguage links are used to link the same concepts between two Wikipedia language editions and this information can be extracted directly from wikitext.

DBpedia's resource names are just taken from the URLs of Wikipedia articles, linking DBpedia and wikipedia dataset in Zhishi.me is straightforward. Likewise, resources in DBpedia and the whole Zhishi.me can be connected based on the transitive properties of <owl:sameAs>. In total, 192,840 links are found between CLOD and LOD.

---

[7] A large lexical database of English. http://wordnet.princeton.edu/
[8] A Chinese common-sense knowledge base. http://www.keenage.com/

4   Web Access to Zhishi.me

Since Zhishi.me is an open knowledge base, we provide several mechanisms for Web accessing. Not only client applications can access it freely, but also human users can lookup and get the integrated knowledge conveniently.

4.1   Linked Data

According to the Linked Data principles[9], Zhishi.me create URIs for all resources and provide sufficient information when someone looks up a URI by the HTTP protocol. We have to mention that in practice, we use IRIs (Internationalized Resource Identifiers) instead of URIs. This will be discussed in detail next.

Since Zhishi.me contains three different data sources, we design three IRI patterns to indicate where a resource comes from (Table 4). The Chinese characters are non-ASCII, so we choose IRIs, the complement to the URIs[6], for Web access. But IRIs are incompatible with HTML 4[16], we have to encode non-ASCII characters with the URI escaping mechanism to generate legal URIs as "href" values for common Web browsing.

Table 4. IRI Patterns

Sources	IRI Patterns
Baidu Baike	`http://zhishi.me/baidubaike/resource/[Label]`
Hudong Baike	`http://zhishi.me/hudongbaike/resource/[Label]`
Wikipedia Chinese version	`http://zhishi.me/zhwiki/resource/[Label]`

Data is published according to the best practice recipes for publishing RDF vocabularies [2]. When Semantic Web agents that accept "application/rdf+xml" content type access our server, resource descriptions in RDF format will be returned. We try our best to avoid common errors in RDF publishing to improve the quality of the open data published on the Web. This issue has been discussed in [8].

In order to encourage non Semantic Web community users to browse our integrated data, we merge descriptions of the same instance and design an easy-to-read layout template to provide all corresponding contents. Figure 4 shows a sample page that displays the integrated data.

There are two ways to browse the integrated resources: via the lookup service, which will be introduced in the next section, or via "`<owl:sameAs>` box" at the upper right corner of a resource displaying page. In "`<owl:sameAs>` box", all resources have `<owl:sameAs>` relation with currently displaying resources are listed and users can tick resources they want to merge.

Detailed descriptions listed in the view page are all statements with subjects being currently displaying resources, so all distinct subjects are assigned different colors which can help users to recognize where a description comes from.

---

[9] `http://www.w3.org/DesignIssues/LinkedData.html`

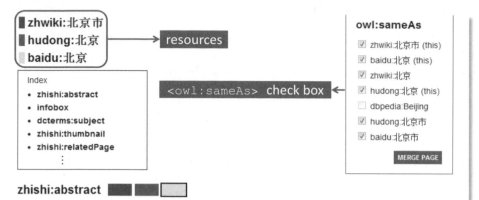

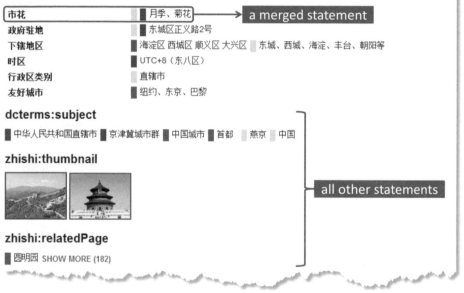

Fig. 4. An Example of Integrated Resources Page

Triples extracted from infoboxes are still presented together. If some statements are sharing the same predicate (property), we will merge the objects (values) but remain the identity colors. The comprehensive property-value pairs rely upon linking heterogeneous data sources.

Other statements grouped by their predicates, such as subjects, thumbnails, relatedPages, etc. are listed subsequently. Users may click on "SHOW MORE" if they want to know more information.

4.2   Lookup Service

For users who do not know the exact IRI of a given resource name, we provide a lookup service to help them. This service is available at `http://zhishi.me/lookup/`. Our index is constructed using four matching strategies presently:

- Returning all resources whose labels exactly match the user's query.
- Using `<owl:sameAs>` links to provide the co-references.
- Some known synonymous are used to provide as many as possible similar resources.
- If a given name has several different meanings (which reflected in a disambiguation page), all corresponding resources of these meanings will also be returned.

Figure 5 gives a sample query. If we search for "Pacific Ocean", it returns not only resources whose labels are exactly "Pacific Ocean" but also a TV miniseries named "The Pacific" (quoted by guillemets) and two disambiguation resources.

As mentioned above, the lookup service is also an entrance to integrate inter-relate resources.

Fig. 5. A Sample Query to Lookup Service

4.3   SPARQL Endpoint

We also provide a simple SPARQL Endpoint for professional users to customize queries at `http://zhishi.me/sparql/`. We use AllegroGraph RDFStore[10] to store the extracted triples and provide querying capabilities.

5   Conclusions and Future Work

Zhishi.me, the first effort to build Chinese Linking Open Data, currently covers three largest Chinese encyclopedias: Baidu Baike, Hudong Baike and Chinese Wikipedia. We extracted semantic data from these Web-based free-editable encyclopedias and integrate them as a whole so that Zhishi.me has a quite wide coverage of many domains. Observations on these independent data sources reveal their heterogeneity and their preferences for describing entities. Then, three

---

[10] `http://www.franz.com/agraph/allegrograph/`

heuristic strategies were adopted to discover `<owl:sameAs>` links between equivalent resources. The equivalence relation leads to about 1.6 million original resources being merged finally.

We provided Web access entries to our knowledge base for both professional and non Semantic Web community users. For people who are familiar with Linked Data, Zhishi.me supports standard URIs (IRIs) de-referencing and provides useful information written in RDF/XML format. Advanced users can also build customized queries by SPARQL endpoint. Casual users are recommended to get well-designed views on the data when they use Web browsers. Both lookup service and data integration operations are visualized.

It is the first crack at building pure Chinese LOD and several specific difficulties (Chinese characters comparing and Web accessing for example) have been bridged over. Furthermore, we have a long-term plan on improving and expanding present CLOD:

- Firstly, several Chinese non-encyclopedia data sources will be accommodated in our knowledge. Wide domain coverage is the advantage of encyclopedia, but some domain-specific knowledge base, such as 360buy[11], Taobao[12] and Douban[13], can supplement more accurate descriptions. A blueprint of Chinese Linking Open Data is illustrated in Figure 6[14].
- The second direction we are considering is improving instance matching strategies. Not only boosting precision and recall of mapping discovering within CLOD, but also augmenting the high-quality entity dictionary to link more Chinese resources to the English ones within Linking Open Data. Meanwhile, necessary evaluations of matching quality will be provided. When matching quality is satisfactory enough, we will use a single constant identifier scheme instead of current source-oriented ones.

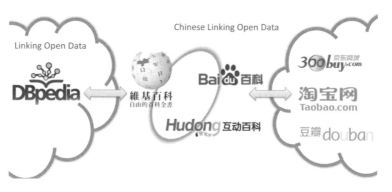

Fig. 6. A Blueprint for Chinese Linking Open Data

---

[11] A Chinese language B2C e-Business site, http://www.360buy.com/
[12] A Chinese language C2C e-Business site, http://www.taobao.com/
[13] The largest Chinese online movie and book database, http://www.douban.com/
[14] All sites mentioned in this figure have rights and marks held by their respective owners.

– Another challenge is refining extracted properties and building a general but consistent ontology automatically. This is an iterative process: initial refined properties are used for ontology learning, and the learned preliminary ontology can help abandon inaccurate properties in return. This iteration will reach the termination condition if results are convergent.

References

1. Auer, S., Lehmann, J.: What Have Innsbruck and Leipzig in Common? Extracting Semantics from Wiki Content. In: Franconi, E., Kifer, M., May, W. (eds.) ESWC 2007. LNCS, vol. 4519, pp. 503–517. Springer, Heidelberg (2007)
2. Berrueta, D., Phipps, J.: Best Practice Recipes for Publishing RDF Vocabularies. W3C Working Group Note (August 2008), http://www.w3.org/TR/2008/NOTE-swbp-vocab-pub-20080828/
3. Bizer, C., Heath, T., Berners-Lee, T.: Linked Data - The Story So Far. Int. J. Semantic Web Inf. Syst. 5(3), 1–22 (2009)
4. Bizer, C., Lehmann, J., Kobilarov, G., Auer, S., Becker, C., Cyganiak, R., Hellmann, S.: DBpedia - A crystallization point for the Web of Data. J. Web Sem. 7(3), 154–165 (2009)
5. Dean, J., Ghemawat, S.: MapReduce: Simplified Data Processing on Large Clusters. In: OSDI, pp. 137–150 (2004)
6. Duerst, M., Suignard, M.: Internationalized Resource Identifiers (IRIs). proposed standard 3987 (January 2005)
7. Fu, B., Brennan, R., O'Sullivan, D.: Cross-Lingual Ontology Mapping – An Investigation of the Impact of Machine Translation. In: Gómez-Pérez, A., Yu, Y., Ding, Y. (eds.) ASWC 2009. LNCS, vol. 5926, pp. 1–15. Springer, Heidelberg (2009)
8. Hogan, A., Harth, A., Passant, A., Decker, S., Polleres, A.: Weaving the pedantic web. In: 3rd International Workshop on Linked Data on the Web, LDOW 2010 (2010)
9. Jain, P., Hitzler, P., Sheth, A.P., Verma, K., Yeh, P.Z.: Ontology Alignment for Linked Open Data. In: Patel-Schneider, et al. (eds.) [15], pp. 402–417
10. Levenshtein, V.: Binary codes capable of correcting deletions, insertions, and reversals. Soviet Physics Doklady 10(8), 707–710 (1966)
11. de Melo, G., Weikum, G.: Towards a universal wordnet by learning from combined evidence. In: Cheung, D.W.L., Song, I.Y., Chu, W.W., Hu, X., Lin, J.J. (eds.) CIKM, pp. 513–522. ACM (2009)
12. Ngai, G., Carpuat, M., Fung, P.: Identifying Concepts Across Languages: A First Step towards a Corpus-based Approach to Automatic Ontology Alignment. In: COLING (2002)
13. Nikolov, A., Uren, V.S., Motta, E., De Roeck, A.: Integration of Semantically Annotated Data by the KnoFuss Architecture. In: Gangemi, A., Euzenat, J. (eds.) EKAW 2008. LNCS (LNAI), vol. 5268, pp. 265–274. Springer, Heidelberg (2008)
14. Parundekar, R., Knoblock, C.A., Ambite, J.L.: Linking and building ontologies of linked data. In: Patel-Schneider, et al. (eds.) [15], pp. 598–614
15. Patel-Schneider, P.F., Pan, Y., Hitzler, P., Mika, P., Zhang, L., Pan, J.Z., Horrocks, I., Glimm, B. (eds.): ISWC 2010, Part I. LNCS, vol. 6496. Springer, Heidelberg (2010)
16. Raggett, D., Hors, A.L., Jacobs, I.: HTML 4.01 Specification - Appendix B: Performance, Implementation, and Design Notes. W3C Recommendation (December 1999), http://www.w3.org/TR/html4/appendix/notes.html

17. Raimond, Y., Sutton, C., Sandler, M.: Automatic interlinking of music datasets on the semantic web. In: Proceedings of the 1st Workshop about Linked Data on the Web, LDOW 2008 (2008)
18. Suchanek, F.M., Kasneci, G., Weikum, G.: Yago: a core of semantic knowledge. In: Williamson, C.L., Zurko, M.E., Patel-Schneider, P.F., Shenoy, P.J. (eds.) WWW, pp. 697–706. ACM (2007)
19. Volz, J., Bizer, C., Gaedke, M., Kobilarov, G.: Discovering and Maintaining Links on the Web of Data. In: Bernstein, A., Karger, D.R., Heath, T., Feigenbaum, L., Maynard, D., Motta, E., Thirunarayan, K. (eds.) ISWC 2009. LNCS, vol. 5823, pp. 650–665. Springer, Heidelberg (2009)
20. Zhao, J.: Publishing Chinese medicine knowledge as Linked Data on the Web. Chinese Medicine 5(1), 1–12 (2010)

An Implementation of a Semantic, Web-Based Virtual Machine Laboratory Prototyping Environment

Jaakko Salonen[1], Ossi Nykänen[1], Pekka Ranta[1], Juha Nurmi[1],
Matti Helminen[2], Markus Rokala[2], Tuija Palonen[2], Vänni Alarotu[2],
Kari Koskinen[2], and Seppo Pohjolainen[1]

[1] Tampere University of Technology, Hypermedia Laboratory, 33101 Tampere, Finland
{jaakko.salonen,ossi.nykanen,pekka.a.ranta,juha.t.nurmi,
seppo.pohjolainen}@tut.fi
[2] Tampere University of Technology, Department of Intelligent Hydraulics and Automation,
33101 Tampere, Finland
{matti.helminen,markus.rokala,tuija.palonen,vanni.alarotu,
kari.koskinen}@tut.fi

Abstract. Creation of virtual machine laboratories – simulated planning and learning environments demonstrating function and structure of working machines – often involve a lot of manual labor. A notable source of the labor is the programming required due to changes in structural and functional models of a system. As a result, rapid prototyping of a virtual machine laboratory becomes difficult, if not impossible. We argue that by using a combination of semantic modeling and prototyping with a web-based system, more rapid development of virtual machine laboratories can be achieved. In this paper, we present the design and implementation of a semantic, web-based virtual machine laboratory prototyping environment. Application of the environment to a case example is also described and discussed.

Keywords: Semantic Web, Resource Description Framework, Web Ontology Language, Prototyping, Virtual Laboratory.

1 Introduction

Virtual machine laboratories (compare [1]) are simulated planning and learning environments demonstrating function and structure of working machines. The creation of such environments has traditionally involved a lot of manual labor. In essence, a developer needs to understand how the machine has been designed and what kinds of planning information and formats are used. In this task, information from various design documents such as CAD drawings and models need to be obtained and integrated. Especially in the context of working machines, it is often desirable to use mathematical modeling based real-time simulations for added interactivity and realistic behavior of the machine, adding a next level of challenge to the creation process.

Since these design documents are primarily created for purposes of the machine's implementation, they often provide less semantic information than what is required to

L. Aroyo et al. (Eds.): ISWC 2011, Part II, LNCS 7032, pp. 221–236, 2011.
© Springer-Verlag Berlin Heidelberg 2011

create functional prototypes in the form of virtual machine laboratories. Especially semantic information connecting various design domains – such hydraulic and mechanical designs – is often informally or implicitly documented, since such pieces of information are more rarely needed for purposes of manufacturing.

Another aspect of this problem is that many perspectives of the data required for this task are largely missing in the designs. For instance, detailed system parameters required for simulation model generation are often missing. Similarly, while the functional model of the system is often implicitly understood by its designers, it may not be explicitly documented. As a consequence, the information required for the virtual prototypes, may need to be manually re-written by a virtual machine laboratory developer. In the worst case, the developer may need to re-engineer or re-design some parts of the system. As such, this process is potentially laborious, lengthy, rigid as well as prone to errors.

In Semogen research project (Phase I during 2010-2011), we have taken an alternative approach to the virtual machine laboratory generation in seek of a more rapid development model. Since most of the problems could be avoided by making sure the design data produced by the primary design activities is complete, we have defined a semantic process for tracking data requirements and the related information objects [2]. Instead of re-engineering and re-designing, we have focused on improving the machine-readability, i.e. the semantic quality of the primary design documents, by using semantic web technologies.

On a general level, this approach is not itself a novel idea; semantic web technologies have readily been applied to other and related domains. For instance in neuromedicine, an ontology for semantic web applications has been readily defined [3]. Benefits for applying various knowledge representation languages to systems and software engineering practices has also been outlined [4]. Very related to our work is a recent effort of product modeling using semantic web technologies [5], as well as a work towards describing linked datasets with an RDF Schema based vocabulary [6].

Also within our domain of application, the need for rapid prototyping and more semantic design data has been also recognized in other ventures. In Simantics project, an open cross-domain modeling and simulation platform was implemented [7]. With an Eclipse Platform -based infrastructure, various simulation and visualization plug-ins were integrated together, resulting in a toolkit for ontology based modeling and simulation. In an another project, TIKOSU, a data model, workflow and a prototype of database-based system was designed [8].

As according to the current machine and virtual laboratory design process, various aspects of the system are designed with a multitude of modeling applications and formats, both of which are often proprietary. Instead of changing these applications, our semantic process provides a model according to which information objects encoded to design documents can be tracked [2].

Our approach to extracting information from the design documents is based on the concept of adapters. For each individual modeling format and/or tool, a specific adapter software is written. This adapter accesses the raw design data and outputs a machine-readable presentation of this information. The information is then integrated together into a semantic model that comprehensively defines the modeled machine. Based on this integrated semantic model, various aspects of the virtual machine laboratory can then be generated.

Generating a virtual machine laboratory with our reference technology required us to generate application-specific configuration documents from the semantic model [9]. In an ideal case, a virtual machine laboratory generation could be approached as a configuration management problem. However, in practice creating a functioning prototype has previously required us to make changes to the underlying software itself. For instance, introducing a new attribute type to a component requires programming. As a result, even with perfected source data, the automated virtual machine laboratory generation could not be realized in many cases.

In order to support more rapid prototyping, a new environment was designed. The fundamental idea was to allow a developer to generate a more light-weight virtual machine laboratory, directly from a semantic model. Technologies were chosen so that they would support easily adding new features to the system with no programming labor whenever possible.

In this article, we present the design and implementation of this light-weight, virtual machine laboratory prototyping environment. The article is organized as follows: In chapter 2, use cases and user requirements for the environment are presented. Additionally a case example used during prototype creation is presented. In chapter 3 the design and the implementation of our prototyping environment is presented. In chapter 4 we present and discuss the results of applying our prototyping environment to the case example. Finally, in chapter 5, we conclude our work.

2   Use Cases, User Requirements and a Case Example

2.1   Use Cases and User Requirements

By definition, we consider *virtual machine laboratory (VML)* to be any environment that can be used to demonstrate a function and structure of a working machine. Different kinds of VMLs may be built for different purposes. Especially three use cases have been identified:

- **Design support.** VMLs to support designers (hydraulics, mechanics, etc.) and collaboration between design areas of the system, by providing a real-time simulation based functional prototype.
- **Educational use.** VMLs for educational purposes are used for providing understanding of the phenomena and function chains of mechatronic machine systems. In additional, educational VMLs may need to consider integration to virtual learning environments as well as assignment of various learning tasks.
- **User guide.** VMLs that provide interactive maintenance and spare parts guides

While details of the systems may change depending on its use case, they come to share many features. Several of our latest VMLs, including a harvester simulator for educational purposes have been developed using our M1 technology [9]. As such, we consider these prototypes and this technology as our reference and as a source for our user requirements. However, regardless of the use case in question, similar features are often requested.

Support for Rapid Prototyping. The ability to make changes to the underlying design and quickly apply these changes to the related virtual environment.

Dynamic Real-Time Simulation. Running real-time simulation of the system (or specific part of the system) in action, including interacting with the model (using controls).

Web Browser Based User Interface. Using the system running either locally or remotely with a web browser (device independence, effortless launch and use with software as a service).

Semantic Search. Locating resources, especially components based on given search criteria (for instance: "find components with a specific material" or "find all components of a given size range").

Ability to Use Data with Potentially Complex and Evolving Schemata. Since data from multiple design domains needs to be integrated, the underlying schema for a VML is potentially very complex. Additionally, since new design domains may be introduced, we should be able to handle some schema evolution with minimal additional work.

Support for Dynamic 2D and 3D Visualizations. The laboratory should be able to represent designs as 2- and 3-dimensional visualizations. In addition, dynamic simulation data is often desired for visualizing function in the designs.

Functional and Integrated Views. Views that representing how various design domains integrate together, for instance to form a chain of actions, are often requested. As machine designs are potentially very complex, provide functional representations a valuable abstraction.

Measurements. Ability to measure and analyze both static (size, width, height, diameter) and dynamic (position, velocity, pressure) properties of a system and its components. Especially for measuring dynamic properties, measurement tools recording history of changes is potentially very useful.

Support for Controllers and Hardware-in-the-Loop. In order to interact with the dynamic system model, at least some controller support is required. Minimally controller devices can be simulated with a graphical interface. Optionally various peripheral devices such as joysticks can be used to emulate controllers. For the scenarios requiring genuine hardware, the system should be able to provide support for hardware-in-the-loop (with other parts of the system being simulated).

Simulation Controls. While continuously playing real-time simulation is sufficient for many uses, it may be useful to be able to control the simulations. For this task, simulation controls including changing simulation speed as well as recording and playing back are potentially useful features.

Our reference technology, M1, already provides many of these features. However, it notably meets only partially the following requirements: 1) support for rapid prototyping, 2) semantic search, 3) web browser based user interface (currently only partial), 4) ability to use data with potentially complex and evolving schemata, 5) support functional and integrated views. Since developing a full-scale alternative to our current technology would require a lot work, it would make sense to aiming at creating a prototype specifically addressing these lacking features.

2.2 Case Example

In order to create a case example, we received real design information and expertise from the industry partners of the Semogen project.

In an attempt to understand and formalize how working machines are designed, we have analyzed the design process. As a result of this work, we have defined a model and methods for analyzing and designing semantic processes and manipulating design information through validation, transformations, and generation applications [10].

As according to this semantic process model, we have identified individual *design activities* and *information requirements* as dependencies between given activities. Each of the design activities may produce design data using *a designing application* often specific for the given design activity. Design applications may use any *data formats* as containers for the *design information*.

In an optimal solution, these data formats are based on open and standard specifications for which processing tools are readily available. Especially extensible markup language (XML) based formats are therefore favorable. As a second to best option, the designing application's export capabilities can be used to export the design data. As a last resort, the designing application's application programming interface (API) capabilities can be utilized in order to create an export format to capture design information.

A subset of the design data and process was chosen as a case example. As a scope of the example, we chose to study a single functionality ("boom lift") within the studied machine. By doing so enabled us to include a heterogeneous and a covering sample of various design activities and materials that potentially are linked by the case functionality in the level of design information. A listing of design activities, their outputs and designing applications used in the case example are provided in table 1.

Table 1. Design activities in the case example

Activity	Output(s)	Designing application(s)
Conceptual design	Conceptual and requirements design documentation.	PDF and Word documents
Hydraulic design	Hydraulic circuit diagrams (2D)	Vertex HD
Mechanical design	Mechanical models (3D)	Vertex G4, SolidWorks
Controller area network (CAN) design	Network and object models	Vector ProCANopen
Simulation design	Simulation models	Simulink

Also within the scope of the case example is some supplementary material. This includes some spare parts documentation and component information data sheets in Portable Document Format (PDF). It must be also noted that simulation models were not directly received from industrial partners, but instead were designed in-house.

3   Environment Design and Implementation

In this section, we will describe the design and implementation of our prototyping environment. Firstly, the semantic process and model are described. Secondly, we will describe details of implementation of the associated semantic viewer application.

3.1 Semantic Process

As described, our approach is based on empowering designers to use the tools they are familiar with. In order to support generation of machine-readable, semantic data, we formalize design activities into a semantic process that produces a semantic model that captures created design information.

In a high level overview, the design activities are formalized into a process. The design activities in our case example includes: concept design, hydraulic design, mechanical design, CANopen design, simulation design and system design. As an example of this process, let us consider the hydraulic design activity from our case example (Fig. 1).

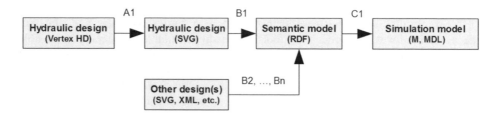

Fig. 1. Semantic process implementation example

Hydraulic diagrams were designed using a domain-specific application, Vertex HD. Following the design similar to data processing and visualization pipelines [11], we can now process the design data in various ways. In the first step (*A1*), the diagrams are then exported to an easily accessible format (SVG; Scalable Vector Graphics). Use of SVG enables us to further process the data with standard XML tools.

In the second step (*B1*), semantic data from the hydraulics SVG can then be extracted. In our case example we realized this by creating a custom XSL transformation. With the transformation, the semantic data encoded into the SVG file was captured and modeled in RDF [12] with a canonical XML serialization [13]. Use of a canonical serialization enabled us to process and validate the data with standard XML tools. For instance, rudimentary input data validators could be written with Schematron (http://www.schematron.com/). Note that as an input to this step, any SVG file containing required information may be used. Thus, other design applications producing conforming SVG can used in the making of hydraulics designs as well.

Similar to hydraulic design, other design activities may also produce input data (steps *B2, ..., Bn*). The semantic model of a target system can then be simply formed by collecting together the RDF documents generated by the pipeline. RDF data model enables us to trivially aggregate these documents together to form a cross design-domain semantic model of the target system. By convention, we have chosen to store this aggregated instance data in a single document (`system.rdf`). For convenience, the definition of a related, domain ontology is passed along as well (`schema.rdf`).

Once the semantic model is generated as a result from running the data processing pipeline, it can be used to generate various portions of a VML. One important use case for the semantic model is generation of simulation models and their templates (step *C1*). Simulation models are required in order to provide VMLs with real-time simulations that enable various dynamic visualizations. While details of simulation model generation are outside the scope of this paper, they are discussed in our other works ([10], [2]).

3.2 Semantic Model and Ontology

The semantic model forms the core of our prototyping environment. The semantic model integrates together design information from various activities. Very importantly, it provides us a standardized mechanism for not only creating machine-readable representations of design data inside the design domains, but also to connect these domains together. We strive at creating a comprehensive semantic model that could be used to generate any aspects of a VML as well as provide basis of integrating design data from many of the disciplines participant machine design. With a schema, we can also provide validation and integrity checks to the data.

Initially, the semantic model was created as an *ad hoc* aggregation of various RDF files from design data adapters. A semi-formal schema was written with SKOS (Simple Knowledge Organization System; [14]). This approach was sufficient for environment bootstrapping as well as for some rudimentary use cases such as using SKOS broader concept to interconnect design domains.

In order to effectively manage instance data, features of RDF Schema [15] were used. We added `rdfs:Class`, `rdfs:subClassOf`, `rdf:Property`, `rdfs:domain` and `rdfs:range` assertions to create classes and their hierarchies as well as properties associated to classes. Adding these assertions was important to provide the semantic model with rudimentary schema-based validation capabilities, including checking for valid properties and valid property values.

Our desire was to provide engineers with a graphical user interface for schema development. While modeling-wise RDFS would have provided us with most of the features required for schema writing, we could not find suitable tools for easily managing an RDFS-based schema. The most suitable software for this purpose was Protégé-OWL ontology editor (http://protege.stanford.edu/). Thus for mostly practical reasons, the original schema file was maintained in Web Ontology Language (OWL) format [16].

3.3 Environment

In terms of a practical implementation, we used a set of tools on Eclipse Platform [17] along with some specific conventions.

In order to simulate a "real-world" engineering process, our multidisciplinary team of researchers used the platform for collaboration through Subversion (http://subversion.tigris.org/). Since we run our environment as "stand-alone", a version control system was necessary to simulate rudimentary product data management (PDM) system functionality (See e.g. [18]).

A new Eclipse project was used to represent an individual VML prototype. Individual design activities were modeled as folders. For each of the activities we further defined (information) requirements, and resources. The resources managed outside the prototyping environment (original CAD drawings) were labeled as external, while resources generated due to data processing were separated under a folder labeled as generated.

We implemented data processing pipelines using Apache Ant (http://ant.apache.org/). Each individual processing step was designed as a new target in Ant. By specifying dependencies between these targets, a pipeline of activities could then be executed in the environment. In addition to XSL transformations that Eclipse supports out-of-the-box, we also configured the environment to enable adapter development with Python (http://www.python.org/).

The rest of the project was organized to following subsections (folders): 1) various tools and adapters required by the pipeline, 2) design application-specific libraries, 3) schema specification. We separated these items from the design activities since they all can be potentially shared between multiple projects.

3.4 Semantic Viewer Application

As a core component of the prototyping environment we implemented a semantic viewer application ("Semogen Player"). Key idea in the viewer application was to make it possible to directly use the semantic model, with no further data processing required for viewing and running a specified machine model.

After reviewing some suitable technologies, a stack consisting of open source components written in Python was chosen. The application architecture was based on the common Model-view-Controller pattern.

Web server was implemented with Tornado (http://www.tornadoweb.org/). A key requirement that lead us to choose Tornado was the need for real-time simulation support. Sufficiently low-latency (10-100 ms) for the simulation interface could be achieved with WebSocket protocol [19], a technology which Tornado was found to readily support.

As for model, we decided to use RDFLib 3 (http://code.google.com/p/rdflib/) along with SuRF (http://code.google.com/p/surfrdf/). With SuRF, RDF triples can be accessed as resources representing classes, properties and their instances. The library also provides various methods for locating and accessing these resources, including a SPARQL [20] interface. SuRF also enables us to support several different RDF triplestores including Sesame 2 (http://www.openrdf.org/) which may need to be used in larger data models. Thus, while we now chose to use RDFLib for practical reasons, other more efficient datastores could be used as well.

New functionality of the system was encapsulated in several Python modules. Framework-like features of the system were placed as part of Semolab model. These features included semantic model bindings, real-time simulation interface as well as application model and various utilities. The main module (*semoplayer*) was used for providing controllers to various views of the system. For view generation, we used Tornado's built-in template engines (HTML with embedded Python code blocks).

User interface components were designed with JavaScript. For interactivity and Ajax handling, jQuery (http://jquery.com/) was used. In order to create a more desktop-like user interface, we used jQuery UI (http://jqueryui.com/) and jQuery UI.Layout Plug-in (http://layout.jquery-dev.net/) as well as several other jQuery plugins. For 2D diagrams, SVG was used. 3D views were implemented using X3DOM (http://www.x3dom.org/).

4 Virtual Machine Laboratory Prototype

Based on the source design materials we received, the defined prototyping environment was used for adapting the source materials to the semantic model. As a result, various virtual machine laboratory views were generated from the design data. In this section we will describe how the semantic modeling and data processing was performed for various design materials, as well as how these materials were connected together. As examples of the design domains, we have included hydraulics and CANopen network designs. The described approach is similarly applied to other design domains such as mechanics, but for brevity are not covered in this paper.

4.1 Hydraulics

For the case example, a simplified hydraulic diagram of the boom lift functionality was drawn with Vertex HD. The diagram was carefully designed for machine-readability (Fig. 2) as follow: 1) for each hydraulic component, a title as well as model name was specified, 2) hydraulic ports were associated with individual components, 3) hydraulic pipes were connected semantically into various ports in the components.

Unique identifiers for hydraulic resources (components, ports and hydraulic pipes) were readily available in the exported design data. Only local uniqueness (per diagram) of the identifiers was guaranteed. In order to generate globally unique identifiers, a document specific prefix (X) was generated by combining a predefined project URI with the filename. Thus, for instance the full URI of the cylinder presented in figure 2, would resolve to `http://project-url/file/#comp-3_21`. In addition to URIs, the resources were identified with Dublin Core identifiers [21] containing local ID in textual format.

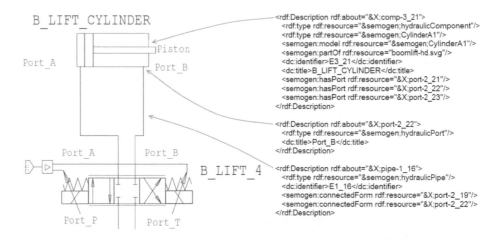

Fig. 2. Excerpt from hydraulic diagram and extracted RDF content

In terms of references, we also encoded model names of various hydraulic components into the design data. Any URI can be specified as a reference to a model name. The component model identifier specified in the source data resolves to two assertions in RDF 1) as an `rdf:type` assertion as well 2) as a `semogen:model` assertion. While model properties could be obtained via the type assertion, the latter was found more practical, since – due to type inference – a component may contain multiple type definitions, from which usually only one refers to a component's model.

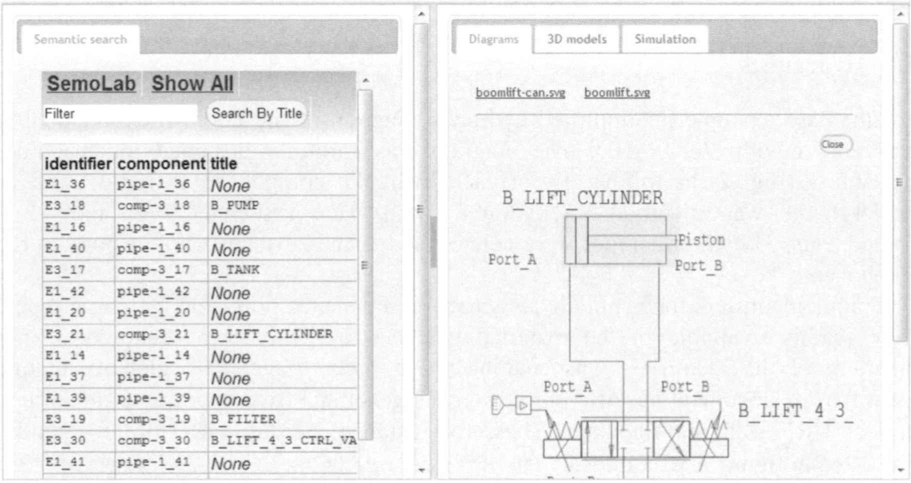

Fig. 3. Semogen Player user interface for hydraulic diagram with semantic search

Based on the exported SVG diagram and extracted RDF data, the data could be then viewed in our viewer application (Fig. 3). The viewer readily detects hydraulic diagrams from the input data (top-right corner). A semantic view (left side) can be used to list all components and pipes as well as their properties. Finally an interactive hydraulic diagram can be presented (right side). A component can be selected by hovering over it with a mouse providing visual linking between the search and the diagram visualization. In addition, any component can be clicked to display all the available RDF data for it.

4.2   CANopen Design

CAN design for the boom (lift) functionality consists of three different CAN buses. These buses were planned using ProCANopen software [22]. Each bus design includes all CAN nodes and their signal routing [23, 24]. This information was then converted from software's native output format (DCF) to an RDF model (Fig. 4). The designed RDF model includes buses, nodes, object and signals.

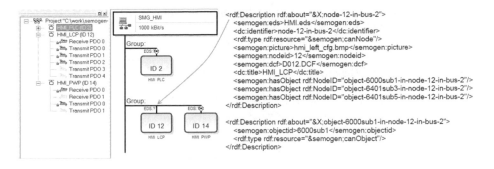

Fig. 4. A network in ProCANopen user interface annotated with related RDF data

In terms of identifiers, original design readily had unique identifiers for CAN buses. Further, each CAN node could be identified by its *NodeID* (`semogen:nodeid`) similarly made available in the original design. As a *NodeID* was only locally unique, a project specific prefix (X) was combined with *busID* and *nodeID* to create globally unique node identifier.As of nodes, each of them included object dictionaries, storing related data objects [23]. These objects are identified by a hexadecimal index number. CAN signals carries these object values to other nodes [23, 24]. So we generated these signals to RDF model and linked them to corresponding CAN objects.

The resulting RDF model was used to generate an SVG-based view of the CAN bus design (Fig. 5). Each component (node and bus) in the SVG contained an element which refers to the RDF model, so the user interface can be used to fetch more information and link to an RDF model on any component in the SVG diagram.

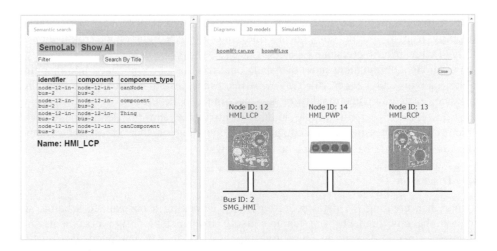

Fig. 5. CAN network visualization with component (node) selection

The RDF data was also used in generation of simulation model. Generator creates Simulink model with CAN nodes and buses. In the future, objects and signals can be used to parametrize these simulated nodes as well.

4.3 Connecting the Design Domains

While designs with various domains are typically created as "self-contained", there are several reasons why cross-references connecting the design domains may be needed. Firstly, an individual physical component such as *cylinder* may appear in multiple designs, potentially in different roles. Secondly, we may need to describe function chains – processes that describe how the various parts of the design work together to implement a functional – for which connections between various designs need to be provided. Finally, even additional cross-references may need to be provided especially for the generation of an integration simulation model.

In some of the current industry practices, engineers use arbitrary mapping tables, for instance implemented as Excel tables (or lists) of references. In the simplest case, these tables contain rows that provide mappings between domain-local identifiers as well as optionally define a global identifier. Additionally the mapping table may encode design information that either is not defined elsewhere or is scattered or otherwise hard to locate from the other design materials.

In order to provide similar mappings within our semantic model, a more formalized approach was needed: the cross-references need to be defined by using full identifiers (URIs). For a rudimentary approach to this mapping, we added an RDF file containing mapping assertions (`mappings.rdf`).

As an example, let us consider mapping various roles of a physical component, a hydraulic cylinder present in examples of sections 4.1 and 4.2. The cylinder appears both in hydraulic design diagram (`&X;comp-3_21`) as well as in CANopen design (`&X;node-1-in-bus-1`) resulting in two different URIs describing the same component. If a mapping assertion between these identifiers is to be defined as an RDF triple, its predicate depends on which relationship it describes. For instance, mapping from hydraulic component (role) into a CAN component (role), can be defined with specific predicate (`semogen:hasACANRole`).

Functions, abbreviations and other meta data that may be embedded into mapping tables can instead be described as new resources inside the mappings file. For instance, each new functionality of the machine can be created as an instance of `semogen:function` for which additional attributes can be defined. For each function, references to various components depending on their roles (for example `semogen:isControlledBy`) can be added as well.

4.4 Discussion

Rather than creating the data model by firstly designing a normalized schema, we created the environment from features rising from actual design documents and processes. In this design, an interactive process model was employed. During this process, various adapters, semantic models as well as features of the viewer application were iteratively developed.

In terms of VML features, our environment is already able to provide the rudimentary aspects of the required features. Firstly, it has proven to support rapid prototyping with the ability to use data with potentially complex and evolving schemata as somewhat demonstrated with the implementation examples. Secondly, the environment provides limited support for semantic search, as well as functional and integrated views to the data (mapping table view). Thirdly, while not demonstrated in this paper, rudimentary support for real-time simulations as well as measurement views have been implemented in the viewer, covering all of the most requested features for a virtual machine laboratory.

In terms of scaling our approach to full machine models and genuine engineering environments, some further work needs to be done. Especially data integration between various domains is seen as a challenge. In principle use of globally unique identifiers allows us to create cross-references between any designs. In practice, manual creation of cross-references is potentially cumbersome and thus impractical, requiring us to look into more automated solutions for scalability. For instance by providing a graphical user interface for the mapping generation, technical details of the RDF data model could be hidden from the system's designer. In addition, to reduce the manual work required for the mappings, the designer could be allowed to provide some general purpose rules for the mappings (for instance: "map together all hydraulic and CAN components that have identical titles").

Another problem with our approach is that in order to integrate new design domains, new adapter needs to be written per new input data format. However, since our data model is based on use of standard RDF, a sophisticated designing application could overcome this problem by readily providing RDF export capabilities, thus leaving only the challenge integrating of the various data models.

Some important lessons were learned during the implementation process. Firstly, it seems that technically semantic web modeling and implementation tools are readily usable and mostly mature. What came as a partial surprise is the lack of well-defined, open vocabularies, schemata and ontologies for the modeling domains in question. For a wider application of semantic modeling, these definitions would be clearly needed. Finally, while some best practices and design patterns for semantic modeling especially in the domain of product modeling, have been defined [5], they are needed in a more wider deployment.

We applied our semantic modeling and process approach to the domain of virtual machine laboratory. As a direction of extension, we see that this process and the tools could be well generalized for other design and engineering practices as well. Especially a similar approach could be applied to other production modeling and engineering domains.

5 Conclusions

In this article, we presented a design and implementation of a virtual machine laboratory prototyping environment. The key approach in the presented environment was the use of understanding machine's and its virtual laboratory's design as a semantic process with defined information objects providing semantic links between various design activities. In the core of our approach was a semantic model implemented in RDF, RDFS and OWL as well as a viewer application for a using conforming data.

In terms of our case example, the given technologies were found as a feasible way to model the underlying data. We see that this semantic modeling approach provided us with some fundamental benefits. Firstly, by using a design of adapters and integrating model, we were able to manage the design data in various domain-specific designing applications. Secondly, it enabled us to create a global, comprehensive representation of the all design data with references between various resources regardless of their design domain. By doing so, we were – in overall – able to understand and build formalized models of how various engineers understand the designs. Thirdly, we recognized that by using an RDFS/OWL based data model, we can fairly easily and quickly adapt the prototype to changes in a data schema. For instance, introducing new properties or classes can be done trivially, with no additional programming work required. Finally, we see that the use general-purpose semantic modeling maximizes data re-usability. For instance, simulation models as well as various aspects of a virtual machine laboratory, can both be generated from the same model.

A notable challenge in our approach is the management of the complexity arising from the heterogeneous process and modeling environment. As virtually any design tools or activities can be introduced to the semantic process, it can potentially become very complex. For instance, while implementation and use of various adapters for integrating data from the tools is often required, can exhaustive use of them result in unmanageable complexity in software design and maintenance. Similarly the design process may become overly complex, resulting in inefficiency in form of poorly manageable structures.

We see that the key in successfully applying this semantic process approach to machine design, requires identification and understanding of the concerns that cross-cut through various design activities. These concerns include, but are not limited to management of global identifiers as well as representation of functional, machine-level models. Without a systematic, process-oriented approach, the risk is that instead of integrating design information, we fall back to *ad hoc* methods of encoding the design data resulting in scattered, unconnected blocks of design data that hinders the re-usability of the data and scalability of the methods.

A lesson learned in the context of semantic web technology application is that while the technologies themselves can be considered mature, some known methods for efficient semantic modeling would have potentially proven to be valuable. We see further room for improvement for instance in introducing an efficient method for managing and integrating local and global identifiers. For industrial deployment of the approach, having more standard, domain vocabularies would be crucially important. Also even though best practices and patterns have already been somewhat extensively recognized and described, we see a room for further studies. Especially since several parts of the modeling technologies (update language, rule languages) are still under standardization and development, a potential of benefit exists from having definitions of more general purpose design patterns.

Individual engineering organizations may not find the motivation for developing domain vocabularies and ontologies. As such, the whole industry would likely benefit from utilizing a more open, collaborative process in the development of these artifacts that are often a requirement for semantic model utilization. In our visions this work could lead to development of an open entity graph within the industry, similar to collaborative databases like Freebase (http://www.freebase.com/).

Within the working machine industry, we see that successfully applying integrated semantic modeling would open up entirely new possibilities for organizing design work. By introducing a semantic modeling-based generation of virtual machine laboratories for design, more agile research and development could be potentially realized. Especially a simulation-driven virtual prototyping process could lead to a new level of efficiency in machine design processes.

References

1. Grimaldia, D., Rapuanob, S.: Hardware and software to design virtual laboratory for education in instrumentation and measurement. Measurement 42(4), 485–493 (2009) ISSN 0263-2241
2. Nykänen, O., Salonen, J., Markkula, M., Ranta, P., Rokala, M., Helminen, M., Alarotu, V., Nurmi, J., Palonen, T., Koskinen, K., Pohjolainen, S.: What Do Information Reuse and Automated Processing Require in Engineering Design? Semantic Process. Journal of Industrial Engineering and Management (2011) (in Review)
3. W3C (2009) Semantic Web Applications in Neuromedicine (SWAN) Ontology. W3C Interest Group Note 20 (October 2009), http://www.w3.org/TR/hcls-swan/ (accessed August 26, 2011)
4. W3C (2006) Ontology Driven Architectures and Potential Uses of the Semantic Web in Systems and Software Engineering. Editor's Draft (Fubruary 11, 2006), http://www.w3.org/2001/sw/BestPractices/SE/ODA/ (accessed August 26, 2011)
5. W3C (2009) Product Modelling using Semantic Web Technologies. W3C Incubator Group Report (October 8, 2009), http://www.w3.org/2005/Incubator/w3pm/XGR-w3pm-20091008/ (accessed August 28, 2011)
6. W3C (2011) Describing Linked Datasets with the VoID Vocabulary. W3C Interest Group Note (March 3, 2011), http://www.w3.org/TR/void/ (accessed August 28, 2011)
7. Simantics (2010) Simatics Platform - Details about Simatics platform, the software architecture, and its applications, http://www.simantics.org/simantics/about-simantics/simantics-platform (accessed June 23, 2011)
8. VTT (2011) TIKOSU - Tietokantakeskeinen koneenohjausjärjestelmän suunnittelu ja toteutus, http://www.hermia.fi/fima/tutkimus/tikosu/ (accessed June 23, 2011)
9. Helminen, M., Palonen, T., Rokala, M., Ranta, P., Mäkelä, T., Koskinen, T.K.: Virtual Machine Laboratory based on M1-technology. In: Proceedings of the Twelfth Scandinavian International Conference on Fluid Power, Tampere, Finland, May 18-20, vol. 1, pp. 321–334 (2011)
10. Markkula, M., Rokala, M., Palonen, T., Alarotu, V., Helminen, M., Koskinen, K.T., Ranta, P., Nykänen, O., Salonen, J.: Utilization of the Hydraulic Engineering Design Information for Semi-Automatic Simulation Model Generation. In: Proceedings of the Twelfth Scandinavian International Conference on Fluid Power, Tampere, Finland, May 18-20, vol. 3, pp. 443–457 (2011)

11. Nykänen, O., Mannio, M., Huhtamäki, J., Salonen, J.: A Socio-technical Framework for Visualising an Open Knowledge Space. In: Proceedings of the International IADIS WWW/Internet 2007 Conference, Vila Real, Portugal, October 5-8, pp. 137–144 (2007)
12. W3C (2011) Resource Description Framework (RDF), http://www.w3.org/RDF/ (accessed June 23, 2011)
13. Nykänen, O.: (2011) RDF in Canonical XML (RDF/cXML), http://wiki.tut.fi/Wille/RDFcXML (accessed June 26, 2011)
14. W3C (2011) SKOS Simple Knowledge Organization System – Home Page, http://www.w3.org/2004/02/skos/ (accessed June 23, 2011)
15. W3C (2004) RDF Vocabulary Description Language 1.0: RDF Schema. W3C Recommendation (February 10, 2004)
16. W3C Web Ontology Language (OWL), http://www.w3.org/2004/OWL/ (accessed June 23, 2011)
17. The Ecllipse Foundation (2011) Eclipse Platform, http://www.eclipse.org/platform/ (accessed June 23, 2011)
18. Bilgic, T., Rock, D.: Product Data Management Systems: State-Of-The-Art And The Future. In: Proceedings of DETC 1997, ASME Design Engineering Technical Conferences, Sacramento, California, September 14-17 (1997)
19. W3C (2011) The Websocket API. Editor's Draft (June 21, 2011), http://dev.w3.org/html5/websockets/ (accessed June 23, 2011)
20. W3C SPARQL (2008) SPARQL Query Language for RDF. W3C Recommendation (January 15, 2008)
21. Dublin Core Metadata Initiative (2010) DCMI Metadata Terms. DCMI Recommendation, http://dublincore.org/documents/dcmi-terms/ (accessed June 23, 2011)
22. Vector (2011) ProCANopen – Project Planning Tool for CANopen Networks, http://www.canopen-solutions.com/canopen_procanopen_en.html (accessed June 23, 2011)
23. IXXAT Automation GmbH (2011) Process data exchange with PDOs ("Process Data Objects"), http://www.canopensolutions.com/english/about_canopen/pdo.shtml (accessed June 23, 2011)
24. Vector (2011) CANopen Fundamentals – The CANopen Standard, http://www.canopen-solutions.com/canopen_fundamentals_en.html (accessed June 23, 2011)

Rule-Based OWL Reasoning for Specific Embedded Devices

Christian Seitz and René Schönfelder

Siemens AG, Corporate Technology
Intelligent Systems and Control
81739 Munich, Germany
{ch.seitz,rene.schoenfelder}@siemens.com

Abstract. Ontologies have been used for formal representation of knowledge for many years now. One possible knowledge representation language for ontologies is the OWL 2 Web Ontology Language, informally OWL 2. The OWL specification includes the definition of variants of OWL, with different levels of expressiveness. OWL DL and OWL Lite are based on Description Logics, for which sound and complete reasoners exits. Unfortunately, all these reasoners are too complex for embedded systems. But since evaluation of ontologies on these resource constrained devices becomes more and more necessary (e.g. for diagnostics) we developed an OWL reasoner for embedded devices. We use the OWL 2 sub language OWL 2 RL, which can be implemented using rule-based reasoning engines. In this paper we present our used embedded hardware, the implemented reasoning component, and results regarding performance and memory consumption.

1 Introduction

Ontologies have been used for formal representation of knowledge for many years now. An ontology is an engineering artifact consisting of a vocabulary used to describe some domain. Additional constraints capture additional knowledge about the domain. One possible knowledge representation language for ontologies is the OWL 2 Web Ontology Language, informally OWL 2. The OWL specification includes the definition of variants of OWL, with different levels of expressiveness. To answer queries over ontology classes and instances some reasoning mechanism is needed. OWL DL and OWL Lite are based on Description Logics, for which sound and complete reasoners exits. Unfortunately, all these reasoners are too complex for embedded systems. But since evaluation of ontologies on these resource constrained devices becomes more and more necessary (e.g. for diagnostics) we developed an OWL reasoner for embedded devices.

A possible use case for embedded reasoning is industrial diagnosis, for example in a car. A car has several system elements like the engine and tires. These elements have physical characteristics, e.g. a tire has a pressure and the engine has a temperature and specific fuel efficiency. Sensor nodes can measure these physical characteristics and deliver the measurement values to a control unit. An ontology specifies all known problems and how to detect them. To create such an ontology, a predefined language is used. With the help of the ontology and the measured data, the software can find the root

L. Aroyo et al. (Eds.): ISWC 2011, Part II, LNCS 7032, pp. 237–252, 2011.
© Springer-Verlag Berlin Heidelberg 2011

cause of the error and may give suggestions how to fix it. To realize these possibilities, in many cases it is important that the software is executable on limited hardware like an embedded system. This work answers the questions of whether reasoning on embedded hardware is possible and how it can be implemented.

The paper is organized as follows. The next section presents already existing approaches for embedded reasoning. This is followed by a detailed explanation of our activities to use an existing DL reasoner on embedded hardware. After this, our rule-based approach is introduced. We present an architecture and the reasoning process in the next section. In the evaluation section, we show the results of the evaluation process. The paper concludes with a summary and a future outlook.

2   Related Work

Various implementations of OWL reasoners exist, e.g. Pellet, FaCT++ and RacerPro. But the memory requirements for installation and at runtime are quite high. Some of them are limited to use on desktop systems or servers only. In the following, we present some approaches that focus on embedded hardware.

SweetRules [5] pioneered rule-based implementation of DL with rule engines. Sweet-Rules does not implement OWL 2, but it supports inferencing in Description Logic Programs subset of DL via translation of first DAML, then OWL 1, into rule engines (Jess/CLIPS).

Bossam[15] is a RETE-based example of a DL reasoner. Bossam is based on a forward chaining production rule engine, which only needs 750Kb runtime memory. In 2007 Bossam has been performance-tuned and released in a new version. The meta-reasoning approach of Bossam was to be changed to a translation-based approach. The results of this work were never released.

One further embedded reasoner is Pocket KRHyper[9]. The core of the system is a first order theorem prover and model generator based on the hyper tableaux calculus. But the development of Pocket KRHyper stopped years ago. This reasoner is not up to date and no support is provided. Documentation on how exactly KRHyper was designed and implemented is also not available.

Another embedded reasoner is μOR[1]. It is a lightweight OWL description logic reasoner for Biomedical Engineering. It was developed for resource-constrained devices in order to enrich them with knowledge processing and reasoning capabilities. To express semantic queries efficiently, the team of μOR has developed SCENT, which is a Semantic Device Language for N-Triples. This approach is similar to but different from ours. μOR is more complex and comes with more overhead. Also it is only designed for OWL Lite and not for OWL 2. For this reason, only a few parts are further pursued in this work.

The first popular approach to emulate a reasoner using CLIPS is described in [13]. Here the object oriented extension of CLIPS, called COOL, was used to build a reasoner for the OWL1 Lite[23] sub-language. This reasoner, called O-DEVICE, is a knowledge base system for reasoning and querying OWL ontologies by implementing RDF/OWL entailments in the form of production rules in order to apply the formal semantics of the language. O-Device is an OWL 1 reasoner which is very powerful and has a lot of features. Its complexity makes it hard to optimize this reasoner for the use on embedded hardware.

The authors in [14] describe an OWL 2 RL reasoner based on Jena and Pellet, both of which are based on Java.

As described in [12], another approach exists to dynamically perform reasoning depending on the specified query. The authors create rules dynamically for the given ontology. In the case of the deterministic end state of OWL 2 RL reasoning, the complexity of this approach resembles our approach. Therefore, the worst-case memory usage is equal to the first approach because in a specially designed case it can be possible to draw every possible conclusion of the ontology.

3 Reasoning on an Embedded Device with Standard DL Reasoners

There are already a lot of implemented DL reasoners. In this section, we analyze whether the most used reasoners can be executed on embedded hardware. As an embedded system, we use a Gumstix Verdex Pro [6], as shown in Figure 1. We opt for the Gumstix because of its modular hardware architecture and its compact size and weight. Additionally, it is best suited for industrial applications because of its supported operating temperature up to $85\,°C$. The Gumstix can also be used in vibrating environments which often occurs if the embedded device is e.g. attached to motors or moving machines. The key features of this stick are: Marvell PXA270 CPU with XScale @ 400 MHz, 64 MB RAM, and 16 MB Flash. The Gumstix was designed to run with a stand alone Linux root image. Additional software can be downloaded or compiled using the OpenEmbedded framework.

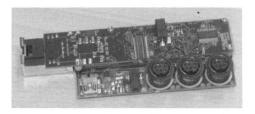

Fig. 1. Gumstix Verdex Pro

In the following, the compatibility of the DL reasoners Pellet, Fact++, and CEL with the Gumstix is analyzed.

3.1 Pellet

Pellet is an open source OWL 2 reasoner for Java. Pellet supports OWL 2 profiles including OWL 2 EL. It incorporates optimizations for nominals, conjunctive query answering, and incremental reasoning [16]. For these tests, Pellet 2.0.2 was used. Pellet is a very powerfully reasoner which uses the tableau reasoning algorithm[7], supports consistency checking, SWRL[17], and DL Safe Rules.

Because of its complexity, Pellet is rather resource intensive. It needs at least 512 MB of RAM at startup plus memory for the Java environment. Anyhow we tried to adapt Pellet to our embedded system. The startup was modified so that Pellet only may use 40 MB of RAM at startup. Unfortunately, the Java virtual machine on the Gumstix is not fully compatible with current Sun Java and Pellet also needs these unsupported functions. Thus, at the moment there is no possibility to reason with the Pellet reasoner on the Gumstix Verdex Pro.

There are other Java DL reasoner, like HOOLET, which cannot be executed on our hardware for the same reasons like Pellet.

3.2   FaCT++

FaCT++ is an open source OWL-DL reasoner written in C++ [20]. For this evaluation the FaCT++ version 1.3.0 is used. After compiling FaCT++ on the embedded hardware three possibilities exist to interact with FaCT++: *(i)* FaCT++ as an HTTP DIG reasoner, *(ii)* FaCT++ as an OWL reasoner with HTTP interface, *(iii)* Standalone FaCT++ with a Lisp-like interface.

First we describe the connection via DIG interface [3], which is a standardized XML interface to Description Logic systems. The DIG language is an XML based representation of ontological entities such as classes, properties, and individuals, and also axioms such as subclass axioms, disjoint axioms, and equivalent class axioms. Unfortunately, there is no useful tool for creating these files for the embedded hardware platform. For that reason the DIG Interface is not considered for our approach.

The next possibility to connect with FaCT++ is the OWL API[8]. The OWL API is a Java API for creating, manipulating and serializing OWL ontologies. Since the OWL API is implemented in Java, the Java native interface (JNI) is necessary. But Java is not completely supported, and the OWL API is not applicable.

The third way to use FaCT++ is as a stand-alone version. For this, an ontology in a special format has to be created, which can be done by an online OWL Ontology converter, e.g. [22]. This procedure works well on a normal x86 computer, on the embedded hardware there are I/O problems while reading the input data. The problem is caused by the ARM architecture.

Since none of the three approaches work, FaCT++ cannot be used for reasoning on embedded systems.

3.3   CEL

CEL [2] is a polynomial-time classifier written in Allegro Common Lisp, which is a closed source, commercial Common Lisp development system and not available for the ARM architecture. On that account CEL unfortunately cannot run on the Gumstix hardware until a version for a free common lisp implementation is published.

4 Rule-Based OWL Reasoning

In the last section we analyzed various reasoners. Unfortunately, non of the above mentioned reasoner can currently be executed on the chosen embedded hardware. One reason is the complexity of the underlying description logic. Therefore, we decided to concentrate on an OWL DL subset to make it tractable.

Fortunately, with the advent of OWL 2 new profiles were introduced by the W3C. Currently the following profiles are defined: OWL 2 EL, OWL 2 QL and OWL 2 RL. OWL 2 EL serves polynomial time algorithms for standard reasoning tasks especially for applications which need very large ontologies. It is based on the description logic EL++. All known approaches for EL reasoning are $O(n^4)$[11]. OWL 2 QL handles conjunctive queries to be answered in LogSpace. It is designed for lightweight ontologies with a large number of individuals. OWL 2 RL is the profile which is used for the reasoner in this paper. It enables polynomial time reasoning using rule engines and operating directly on RDF triples[24]. The complexity of the RL sub language is $O(n^2)$ for standard reasoning. The mathematical proof for that can be done similar to [10]. Therefore OWL 2 RL is the best documented profile. Since it is processable with a standard rule engine, we decided to implement this subset for embedded reasoning.

4.1 CLIPS

In this paper a rule-based system for inferencing and querying OWL ontologies is considered. For this, CLIPS [19] the well-known production rule engine is used because of the small size of CLIPS and its programming language which makes it possible to port it to almost any embedded system. Additionally, it is fast, efficient, and open source.

CLIPS is based on a fact database and production rules. When the conditions of a rule match the existing facts, the rule is placed in the conflict set. A conflict resolution mechanism selects a rule for firing its action which may alter the fact database. Rule condition matching is performed incrementally using the RETE[4] algorithm.

4.2 Rule-Based Reasoning

The concept of our rule-based reasoner is based on a concept-ontology and an instance-ontology which should be reasoned with and a query which should be answered. The rules for the reasoning process can be found in [24]. They are subdivided into 6 parts: *i* the Semantics of Equality, *ii* the Semantics of Axioms about Properties, *iii* the Semantics of Classes, *iv* the Semantics of Class Axioms, *v* the Semantics of Data types, and *vi* the Semantics of Schema Vocabulary. The rules are described by the W3C in the following style:

$Name$	If	$Then$
eq-sym	$T(?x, owl : sameAs, ?y)$	$T(?y, owl : sameAs, ?x)$

Every rule has a name, some if-conditions, and some then-parts. It is also possible to have no if-conditions. This means that the rule should be executed at program start. For the embedded reasoning process, we transformed all of the 80 rules of the W3C to the CLIPS syntax. In CLIPS the rule from above is expressed with the following syntax:

```
(defrule eq-sym
    (. ?x owl:sameAs ?y)
=>
    (assert (. ?y owl:sameAs ?x))
)
```

4.3 Reasoning Component

We use CLIPS as a key element for our embedded reasoning component. Nevertheless, additional modules are necessary. The complete architecture of the reasoning component is shown in Figure 2. The reasoning component is written in C++ which uses CLIPS functions internally. In order to get the OWL 2 RL functions into our system, the rules defined in [24] by the W3C were converted into CLIPS rules and are stored in the file Rules.clp.

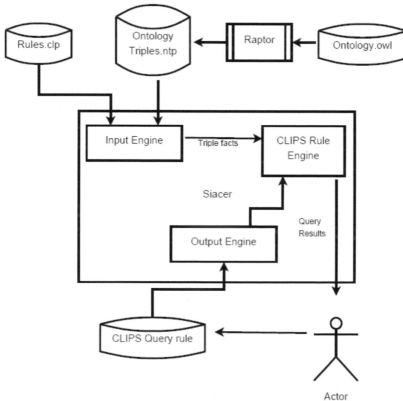

Fig. 2. System architecture

Another important element which must be processed is the ontology. We assume the OWL ontology is available in an XML based format. This OWL file is inserted in the tool Raptor[18]. In this step the ontology is checked for correctness and converted into a format called ntriples[26]. In this format facts are saved as collections of triples with subject, predicate, and object. Together with the rules and the facts CLIPS performs the reasoning. This means that the facts are input data for the OWL RL rules. When all possible facts are added, the reasoning process is finished and a complete materialized knowledge base is created. This means that no new information can be drawn of the existing information and the reasoning process is finished. This knowledge base is based on OWL 2 RL which is included in the rules. The only exception is the semantics of data types. This information causes massive performance problems, because these rules generate too many facts, which finally overflows the memory. Therefore, we only specify which literal is of which data type. This is done by a separate rule.

The test whether a literal is defined from a type which is not contained in the value space is also performed during parsing the OWL RDF file to an ntriples file using the tool Raptor. If there is any syntactical data type error in the ontology, raptor will find it at parsing time. To interact with this knowledge base an interface is necessary to enable the insertion of queries. Executing queries is only possible using a further rule. This rule must be manually created and contains CLIPS code. The syntax of the query is not conform to other knowledge based query languages like SPARQL[25]. Furthermore the rule which contains the query must exist before the reasoning begins.

4.4   Queries

To send a query to the system the simplest way is to create a new rule in CLIPS which fires if the query conditions are passed. For example to identify which student takes the math course the query rule looks like:

```
(defrule abfrage
    (. ?x rdf:type Student)
    (. ?x takesCourse math)
=>
    (printout t "Query 1: " ?x    crlf)
)
```

That prints out appropriate facts that pass the conditions. We are currently working on a SPARQL - CLIPS transformation that allows specifying standard SPARQL queries to our system.

4.5   Formal Analysis

When reasoning is performed it is important to observe the worst-case memory usage. Embedded systems do not have something like swap space. Therefore the complete calculation has to fit in the RAM. For example the reasoner needs 80 MB of memory for a full materialization and only 30 MB for a dynamical result calculation, for an example query. In this case it is possible to run this calculation dynamically on a system with only 60 MB of free memory. But if the calculated results are cached the size of memory will raise up the 80 MB because the possible answers are the same as in a fully materialized knowledge base and it cannot be assumed that some requests will never be part of the knowledge base. Therefore an approach with caching the data comes with no benefit, although it can save memory. But this memory cannot be used for anything because it might be needed for further requests. The situation is different if the results are not cached after calculating. Here only the worst case memory usage in a single query is important.

To compare the first approach in which all calculations are performed a priori and the second approach which calculates the results only when they are requested, a theoretical analysis is necessary.

When looking at the first approach, the worst case memory usage depends on the given ontology. Since the ontology is completely materialized the memory consumption is constantly maximal. There are no relevant differences in memory consumption depending on the requests. In the second approach the worst-case memory usage after the reasoning process is the same as in the first approach. This means that there is no difference for memory usage if the reasoning is performed a priori or dynamically after a request.

To prove this hypothesis we apply graph theory and use a strongly connected graph. A directed graph is called strongly connected if there is a path from each vertex in the graph to every other vertex. We create the graph from the rules provided by the W3C [24]. The vertices of the graph are OWL 2 RL statements like `subClassOf`, `sameAs`, or `range`. The edges are defined by the rules. If a rule transforms an OWL statement in another an edge in the graph is drawn. To keep the graph as simple as possible only statements are considered which are in the `If` and `Else` part of rules. If a statement appears only in one side it cannot trigger other statements and it is not added by other statements and can be ignored. Statements with the same expression in the `If` and `Else` condition in one rule, are also ignored because this does not affect the connectivity of the graph. For illustration, a sample part of the graph, defined by the rule $prp - rng$:

Name	If	Then
prp-rng	$T(?p, rdfs : range, ?c)$ $T(?x, ?p, ?y)$	$T(?y, rdf : type, ?c)$

This rule defines the following node relationship in a graph:

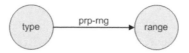

Fig. 3. Representation of the OWL 2 RL rules in a graph

For all OWL rules, the graph is successively built. The resulting graph is strongly connected, which means no information is insignificant for answering a query. Therefore for answering a request nearly all rules and facts are needed. Exceptions are rules which check the consistency of the ontology. The else statement of these rules contains only *false*. To get a correct answer these rules can be ignored theoretically but it is not recommended because they ensure that the ontology is correct. This shows that the second approach needs in the long run generally the same memory amount as the first approach.

There is only a theoretical case in which the worst-case memory usage of the second approach is smaller. When it occurs that the graph is subdivided into two or more subgraphs the worst case memory usage for a given request is the fully materialized subgraph. In this case it can be assured that in no case more memory than the largest fully materialized subgraph is needed and the remaining memory could be used. It is a very unlikely case in which the efficiency of the reasoning could be increased.

5   Evaluation

In this section the evaluation of our rule-based reasoning approach is presented. At first the evaluation environment is introduced. After this, the evaluation results are shown. Generally, we focused on performance and memory usage of the embedded system in contrast to implementing the solution on a system without hardware restrictions.

5.1   Evaluation Environment

The performance of our system has been tested with the Lehigh University Benchmark (LUBM). We use LUBM, because the results can be compared to the results of other reasoners. It is a benchmark developed to evaluate the performance of several Semantic Web repositories in a standardized way. LUBM contains a university domain ontology (Univ-Bench), customizable synthetic data, and a set of test queries. This ontology contains 309 triples and an additional data generator called UBA exists. This tool generates syntactic OWL data based on the Univ-Bench ontology. We only consider the OWL 2 RL part of the ontology for our evaluation purposes.

For our use cases we identified a number of data sets from a few hundred up to a maximum of 25.000 and we started tests with three different benchmark files. Every benchmark contains 309 concept triples. After the reasoning step, some queries from [21] are tested on the data set to determine if the reasoning was correct.

During the tests, the performance of the systems, the memory from the CLIPS reasoner is observed. The unit for the CPU time is mT [million Ticks]. One million ticks are equal to one second calculation time in the cpu and only the time when the CPU works on this process is measured. The time information in the diagrams in the following section is specified in CPU-seconds.

The situation for the memory measurement is similar. Analyzing the memory usage of a reasoning process with and without calculating the CLIPS memory usage shows that the measurement itself needs less than 10KB of memory.

As primary benchmark setup the University0_0 is taken. This is called "bench1" and contains 1657 class instances and 6896 property instances. Together with University0_1 there are 2984 class instances and 12260 property instances, which is called "bench2". This and University1_1 have 4430 class instances and 18246 property instances and called "bench3".

After the reasoning step, the queries 1, 2 and 5 from [21] tested on the data set.

For comparing our results with PC based reasoning, we use an Intel Pentium 4 @ 3200 Mhz and hyper threading, 1 GB of RAM, 4 GB of swap space and Linux (Ubuntu 9.10).

5.2   Evaluation Results

In the following, the performance and memory analysis is presented.

Performance Analysis. Figure 4 shows the complete runtime of the reasoner at the workstation (t_W) and the Gumstix (t_G). The graphs are nearly linear. In fact the runtime can be approximated on the workstation by

$$t_W(x) = 1.9 \cdot 10^{-6}x^2 - 0.016x + 100,$$

in which t_W is the runtime and x the number of triples. On the Gumstix the runtime t_G can be approximated by

$$t_G(x) = 3.4 \cdot 10^{-5}x^2 - 0.4x + 2279.$$

The complexity of the reasoner is polynomial. Therefore the approximated runtime on the Gumstix can be calculated from the time on the workstation by

$$t_G(t_W) = 2.8 \cdot 10^{-3} \cdot t_W^2 + 13 \cdot t_W - 9.8.$$

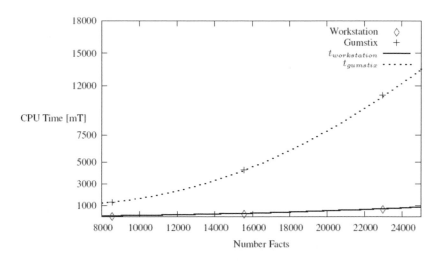

Fig. 4. CPU Time of the rule-based reasoning approach

An execution on the workstation is therefore approximately 12 to 14 times faster than on the Gumstix. This is acceptable considering that the workstation has an eight times higher clock frequency.

For the further evaluation five different measuring points are defined: (1) after all facts are loaded, (2) after the processing, (3) after the first query, (4) after the second query, and (5) after the third query. These points are depicted in the Figures. Most time is needed for the processing, see Figure 5. Here all facts are calculated and all rules have fired. The executions of the queries normally does not need much time, the first query nevertheless needs still a long time. The explanation for this is that CLIPS internally builds some kind of trees or hash tables to find the facts quicker.

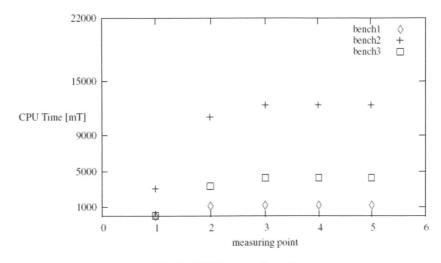

Fig. 5. CPU-Time on Gumstix

Memory Analysis. More important than the performance evaluation is the memory consumption of our approach. The amount of time, the reasoning process takes plays often only a minor role, because it can be done in the background and real time processing is currently not our focus. The memory consumption is therefore more important because it is a criterion for exclusion. If the ontology is too large, it cannot be processed at all. Therefore we need to know how our approach behaves.

In order to reduce the memory usage of CLIPS its memory allocation system was inspected. CLIPS has an integrated garbage collection. This allocates and de-allocates numerous types of data structures during runtime and only reserves new memory if it is actually needed.

Additionally, CLIPS has a special function which tries explicitly to release all memory which is currently not needed. When this function is called after the reasoning process, it can release some memory.

Therefore CLIPS can be forced to use memory economically. This is activated by a flag. If it is enabled, CLIPS will not save information about the pretty printing of facts.

This disables only formatting and can be deactivated without problems. Activating the memory release function saves about 5.8% of memory for all facts and improves our system additionally.

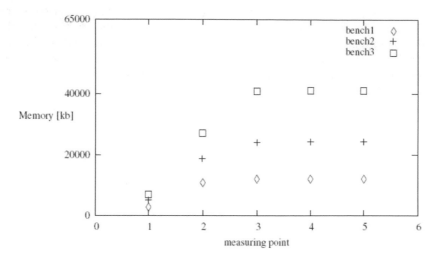

Fig. 6. Memory of rule-based reasoning approach

In Figure 6 the memory usage of the Gumstix can be seen. To calculate the approximate memory usage the formula

$$mem_{poly}(x) = 4.2 \cdot 10^{-5}x^2 + 0.69x + 3459,$$

or for only a few data sets:

$$mem_{lin}(x) = 2x - 5565,$$

in which mem is the size of needed memory in kilo bytes and x the number of triples. It is possible to calculate how many triples can be reasoned with a specified memory size. For this approximation the following formula holds:

$$triples(x) = -5.3 \cdot 10^{-6}x^2 + 0.79x - 357.$$

This is very important for us, since we plan to implement various industrial applications and with this information we can tailor our needed hardware accordingly.

5.3   Evaluation Summary

In the end of this section we provide some detailed facts of our evaluation. This is summarized in Table 1.

Table 1. Detailed evaluation results

Number of facts	CPU time System II	Mem System II	CPU time System I	Mem System I
8864	8.5 mT	7470 kb	80.24 mT	11194 kb
15557	27.54 mT	12686 kb	214.46 mT	18932 kb
22991	59.24 mT	18463 kb	437.89 mT	27482 kb

On the right hand side the data for our System I is shown. We made some additional performance tuning, e.g. compiler options, rule rewriting. This data is shown on the left side, entitled with System II. The performance gain is considerable. The calculation time has been reduced to about 12% from the original calculation time, the memory usage is reduced to about 67%.

Finally, we compared the embedded with a desktop approach. A test ontology took us 8.5 seconds on the embedded hardware. If we use Fact++, the ontology takes about 6 seconds to process the ontology. Thus, the implementation is only a bit slower on the embedded hardware, which is a really good result.

6   Embedded Reasoning Applications

We apply the above presented reasoning approach in some sample applications. Key concept is a digital product memory, which stores product relevant data of the complete product life cycle. With the reasoning approach we work with the stored data, analyze them, infer new date and perform appropriate actions.

6.1   Predictive Maintenance

A possible benefit of a product memory is the detection of technical issues when the product is already deployed in the field. Since the product memory contains a lot of sensor data, an analysis is useful for diagnostics and predictive maintenance.

An example is an industrial robot that moves a work piece from one machine to another. The robot is equipped with an acceleration sensor. While the robot moves the work pieces, the sensor values are continuously (every 20 ms) recorded in the digital product memory of the robot. On basis of this data, an evaluation of the robots product memory can detect, if the robot still works properly or whether an execution has occurred. A more sophisticated method is not only to detect errors if they have occurred, but to detect slight changes in the normal behavior. An anomaly detector should report if new measured data of certain sensors have a different pattern or are beyond the normal sensor values. The basis for an abnormality detector are sensor values and a domain ontology which specifies normal behavior. With the embedded reasoner we try to perform a mapping of sensor value to machine states. On this basis it can be decided, if the current state is an allowed one or whether an alarm must be triggered.

6.2   Decentralized Manufacturing Control

The need of products which are tailored to customers' needs, results in a reduction of the lot size and implies a more flexible production and the associated processes. In the course of an increased diversification the changeover time will be a critical cost factor. This essentially needed flexibility is hard to realize with traditional central control architectures that can be found in nowadays automation systems. One solution is a decentralized production control, done by the product itself. The goal is to operate autonomous working stations and all data that is needed to assemble the product is kept on the product in the digital product memory. If a product enters the vicinity of a working station the necessary machine configuration information is sent from the product memory to the machine and the station accomplishes the necessary tasks.

Thus, if a product is assembled in multiple steps, the necessary data is written to the product memory when the order is entered into the order system. The data contains the description of the single production steps with all its parameters, e.g. the position of bore holes or welds, the used materials, the size or the color. The memory is read by the machine (e. g. via RFID) and the machine is parameterized and set up accordingly. If a production step is finished, the product itself is responsible for the routing to the next station. Depending on its weight and shape either an automated guided vehicle or a conveyer belt can be used. This inference step is done by our embedded reasoning component and is therefore an essential step for next generation automation systems.

6.3   Situation Recognition in Assisted Living

Due to the dramatic growth of elderly population, we additionally aim at research of near-future systems providing elderly people a safe and comfort life during daily living. The people have the possibility to stay either at home or still being mobile and could be relatively healthy or having some physical disabilities or medical liabilities. The diversity and breadth of these scenarios and realistic approaches make this target challenging, assuming the use of various medical devices, different home and mobile systems, heterogeneous and data-rich environments.

A core functionality of such assisted living systems is the conclusion of knowledge about the activities of the user and the current situation in the environment from low-level sensor data and to plan the appropriate short-term and long-term reaction. This reaction on the situations and activities to be recognized are based on situational models that have to keep reasoning system safe for people using it and preserve the relevancy of reactions to the situations respectively. Typical reaction aims at recognizing and/or preventing an urgency situation, defined in form of situation rules within ontology-based situation understanding system.

A small embedded device which can be easily worn on the wrist was chosen. Sensors are needed that monitor the health state of a person. Thus, the product memory becomes a patient memory. It records the vital signs and activities of the senior, does some basic evaluation and executes finally necessary tasks, e.g. informing the patient or sending text messages to doctors.

Additional rules specify the interaction with the environment, because other objects in the smart home environment may be equipped with a product memory as well. This eases the detection of certain situations, e.g. when the temperature of a kettle changes, maybe tea is prepared. Such activities and interdependencies can be optimally expressed with rules.

7 Conclusion

This paper shows that it is currently not possible to use a standard OWL reasoner on the specified embedded system, because existing OWL reasoner are too resource intensive and difficult to port to embedded and source restricted architectures. Therefore, we use a rule-based approach to achieve OWL reasoning on embedded devices. We apply OWL 2 RL rules to the rule engine CLIPS to accomplish the OWL behavior. We integrate the rule engine in an embedded architecture to enable ontology processing in an comfortable and easy way. Several methods for reducing calculation time and memory consumption were reviewed and selected methods were implemented. The created reasoner is compatible with OWL 2 RL with the exception of the semantics of data types which are deleted due to performance reasons. Performance evaluations and the results of our approach are very satisfying. Additionally, we are able to calculate the necessary amount of memory for a given number of facts.

In the future, we will extend our embedded reasoning system. To improve the expressive power of the reasoner, it can be extended for reasoning with OWL 2 EL using the work from [11]. Although this will increase the calculation time and the complexity of reasoning, more complex problems can be addressed.

References

1. Ali, S., Kiefer, S.: μOR – A Micro OWL DL Reasoner for Ambient Intelligent Devices. In: Abdennadher, N., Petcu, D. (eds.) GPC 2009. LNCS, vol. 5529, pp. 305–316. Springer, Heidelberg (2009)
2. Baader, F., Lutz, C., Suntisrivaraporn, B.: CEL — A polynomial-time reasoner for life science ontologies. In: Furbach, U., Shankar, N. (eds.) IJCAR 2006. LNCS (LNAI), vol. 4130, pp. 287–291. Springer, Heidelberg (2006)
3. Bechhofer, S.: The DIG description logic interface: DIG/1.1. Tech. rep., University of Manchester (2003)
4. Forgy, C.L.: Rete: A fast algorithm for the many pattern/many object pattern match problem. Department of Computer Science, Carnegie-Mellon University, Pittsburgh (2003)
5. Grosof, B., Dean, M., Ganjugunte, S., Tabet, S., Neogy, C.: Sweetrules homepage (2005), http://sweetrules.semwebcentral.org/
6. Gumstix: Gumstix website (2010), http://www.gumstix.com
7. Hähnle, R.: Tableaux and Related Methods. Handbook of Automated Reasoning (2001)
8. Horridge, M.: OWL2 API (2010), http://owlapi.sourceforge.net/
9. Kleemann, T., Sinner, A.: KRHyper - in your pocket. In: Nieuwenhuis, R. (ed.) CADE 2005. LNCS (LNAI), vol. 3632, pp. 452–457. Springer, Heidelberg (2005)
10. Krötzsch, M.: Efficient inferencing for OWL EL. In: Janhunen, T., Niemelä, I. (eds.) JELIA 2010. LNCS, vol. 6341, pp. 234–246. Springer, Heidelberg (2010)
11. Krötzsch, M.: Efficient inferencing for the description logic underlying OWL EL. Institut AIFB, KIT, Karlsruhe (2010)
12. Krötzsch, M., ul Mehdi, A., Rudolph, S.: Orel: Database-driven reasoning for OWL 2 profiles. In: Int. Workshop on Description Logics (2010)
13. Meditskos, G., Bassilades, N.: A rule-based object-oriented OWL reasoner. In: IEEE Transactions on Knowledge and Data Engineering (2008)
14. Meditskos, G., Bassiliades, N.: DLEJena: A practical forward-chaining OWL 2 RL reasoner combining Jena and Pellet. Web Semantics 8(1), 89–94 (2010)

15. Jang, M., Sohn, J.-C.: Bossam: An Extended Rule Engine for OWL Inferencing. In: Antoniou, G., Boley, H. (eds.) RuleML 2004. LNCS, vol. 3323, pp. 128–138. Springer, Heidelberg (2004)
16. Parsia, B., Sirin, E.: Pellet: An OWL DL Reasoner. In: Third International Semantic Web Conference-Poster (2004)
17. Parsia, B., Sirin, E., Grau, B.C., Ruckhaus, E., Hewlett, D.: Cautiously approaching SWRL. Tech. rep., University of Maryland (2005)
18. Raptor: Raptor website (2010), http://librdf.org/raptor/
19. Riley, G.: CLIPS (2010), http://clipsrules.sourceforge.net/
20. Tsarkov, D., Horrocks, I.: FaCT++ Description Logic Reasoner: System Description. In: Furbach, U., Shankar, N. (eds.) IJCAR 2006. LNCS (LNAI), vol. 4130, pp. 292–297. Springer, Heidelberg (2006)
21. University, L.: LUBM website (2010),
http://swat.cse.lehigh.edu/projects/lubm/query.htm
22. Volz, R.: FactConverter (2010), http://phoebus.cs.man.ac.uk:9999/OWL/Converter
23. World Wide Web Consortium (W3C): OWL-Lite (2010), http://www.w3.org/TR/2004/REC-owl-features-20040210/#s3
24. World Wide Web Consortium (W3C): OWL profiles (2010), http://www.w3.org/TR/owl2-profiles/
25. World Wide Web Consortium (W3C): SPARQL (2010), http://www.w3.org/TR/rdf-sparql-query/
26. World Wide Web Consortium (W3C) OWL Working Group: nTriples Format (2010),
http://www.w3.org/TR/rdf-testcases/#ntriples

A Semantic Portal for Next Generation Monitoring Systems

Ping Wang[1], Jin Guang Zheng[1], Linyun Fu[1], Evan W. Patton[1], Timothy Lebo[1],
Li Ding[1], Qing Liu[2], Joanne S. Luciano[1], and Deborah L. McGuinness[1]

[1] Tetherless World Constellation, Rensselaer Polytechnic Institute, USA
[2] Tasmanian ICT Centre, CSIRO, Australia
{wangp5,zhengj3,ful2,pattoe,lebot,dingl,jluciano,dlm}@rpi.edu,
Q.Liu@csiro.au

Abstract. We present a semantic technology-based approach to emerging monitoring systems based on our linked data approach in the Tetherless World Constellation Semantic Ecology and Environment Portal (SemantEco). Our integration scheme uses an upper level monitoring ontology and mid-level monitoring-relevant domain ontologies. The initial domain ontologies focus on water and air quality. We then integrate domain data from different authoritative sources and multiple regulation ontologies (capturing federal as well as state guidelines) to enable pollution detection and monitoring. An OWL-based reasoning scheme identifies pollution events relative to user chosen regulations. Our approach captures and leverages provenance to enable transparency. In addition, SemantEco features provenance-based facet generation, query answering, and validation over the integrated data via SPARQL. We introduce the general SemantEco approach, describe the implementation which has been built out substantially in the water domain creating the SemantAqua portal, and highlight some of the potential impacts for the future of semantically-enabled monitoring systems.

Keywords: Environmental Portal, Provenance-Aware Search, Water Quality Monitoring, Pragmatic Considerations for Semantic Environmental Monitoring.

1 Introduction

Concerns over ecological and environmental issues such as biodiversity loss [1], water problems [14], atmospheric pollution [8], and sustainable development [10] have highlighted the need for reliable information systems to support monitoring of ecological and environmental trends, support scientific research and inform citizens. In particular, semantic technologies have been used in environmental monitoring information systems to facilitate domain knowledge integration across multiple sources and support collaborative scientific workflows [17]. Meanwhile, demand has increased for direct and transparent access to ecological and environmental information. For example, after a recent water quality episode in Bristol County, Rhode Island where *E. coli* was reported in the water, residents requested information

L. Aroyo et al. (Eds.): ISWC 2011, Part II, LNCS 7032, pp. 253–268, 2011.
© Springer-Verlag Berlin Heidelberg 2011

concerning when the contamination began, how it happened, and what measures were being taken to monitor and prevent future occurrences.[1]

In this paper, we describe a semantic technology-based approach to ecological and environmental monitoring. We deployed the approach in the Tetherless World Constellation's Semantic Ecology and Environment Portal (SemantEco). SemantEco is an exemplar next generation monitoring portal that provides investigation support for lay people as well as experts while also providing a real world ecological and environmental evaluation testbed for our linked data approach. The portal integrates environmental monitoring and regulation data from multiple sources following Linked Data principles, captures the semantics of domain knowledge using a family of modular simple OWL2 [7] ontologies, preserves provenance metadata using the Proof Markup Language (PML) [11], and infers environment pollution events using OWL2 inference. The web portal delivers environmental information and reasoning results to citizens via a faceted browsing map interface[2].

The contributions of this work are multi-faceted. The overall design provides an operational specification model that may be used for creating ecological and environmental monitoring portals. It includes a simple upper ontology and initial domain ontologies for water and air. We have used this design to develop a water quality portal (SemantAqua) that allows anyone, including those lacking in-depth knowledge of water pollution regulations or water data sources, to explore and monitor water quality in the United States. It is being tested by being used to do a redesign of our air quality portal[3]. It also exposes potential directions for monitoring systems as they may empower citizen scientists and enable dialogue between concerned citizens and professionals. These systems, for example, may be used to integrate data generated by citizen scientists as potential indicators that professional collection and evaluation may be needed in particular areas. Additionally subject matter professionals can use this system to conduct provenance-aware analysis, such as explaining the cause of a water problem and cross-validating water quality data from different data sources with similar contextual provenance parameters (e.g. time and location).

In this paper, section 2 reviews selected challenges in the implementation of the SemantEco design in the SemantAqua portal on real-world data. Section 3 elaborates how semantic web technologies have been used in the portal, including ontology-based domain knowledge modeling, real-world water quality data integration, and provenance-aware computing. Section 4 describes implementation details and section 5 discusses impacts and several highlights. Related work is reviewed in section 6 and section 7 describes future directions.

2 Ecological and Environmental Information Systems Challenges

SemantEco provides an extensible upper ontology for monitoring with an initial focus on supporting environmental pollution monitoring with connections to health impacts.

---

[1] Morgan, T. J. 2009. "Bristol, Warren, Barrington residents told to boil water" Providence Journal, September 8, 2009. http://newsblog.projo.com/2009/09/residents-of-3.html

[2] http://was.tw.rpi.edu/swqp/map.html

[3] http://logd.tw.rpi.edu/demo/clean_air_status_and_trends_-_ozone

Our initial domain area for an in depth dive was water quality. The resulting portal is a publicly accessible semantically-enabled water information system that facilitates discovery of polluted water, polluting facilities, specific contaminants, and health impacts. We are in the process of extending it to include air quality data as well as industrial connections to the operating entities of polluting facilities. We faced a number of challenges during implementation, which we will now discuss.

2.1 Modeling Domain Knowledge for Environmental Monitoring

Environmental monitoring systems must model at least three types of domain knowledge: background environmental knowledge (e.g., water-relevant contaminants, bodies of water), observational data items (e.g., the amount of arsenic in water) collected by sensors and humans, and (preferably authoritative) environmental regulations (e.g., safe drinking water levels for known contaminants). An interoperable model is needed to represent the diverse collection of regulations, observational data, and environmental knowledge from various sources.

Observational data include measurements of environmental characteristics together with corresponding metadata, e.g. the type and unit of the data item, as well as provenance metadata such as sensor locations, observation times, and optionally test methods and devices used to generate the observation. A light-weight extensible domain ontology is ideal to enable reasoning on observational data while limiting ontology development and understanding costs.

A number of ontologies have been developed for modeling environmental domains. Raskin et al. [13] propose the SWEET ontology family for Earth system science. Chen et al. [5] models relationships among water quality datasets. Chau et al. [4] models a specific aspect of water quality. While these ontologies provide support to encode the first two types of domain knowledge, they do not support modeling environmental regulations.

Environmental regulations describe contaminants and their allowable thresholds, e.g. "the Maximum Contaminant Level (MCL) for Arsenic is 0.01 mg/L" according to the National Primary Drinking Water Regulations (NPDWRs)[4] stipulated by the US Environmental Protection Agency (EPA). Water regulations are established both at the federal level and by different state agencies. For instance, the threshold for Antimony is 0.0056 mg/L according to the Rhode Island Department of Environmental Management's Water Quality Regulations[5] while the threshold for Antimony is 0.006 mg/L according to the Drinking Water Protection Program[6] from the New York Department of Health. To capture the diversity of the water regulations, we generated a comparison table[7] (including provenance) of different contaminant thresholds at federal and state levels.

---

[4] http://water.epa.gov/drink/contaminants/index.cfm

[5] http://www.dem.ri.gov/pubs/regs/regs/water/h20q09.pdf

[6] http://www.health.ny.gov/environmental/water/drinking/part5/tables.htm

[7] http://tw.rpi.edu/web/project/TWC-SWQP/compare_five_regulation

2.2   Collecting Environmental Data

Environmental information systems need to integrate data from distributed data sources to enrich the source data and provide data validation. For water quality monitoring, two major U.S. government agencies publish water quality data: the Environmental Protection Agency (EPA) and US Geological Survey (USGS). Both release observational data based on their own independent water quality monitoring systems. Permit compliance and enforcement status of facilities is regulated by the National Pollutant Discharge Elimination System (NPDES[8]) under the Clean Water Act (CWA). The NPDES datasets contain descriptions of the facilities (e.g. name, permit number, and location) and measurements of water contaminants discharged by the facilities for up to five test types per contaminant. USGS publishes data about water monitoring sites and measurements from water samples through the National Water Information System (NWIS)[9].

Although environmental datasets are often organized as data tables, it is not easy to integrate them due to syntactic and semantic differences. In particular, we observe multiple needs for linking data: (i) the same concept may be named differently, e.g., the notion "name of contaminant" is represented by "CharacteristicName" in USGS datasets and "Name" in EPA datasets, (ii) some popular concepts, e.g. name of chemical, may be used in domains other than water quality monitoring, so it would be useful to link to other accepted models, e.g. the ChemML chemical element descriptions and (iii) most observational data are complex data objects. For example, Table 1 shows a fragment from EPA's measurement dataset, where four table cells in the first two columns together yield a complex data object: "C1" refers to one type of water contamination test, "C1_VALUE" and "C1_UNIT" indicate two different attributes for interpreting the cells under them respectively, and the data object reads "the measured concentration of fecal coliform is 34.07 MPN/100mL under test option C1". Effective mechanisms are needed to allow connection of relevant data objects (e.g., the density observations of fecal coliform observed in EPA and USGS datasets) to enable cross-dataset comparisons.

Table 1. For the facility with permit RI0100005, the 469th row for Coliform_fecal_general measurements on 09/30/2010 contains 2 tests

C1_VALUE	C1_UNIT	C2_VALUE	C2_UNIT
34.07	MPN/100ML	53.83	MPN/100ML

2.3   Provenance Tracking and Provenance-Aware Computing

In order to enable transparency and encourage community participation, a public information system should track provenance metadata during data processing and leverage provenance metadata in its computational services. Similarly, an

---

[8] http://www.epa-echo.gov/echo/compliance_report_water_icp.htm
[9] http://waterdata.usgs.gov/nwis

environmental monitoring system that combines data from different sources should maintain and expose data sources on demand. This enables data curators to get credit for their contributions and also allows users to choose data from trusted sources. The data sources are automatically refreshed from the corresponding provenance metadata when the system ingests new data.

Provenance metadata can maintain context information (e.g. when and where an observation was collected), which can be used to determine whether two data objects are comparable. For example, when pH measurements from EPA and USGS are validated, the measurement provenance should be checked: the latitude and longitude of the EPA and USGS sites where the pH values are measured should be very close, the measurement time should be in the same year and month, etc.

3 Semantic Web Approach

We believe that a semantic web approach is well suited to the general problem of monitoring, and explore this approach with a water quality monitoring portal at scale.

3.1 Domain Knowledge Modeling and Reasoning

We use an ontology-based approach to model domain knowledge in environmental information systems. An upper ontology[10] defines the basic terms for environmental monitoring. Domain ontologies extend the upper ontology to model domain specific terms. We also develop regulation ontologies [11] that include terms required for describing compliance and pollution levels. These ontologies leverage OWL inference to reason about the compliance of observations with regulations.

Upper Ontology Design
Existing ontologies do not completely cover all the necessary domain concepts as mentioned in section 2.1. We provide an upper ontology that reuses and is complementary to existing ontologies (e.g. SWEET, FOAF). The ontology models domain objects (e.g. polluted sites) as classes and their relationships (e.g. has Measurement, hasCharacteristic[12]) as properties. A subset of the ontology is illustrated in Figure 1. A polluted site is modeled as something that is both a measurement site and polluted thing, which is something that has at least one measurement that violates a regulation.

This ontology can be extended to different domains by adding domain-specific classes. For example, water measurement is a subclass of measurement, and water site is the intersection of body of water, measurement site and something that has at least one water measurement[13]. Our water quality extension is also shown in Figure 1.

---

[10] http://escience.rpi.edu/ontology/semanteco/2/0/pollution.owl#

[11] e.g., http://purl.org/twc/ontology/swqp/region/ny; others are listed at
http://purl.org/twc/ontology/swqp/region/

[12] Our ontology uses characteristic instead of contaminant based on the consideration that some characteristics measured like pH and temperature are not contaminants.

[13] http://escience.rpi.edu/ontology/2/0/water.owl#

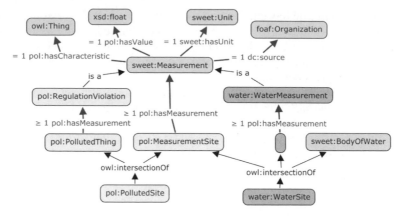

Fig. 1. Portion of the TWC Environment Monitoring Ontology

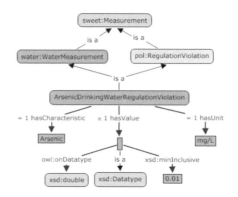

Fig. 2. Portion of the EPA Regulation Ontology

Regulation Ontology Design

Each domain must define its own regulation ontology that maps the rules in regulations to OWL classes. For water quality monitoring, we developed a regulation ontology in which the allowable ranges of regulated characteristics are encoded via numeric range restrictions on datatype properties. The rule-compliance results are reflected by whether an observational data item is a member of the class mapped from the rule. Figure 2 illustrates the OWL representation of one rule from EPA's NPDWRs, i.e. drinking water is polluted if the concentration of Arsenic is more than 0.01 mg/L. In the regulation ontology, ArsenicDrinkingWaterRegulationViolation is a water measurement with value greater than or equal to 0.01 mg/L of the Arsenic characteristic. Regulations in other environment domains can be similarly mapped to OWL2 restrictions if they represent violations as ranges of measured characteristics.

Reasoning Domain Data with Regulations

Combining observational data items collected at water-monitoring sites and the domain and regulation ontologies, an OWL2 reasoner can decide if any sites are polluted. This design provides several benefits. First, the upper ontology is small and easy to maintain; it consists of only 7 classes, 4 object properties, and 10 data properties. Secondly, the ontology design is extensible. The upper ontology can be extended to other domains, e.g. air quality monitoring[14]. Regulation ontologies can be extended to incorporate more regulations as needed. We wrote converters to extract federal and four states' regulation data from web pages and translated them into OWL2 [7] constraints that align with the upper ontology. The same workflow can be used to obtain the remaining state regulations using either our existing converters or potentially new converters if the data are in different forms. The design leads to flexible querying and reasoning: the user can select the regulations to apply to the data and the reasoner will classify using only the ontology for the selected regulations. For example, when Rhode Island regulations are applied to water quality data for zip code 02888 (Warwick, RI), the portal detects 2 polluted water sites and 7 polluting facilities. If the user chooses to apply California regulations to the same region, the portal identifies 15 polluted water sites, including the 2 detected with Rhode Island regulations, and the same 7 polluting facilities. One conclusion is that California regulations are stricter than Rhode Island's (and many other states), and the difference could be of interest to environmental researchers and local residents.

3.2 Data Integration

When integrating real world data from multiple sources, monitoring systems can benefit from adopting the data conversion and organization capabilities enabled by the TWC-LOGD portal [6]. The open source tool **csv2rdf4lod**[15] can be used to convert datasets from heterogeneous sources into Linked Data [9].

Linking to ontological terms: Datasets from different sources can be linked if they reuse common ontological terms, i.e. classes and properties. For instance, we map the property "CharacteristicName" in the USGS dataset and the property "Name" in the EPA dataset to a common property water:hasCharacteristic. Similarly, we map spatial location to an external ontology, i.e. wgs84[16]:lat and wgs84:long.

Aligning instance references: We promote references to characteristic names from literal to URI, e.g. "Arsenic" is promoted to "water:Arsenic", which then can be linked to external resources like "dbpedia:Arsenic" using owl:sameAs. This design is based on the observation that not all instance names can be directly mapped to DBpedia URIs (e.g., "Nitrate/Nitrite" from the Massachusetts water regulations[17] maps two DBpedia URIs), and some instances may not be defined in DBpedia (e.g., "C5-C8" from the Massachusetts water regulations). By linking to DBpedia URIs, we reserve the opportunity to connect to other knowledge base, e.g. disease database.

---

[14] http://escience.rpi.edu/ontology/2/0/air.owl#
[15] http://purl.org/twc/id/software/csv2rdf4lod
[16] http://www.w3.org/2003/01/geo/wgs84_pos
[17] The "2011 Standards & Guidelines for Contaminants in Massachusetts Drinking Water" at http://www.mass.gov/dep/water/drinking/standards/dwstand.htm

Converting complex objects: As discussed in section 2.2, we often need to compose a complex data object from multiple cells in a table. We use the cell-based conversion capability provided by csv2rdf4lod to enhance EPA data by marking each cell value as a subject in a triple and bundling the related cell values with the marked subject. The details can be found in [18].

3.3 Provenance Tracking and Provenance-Aware Computing

SemantEco provenance data come from two sources: (i) provenance metadata embedded in the original datasets, e.g. measurement location and time; (ii) metadata that describe the derivation history of the data. We automatically capture provenance data during the data integration stages and encode them in PML 2 [11] due to the provenance support from csv2rdf4lod. At the retrieval stage, we capture provenance, e.g. data source URL, time, method, and protocol used in data retrieval. We maintain provenance at the conversion stage, e.g. engine performing the conversion, antecedent data, and roles played by those data. At the publication stage, we capture provenance, e.g. agent, time, and context for triple store loads and updates. When we convert the regulations, we capture their provenance programmatically. We reveal these provenance data via pop up dialogs when the user selects a measurement site or facility, and utilize them to enable new applications like dynamic data source (DS) listings and provenance-aware cross validation.

Data Source as Provenance
We utilize data source provenance to support dynamic data source listing as follows:

1. Newly gathered water quality data are loaded into the system as RDF graphs.
2. When new graphs come, the system generates an RDF graph, namely the DS graph, to record the metadata of all the RDF graphs in the system. The DS graph contains information such as the URI, classification and ranking of each RDF graph.
3. The system tells the user what data sources are currently available by executing a SPARQL query on the DS graph to select distinct data source URIs.
4. With the presentation of the data sources on the interface, the user is allowed to select the data sources he/she trusts (see Figure 4). The system would then only return results within the selected sources.

Provenance information can allow the user to customize his/her data retrieval request, e.g. some users may be only interested in data published within a particular time period. The SPARQL queries used in each step are available at [18].

Provenance-Aware Cross-Validation over EPA and USGS Data
Provenance enables our system to compare and cross-validate water quality data originating from different source agencies. Figure 3 shows pH measurements collected at an EPA facility (at 41:59:37N, 71:34:27W) and a USGS site (at 41:59:47N, 71:33:45W) that are less than 1km apart. Note that the pH values measured by USGS fell below the minimum value from EPA often and went above

the maximum value from EPA once. We found two nearby locations using a SPARQL filter:

```
FILTER ( ?facLat < (?siteLat+"+delta+")
&& ?facLat > (?siteLat-"+delta+")
&& ?facLong < (?siteLong+"+delta+")
&& ?facLong > (?siteLong-"+delta+"))
```

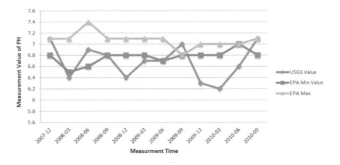

Fig. 3. Data Validation Example

4   SemantAqua: Semantic Water Quality Portal

4.1   System Implementation

Figure 4 shows one example where the semantic water quality portal supports water pollution identification. The user specifies a geographic region of interest by entering a zip code (mark 1). Users can customize queries from multiple facets: data source (mark 3), water regulations (mark 4), water characteristic (mark 6) and health concern (mark 7). After the portal generates the results, it visualizes the results on a Google map using different icons to distinguish between clean and polluted water sources and facilities (mark 5). The user can access more details about a site by clicking on its icon. The information provided in the pop up window (mark 2) include: names of contaminants, measured values, limit values, and time of measurement. The window also provides a link that displays the water quality data as a time series.

The portal retrieves water quality datasets from EPA and USGS and converts the heterogeneous datasets into RDF using csv2rdf4lod. The converted water quality data are loaded into OpenLink Virtuoso 6 open-source edition [18] and retrieved via SPARQL queries. The portal utilizes the Pellet OWL Reasoner [16] together with the Jena Semantic Web Framework [2] to reason over the water quality data and water ontologies in order to identify water pollution events.

The portal models the effective dates of the regulations, but only at the granularity of a set of regulations rather than per contaminant. We use provenance data to generate and maintain the data source facet (mark 3), enabling the user to choose data sources he/she trusts.

---

[18] http://virtuoso.openlinksw.com/dataspace/dav/wiki/Main/

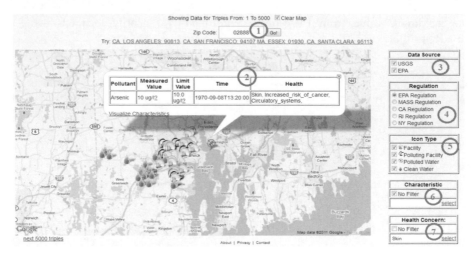

Fig. 4. Water Quality Portal In Action

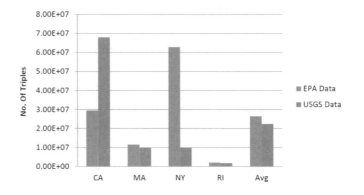

Fig. 5. Triple numbers for our initial four states with average computation

4.2 Scaling Issues

We wanted to test our approach in a realistic setting so we gathered data for an initial set of states to determine scaling issues. We have generated 89.58 million triples for the USGS datasets and 105.99 million triples for the EPA datasets for 4 states, which implies that water data for all 50 states would generate at least a few billion triples. The sizes of the available datasets are summarized in Figure 5. Such size suggests that a triple store cluster should be deployed to host the water data.[19]

Table 2 includes class counts for our initial four state regulations. Our programmed conversion provides a quick and low cost approach for encoding regulations. It took us about 2 person-days to encode hundreds of rules.

---

[19] We have recently obtained the data for the remaining 46 states and are working on the completed US portal.

Table 2. Number of classes converted from regulations

EPA	CA	MA	NY	RI
83	104	139	74	100

5 Discussion

5.1 Linking to a Health Domain

Polluted drinking water can cause acute diseases, such as diarrhea, and chronic health effects such as cancer, liver and kidney damage. For example, water pollution co-occurring with new types of natural gas extraction in Bradford County, Pennsylvania has been reported to generate numerous problems[20, 21]. People reported symptoms ranging from rashes to numbness, tingling, and chemical burn sensations, escalating to more severe symptoms including racing heart and muscle tremors.

In order to help citizens investigate health impacts of water pollution, we are extending our ontologies to include potential health impacts of overexposure to contaminants. These relationships are quite diverse since potential health impacts vary widely. For example, according to NPDWRs, excessive exposure to lead may cause kidney problems and high blood pressure in adults whereas infants and children may experience delays in physical or mental development.

Similar to modeling water regulations, we programmatically extracted the relationships between contaminants and health impacts from a web page[22] and encoded them into OWL classes. We used the object property "hasSymptom" to connect the classes with their symptoms, e.g. health:high_blood_pressure. The classes of health effects are related to the classes of violations, e.g. LeadDrinkingWater RegulationViolation, with the object property hasCause. We can query symptom-based measurements using this SPARQL query fragment:

```
?healthEffect water:hasSymptom health:high_blood_pressure.
?healthEffect rdf:type water:HealthEffect.
?healthEffect water:hasCause ?cause.
?cause owl:intersectionOf ?restrictions.
?restrictions list:member ?restriction.
?restriction owl:onProperty water:hasCharacteristic.
?restriction owl:hasValue ?characteristic.
?measurement water:hasCharacteristic ?characteristic.
```

Based on this modeling, the portal has been extended to begin to address health concerns: (1) the user can specify his/her health concern and the portal will detect only the water pollution that has been correlated the particular health concern; (2) the user can query the possible health effects of each contaminant detected at a polluted

---

[20] http://protectingourwaters.wordpress.com/2011/06/16/black-water-and-brazenness-gas-drilling-disrupts-lives-endangers-health-in-bradford-county-pa/

[21] http://switchboard.nrdc.org/blogs/amall/one_familys_life_in_the_gas_pa.html

[22] As obtained from the NPDWRs at http://water.epa.gov/drink/contaminants/index.cfm

site, which is useful for identifying potential effects of water pollution and for identifying appropriate responses (e.g., boiling water to kill germs, using water only for bathing but not for drinking, etc.)

5.2 Time as Provenance

Temporal considerations were non-trivial in regulation modeling. The thresholds defined in both the NPDWRs' MCLs and state water quality regulations became effective nationally at different times for different contaminants[23]. For example, in the "2011 Standards & Guidelines for Contaminants in Massachusetts Drinking Water", the date that the threshold of each contaminant was developed or last updated can be accessed by clicking on the contaminant's name on the list. The effective time of the regulations has semantic implications: if the collection time of the water measurement is not in the effective time range of the constraint, then the constraint should not be applied to the measurement. In principle, we can use OWL2 RangeRestriction to model time interval constraints as we did on threshold.

5.3 Regulation Mapping and Comparison

The majority of the portal domain knowledge stems from water regulations that stipulate contaminants, pollution thresholds, and contaminant test options. Besides using semantics to clarify the meaning of water regulations and support regulation reasoning, we can also perform analysis on regulations. For example, we compared regulations from five different sources and shows substantial variation.

By modeling regulations as OWL classes, we may also leverage OWL subsumption inference to detect the correlations between thresholds across different regulatory bodies and this knowledge could be further used to speed up reasoning. For example, the California regulations are stricter than the EPA regulations concerning Methoxychlor so we can derive two rules: 1) with respect to Methoxychlor, if a water site is identified as polluted according to the EPA regulations, it is polluted according to the California regulations; and 2) with respect to Methoxychlor, if the available data supports no threshold violation according to the California regulations, it will not exceed thresholds according to the EPA regulations. Since regulations such as these can be subclassed, reasoning efficiencies may be realized when multiple regulations are used to evaluate pollution.

5.4 Scalability

The large number of triples generated during the conversion phase prohibits classifying the entire dataset in real time. We have tried several approaches to improve reasoning speed: organize observation data by state, filter relevant data by zip code (we can derive county using zip code), and reasoning over the relevant data on one (or a small number of) selected regulation(s).

The portal assigns one graph per state to store the integrated data. The triple count at the state level is still quite large: we currently host 29.45 million triples from EPA

---

[23] Personal communication with the   Office of Research and Standards, Massachusetts Department of Environmental Protection.

and 68.03 million triples from USGS for California water quality data. Therefore, we refine the granularity to county level using a CONSTRUCT query (see below). This operation reduces the number of relevant triples to a manageable 10K to 100K size.

```
CONSTRUCT {
  ?s rdf:type water:MeasurementSite.
  ?s water:hasMeasurement ?measurement.
  ?s water:hasStateCode ?state.
  ?s wgs84:lat ?lat.        ?s wgs84:long ?long.
  ?measurement water:hasCharacteristic ?characteristic.
  ?measurement water:hasValue ?value.
  ?measurement water:hasUnit ?unit.
  ?measurement time:inXSDDateTime ?time.
  ?s water:hasCountyCode 085. }
WHERE { GRAPH <http://sparql.tw.rpi.edu/source/usgs-
gov/dataset/national-water-information-system-nwis-
measurements/06>
{ ?s rdf:type water:MeasurementSite.
  ?s water:hasUSGSSiteId ?id.
  ?s water:hasStateCode ?state.
  ?s wgs84:lat ?lat.        ?s wgs84:long ?long.
  ?measurement water:hasUSGSSiteId ?id.
  ?measurement water:hasCharacteristic ?characteristic.
  ?measurement water:hasValue ?value.
  ?measurement water:hasUnit ?unit.
  ?measurement time:inXSDDateTime ?time.
  ?s water:hasCountyCode 085. }}
```

5.5 Maintenance Costs for Data Service Provider

Although government agencies typically publish environmental data on the web and allow citizens to browse and download the data, not all of their information systems are designed to support bulk data queries. In our case, our programmatic queries of the EPA dataset were blocked. From a personal communication with the EPA, we were surprised to find that our previous continuous data queries were impacting their operations budget since they are charged for queries. Consequently, we filed an online form requesting a bulk data transfer from the EPA which has recently been processed. In contrast, the USGS provides web services to facilitate periodic acquisition and processing of their water data via automated means.

5.6 System Evaluation

We provide an online questionnaire [24] to collect feedback from users. In the questionnaire, we ask the users to identify themselves as experts or lay users, then ask them to rate the data quality, responsiveness, and user interface of the portal. The questionnaire also solicits free text comments from users. We will report preliminary results of this ongoing user study at the conference.

---

[24] http://was.tw.rpi.edu/swqp/questionnaire/portal_questionnaire.php

6 Related Work

Three areas of work are considered most relevant to this work, namely knowledge modeling, data integration, and provenance tracking of environmental data.

Knowledge-based approaches have begun in environmental informatics. Chen et al. [5] proposed a prototype system that integrates water quality data from multiple sources and retrieves data using semantic relationships among data. Chau [4] presented an ontology-based knowledge management system (KMS) to enable novice users to find numerical flow and water quality models given a set of constraints. OntoWEDSS [3] is an environmental decision-support system for wastewater management that combines classic rule-based and case-based reasoning with a domain ontology. Scholten et al. [14] developed the MoST system to facilitate the modeling process in the domain of water management. The Registry of EPA Applications, Models and Databases (READ)[25] supports management of information resources. It collects life cycle phase information, how the resource supports environmental statutes, and whether the resource interfaces with other EPA information resources. A comprehensive review of environmental modeling approaches can be found in [17]. SemantEco and SemantAqua differ from these projects since they support provenance-based query and data visualization. Moreover, SemantAqua is built upon standard semantic technologies (e.g. OWL, SPARQL, Pellet, Virtuoso) and thus can be easily replicated or expanded.

Data integration across providers has been studied for decades by database researchers. In the area of ecological and environmental research, shallow integration approaches are taken to store and index metadata of data sources in a centralized database to aid search and discoverability. This approach is applied in systems such as KNB[26] and SEEK[27]. Our integration scheme combines a limited, albeit extensible, set of data sources under a common ontology family. This supports reasoning over the integrated data set and allows for ingest of future data sources.

There also has been a considerable amount of research efforts in semantic provenance, especially in the field of eScience. myGrid [19] proposes the COHSE open hypermedia system that generates, annotates and links provenance data in order to build a web of provenance documents, data, services, and workflows for experiments in biology. The Multi-Scale Chemical Science [12] (CMCS) project develops a general-purpose infrastructure for collaboration across many disciplines. It also contains a provenance subsystem for tracking, viewing and using data provenance. A review of provenance techniques used in eScience projects is presented in [15]. While these eScience projects design their own schemes for modeling provenance, the SemantAqua portal encodes provenance with PML 2, which is a general purpose interlingua for sharing explanations generated by various automated systems. These eScience projects keep provenance for uses like improving data quality, facilitating audits, and data replicability. Our portal demonstrates that provenance also can be used for developing and customizing web applications (e.g. generating the data source facet).

---

[25] http://iaspub.epa.gov/sor_internet/registry/systmreg/home/overview/home.do

[26] Knowledge Network for Biocomplexity Project. http://knb.ecoinformatics.org/index.jsp

[27] The Science Environment for Ecological Knowledge. http://seek.ecoinformatics.org

7 Conclusions and Future Work

We presented a semantic technology-based approach to ecological and environmental monitoring and described our work using this approach in the Tetherless World Constellation SemantEco approach and the SemantAqua Portal. SemantAqua supports both non-expert and expert users in water quality monitoring. We described the overall design and highlighted some benefits from utilizing semantic technologies, including: the design of the ontologies, the methodology used to perform data integration, and the encoding and usage of provenance information generated during data aggregation. The SemantAqua portal demonstrates some benefits and potential of applying semantic web technologies to environmental information systems.

A number of extensions to this portal are in process. First, only four states' regulations have been encoded. We intend to encode the regulations for the remaining states whose regulations differ from the federal regulations. Second, data from other sources, e.g. weather, may yield new ways of identifying pollution events. For example, a contaminant control strategy may fail if heavy rainfall causes flooding, carrying contaminants outside of a prescribed area. It would be possible with real-time sensor data to observe how these weather events impact the portability of water sources in the immediate area. We are also applying this approach to other monitoring topics, e.g. air quality, food safety, and health impacts.

References

1. Batzias, F.A., Siontorou, C.G.: A Knowledge-based Approach to Environmental Biomonitoring. Environmental Monitoring and Assessment 123, 167–197 (2006)
2. Carroll, J.J., Dickinson, I., Dollin, C., Reynolds, D., Seaborne, A., Wilkinson, K.: Jena: Implementing the semantic web recommendations. In: 13th International World Wide Web Conference, pp. 74-83 (2004)
3. Ceccaroni, L., Cortes, U., Sanchez-Marre, M.: OntoWEDSS: augmenting environmental decision-support systems with ontologies. Environmental Modelling & Software 19(9), 785–797 (2004)
4. Chau, K.W.: An Ontology-based knowledge management system for flow and water quality modeling. Advances in Engineering Software 38(3), 172–181 (2007)
5. Chen, Z., Gangopadhyay, A., Holden, S.H., Karabatis, G., McGuire, M.P.: Semantic integration of government data for water quality management. Government Information Quarterly 24(4), 716–735 (2007)
6. Ding, L., Lebo, T., Erickson, J. S., DiFranzo, D., Williams, G. T., Li, X., Michaelis, J., Graves, A., Zheng, J. G., Shangguan, Z., Flores, J., McGuinness, D. L., and Hendler, J.: TWC LOGD: A Portal for Linked Open Government Data Ecosystems. JWS special issue on semantic web challenge 2010 (2010)
7. Hitzler, P., Krotzsch, M., Parsia, B., Patel-Schneider, P., Rudolph, S.: OWL 2 Web Ontology Language Primer, http://www.w3.org/TR/owl2-primer/ (2009)
8. Holland, D.M., Principe, P.P., Vorburger, L.: Rural Ozone: Trends and Exceedances at CASTNet Sites. Environmental Science & Technology 33(1), 43–48 (1999)
9. Lebo, T., Williams, G.T.: Converting governmental datasets into linked data. In: Proceedings of the 6th International Conference on Semantic Systems, I-SEMANTICS 2010, pp. 38:1–38:3 (2010)

10. Liu, Q., Bai, Q., Ding, L., Pho, H., Chen, Kloppers, C., McGuinness, D. L., Lemon, D., Souza, P., Fitch, P. and Fox, P.: Linking Australian Government Data for Sustainability Science - A Case Study. In: Linking Government Data (chapter) (accepted 2011)

11. McGuinness, D.L., Ding, L., Pinheiro da Silva, P., Chang, C.: PML 2: A Modular Explanation Interlingua. In: Workshop on Explanation-aware Computing (2007)

12. Myers, J., Pancerella, C., Lansing, C., Schuchardt, K., Didier, B.: Multi-scale science: Supporting emerging practice with semantically derived provenance. In: ISWC workshop on Semantic Web Technologies for Searching and Retrieving Scientific Data (2003)

13. Raskin, R.G., Pan, M.J.: Knowledge representation in the semantic web for Earth and environmental terminology (SWEET). Computers & Geosciences 31(9), 1119–1125 (2005)

14. Scholten, H., Kassahun, A., Refsgaard, J.C., Kargas, T., Gavardinas, C., Beulens, A.J.M.: A Methodology to Support Multidisciplinary Model-based Water Management. Environmental Modelling and Software 22(5), 743–759 (2007)

15. Simmhan, Y.L., Plale, B., Gannon, D.: A survey of data provenance in e-science. ACM SIGMOD Record 34(3), 31–36 (2005)

16. Sirin, E., Parsia, B., Cuenca-Grau, B., Kalyanpur, A., Katz, Y.: Pellet: A practical OWL-DL reasoner. Journal of Web Semantics 5(2), 51–53 (2007)

17. Villa, F., Athanasiadis, I.N., Rizzoli, A.E.: Modelling with knowledge: A Review of Emerging Semantic Approaches to Environmental Modelling. Environmental Modelling and Software 24(5), 577–587 (2009)

18. Wang, P., Zheng, J.G., Fu, L.Y., Patton E., Lebo, T., Ding, L., Liu, Q., Luciano, J. S., McGuinness, D. L.: TWC-SWQP: A Semantic Portal for Next Generation Environmental Monitoring[28]. Technical Report (2011),
    http://tw.rpi.edu/media/latest/twc-swqp.doc

19. Zhao, J., Goble, C.A., Stevens, R., Bechhofer, S.: Semantically Linking and Browsing Provenance Logs for E-Science. In: Bouzeghoub, M., Goble, C.A., Kashyap, V., Spaccapietra, S. (eds.) ICSNW 2004. LNCS, vol. 3226, pp. 158–176. Springer, Heidelberg (2004)

---

[28] SWQP has been renamed SemantAqua that is one instantiation of the SemantEco design for Ecological and Environmental Monotoring.

DC Proposal: PRISSMA, Towards Mobile Adaptive Presentation of the Web of Data

Luca Costabello

INRIA Edelweiss Team, Sophia Antipolis, France
luca.costabello@inria.fr

Abstract. The Mobile Web is evolving fast and mobile access to the Web of Data is gaining momentum. Interlinked RDF resources consumed from portable devices need proper adaptation to the context in which the action is performed. This paper introduces PRISSMA (Presentation of Resources for Interoperable Semantic and Shareable Mobile Adaptability), a domain-independent vocabulary for displaying Web of Data resources in mobile environments. The vocabulary is the first step towards a declarative framework aimed at sharing and re-using presentation information for context-adaptable user interfaces over RDF data.

Keywords: Mobile Web of Data, RDF Presentation, Adaptive UI.

1 Introduction

The Mobile Web and the Web of Data are strong trends for the next evolution of the web. Ubiquitous access is growing fast, as a result of heterogeneous factors such as new generation access networks, increasing device capabilities, novel interaction paradigms and enhanced user interfaces[1]. At the same time, the Linked Data Initiative[2] contributes to the growing popularity of the world of interlinked, structured and open data on the web. Soon mobile applications will offer novel ways of consuming and contributing to the Web of Data, both with the adoption of novel interaction modalities (e.g. enhanced Mobile Web applications, voice interaction with the web, augmented reality, etc.) and, on the other hand, with a deeper awareness of the surrounding physical environment.

Mobile Web of Data consumption faces the same general issues of classic Web-of-Data-savvy applications, such as access strategy choice, vocabulary mapping, identity resolution, provenance and data quality assessment [10]. Another well-known problem affecting Web of Data applications consists in delivering effective user interfaces for RDF resources. When dealing with mobile devices, the issue of *adaptation* arises: classic Mobile Web experience suggests that the same information is not meant to be represented in the same way on all devices. Moreover, context-awareness research underlines the influence of the surrounding environment on data representation [4,8,12,13]. The same principles and best practices

---

[1] http://pewinternet.org/Reports/2011/Smartphones.aspx
[2] http://linkeddata.org/

L. Aroyo et al. (Eds.): ISWC 2011, Part II, LNCS 7032, pp. 269–276, 2011.
© Springer-Verlag Berlin Heidelberg 2011

apply to RDF resource representation in mobile environments, as shown in the following `foaf:Person`-related scenario.

Alice loves shopping. Today is at the shopping mall, looking for a birthday present for Bob. She needs an idea, so she opens Bob's FOAF profile looking for inspiration. Alice's smartphone is aware of being at the mall, and therefore displays Bob's interests while hiding other properties such as his email or geographical position. As she exits the mall and walks down the street, her smartphone visualizes Bob's profile as a `foaf:Depiction` on a map. In the meanwhile, Bob is jogging. The mobile in his pocket updates him on the status of his FOAF contacts using speech synthesis. Each contact's latest social status update is read out loud. Bob listens the audio reproduction of Alice's last tweet, and decides to reply. As he stops running and takes the device into hands, the smartphone deactivates the audio representation and adopts a visual paradigm listing Alice's phone number, email and social network accounts.

RDF consumption in mobile environments needs *contextual* resource adaptation and heterogeneous dimensions must be involved (e.g. the mobile user profile, the device features and the surrounding physical environment). Representation paradigm independence has to be provided, thus supporting multi-modality. Moreover, tools designed for RDF resource adaptation on mobile devices must not rely on ad-hoc solutions. Re-use must be favoured and encouraged, thus enabling the exchange of context-based knowledge across heterogeneous applications. This paper sets the basis for a Web of Data *mobile presentation framework*, and it focuses on the PRISSMA lightweight vocabulary[3], an early stage contribution providing classes and properties to model core mobile context concepts useful for dynamic adaptation of user interfaces for RDF data.

The remainder of the paper is organized as follows: Section 2 provides a state of the art overview. Section 3 describes with further details the issue of mobile adaptive RDF representation and presents the research plan. Section 4 describes the PRISSMA vocabulary, an early contribution. In Section 5 a discussion of the approach is provided, along with the most relevant perspectives.

2   Related Work

Related work includes proposals from heterogeneous research domains, such as RDF presentation, content adaptation in mobile classic web and context-aware data representation.

Fresnel [14] is a vocabulary modelling core presentation-level concepts for RDF visualization. The assumption is that data and its related schema do not carry sufficient information for representation: further presentation knowledge has to be provided. This information is expressed in RDF as well, in order to ensure homogeneity and favour presentation knowledge sharing among different applications. In Fresnel, *Lens* components select and filter information while *Formats* define how to present data. Designed for static environments, Fresnel leaves contextual adaptation of RDF data to the application logic. Sharing and

---

[3] http://ns.inria.fr/prissma

re-using *contextual conditions* for RDF presentation is therefore not possible. More recent proposals targeting RDF user interfaces have been provided, both of them being RDF template systems [1,2]. In LESS, the authors propose an end-to-end approach to consume linked data based on the Smarty template engine [1]. Dadzie et al. suggest a template-based RDF visualization approach aimed at facilitating Web of Data consumption [2]. Although they move in the direction of easing the development of Web of Data applications, these approaches do not address the specific needs of mobile environments, e.g. context adaptability.

Initiatives such as the now discontinued W3C Model-Based User Interface Incubator Group deal with classic Mobile Web content adaptation (i.e. RDF consumption-related presentations issues are not considered) [6]. The declared goal is to 'evaluate research on model-based user interface design' for the authoring of web applications. The authors propose the Delivery Context Ontology, a modular, fine-grained vocabulary to model mobile platforms [7].

CAMB is a mobile context-aware HTML browser that adapts web pages according to a predetermined set of environmental situations [8]. Other proposals have been made in this direction, e.g. [4,12,13]. None of these address a Web-of-Data-specific scenario.

The approach proposed in the remainder of the paper adapts and extends to the mobile world (static) Web of Data representation best practices, such as those described in Fresnel [14].

3   Towards Mobile Adaptive RDF Representation

Classic web content needs ad-hoc adaptation when displayed on portable devices and its visualization is influenced by the surrounding mobile context as well [8]. Web of Data access is no exception: whenever an RDF-fuelled mobile application needs to display some resources, adaptation (e.g., to the surrounding context) must be performed to deliver better user experience.

A widely-accepted formalization of *context* can be found in [3] and inspires the proposal in [6]. The authors of the latter describe the *mobile context* as an encompassing term, an information space defined as the sum of three different dimensions: the mobile *User* model, the *Device* features and the *Environment* in which the action is performed. This proposal is adopted as operative definition by the present work.

Dealing with Web of Data resources adaptation inside the application logic prevents the sharing of context-related presentation knowledge and intrinsically ties to a fixed representation paradigm. A common and shareable presentation-level solution provides common ground for building mobile adaptive user interfaces, avoids re-inventing the wheel and favours the exchange of context-related presentation knowledge between mobile applications. The need for sharing and reuse contextual presentation is not compatible with hardwired, programmatic approaches: a *declarative* proposition must be adopted to favour presentation data reuse (the latter point being the main Fresnel objective [14]).

As mentioned in Section 1, mobile devices provide heterogeneous *representation paradigms* (not necessarily visual, e.g. voice). A presentation-level approach is independent from the representation paradigm chosen by the application, and it is therefore suitable for mobile applications that need to change their RDF representation paradigm dynamically (e.g. contextually switching from a nested boxes layout à la HTML to a timeline or voice output).

In terms of research plan, achieving contextual content adaptation on mobile devices includes the following steps:

Contextual Presentation-Level Model Definition. The overall approach relies on the PRISSMA vocabulary, used to specify in which context a given representation must be activated. PRISSMA is designed as an extension of Fresnel [14] and it is described as an early contribution in Section 4.

Comparison between Declared and Real Context. Every time a Web of Data resource is requested, PRISSMA-based applications need to compare declared activation contexts and real situations to select the proper visualization. Implementing a rendering solution based on PRISSMA introduces therefore a graph matching problem. On the front-end, this issue can be modeled as a SPARQL ASK query performed against a set of PRISSMA declarations, having the goal of looking for a match between the real, detected context and a declared item in the set (both real context and the declared ones included in the set are modelled with PRISSMA). The intrinsic nature of contextual data and PRISSMA lightweight approach determine the need for an approximate matching solution. In other words, the comparison must take into account the discrepancies between the declared context and the actual situation. The proposed strategy must rely on a compound notion of distance: heterogeneous dimensions need to be considered (e.g. location, time, terminological heterogeneity), therefore different metrics must be chosen, along with a proper composition function. The algorithm needs to be unsupervised and one-pass, in order to enable responsive mobile applications.

Specificity Computation Rules. Conflicts might occur in case more than one representation is suitable for the current situation: a *specificity* computation mechanism has to be provided. This could be implemented giving higher priority to the most detailed and specific PRISSMA declarations.

Presentation Knowledge Distribution. PRISSMA declarations can be published as linked data. A discovery mechanism is needed to enable mobile applications that rely on PRISSMA to potentially adopt representations shared on the web.

4   PRISSMA Vocabulary Overview

The PRISSMA vocabulary provides classes and properties to model core mobile context concepts useful for dynamic adaptation of user interfaces for RDF

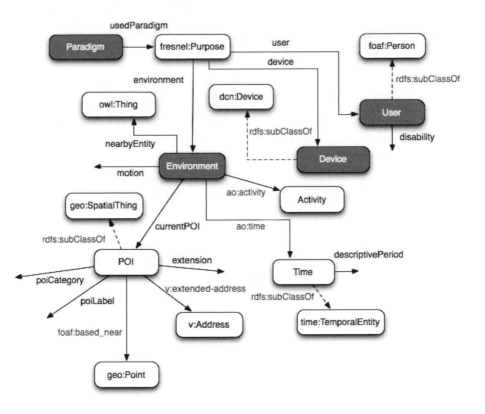

Fig. 1. The PRISSMA vocabulary at a glance. Grey boxes represent core classes.

data. Designed to work in conjunction with the Fresnel presentation vocabulary, PRISSMA[4] specifies *when* (i.e. in which context) Fresnel Lenses and Formats must be activated. The original Fresnel core vocabulary includes the class `fresnel:Purpose` and the property `fresnel:purpose`, used to choose between lenses in conflicting situations, i.e. when more than one lens applies. Predefined instances of `fresnel:Purpose` are provided, e.g. `defaultLens`, `labelLens`. The extended Fresnel vocabulary provides instances such as `screen`, `print` and `projection`, in order to support a basic media-based presentation. PRISSMA re-uses `fresnel:Purpose` to model a *contextual situation* in which an associated Fresnel lens and format must be activated.

PRISSMA is based on the definition of context provided in Section 3: it is not meant to provide yet another mobile contextual model, therefore well-known vocabularies are reused. Moreover, it does not provide an exhaustive set of properties for resource adaptation: (extensions are free to add further functionalities). Fresnel Lenses and Formats are showed according to the following dimensions:

---

[4] From latin *prisma*. In optics, a Fresnel lens can be considered as a series of prisms.

```
:personInterestLens rdf:type fresnel:Lens;
    fresnel:purpose fresnel:shoppingPurpose;
    fresnel:classLensDomain foaf:Person;
    fresnel:showProperties (foaf:name
                            foaf:depiction
                            foaf:interest).

:shoppingPurpose a fresnel:Purpose;
    prissma:user :compulsiveBuyer;
    prissma:environment :shoppingEnv.

:compulsiveBuyer a prissma:User;
    foaf:interest <http://dbpedia.org/resource/Compulsive_shopping>.

:shoppingEnv a prissma:Environment;
    prissma:currentPOI [
      prissma:poiCategory <http://dbpedia.org/resource/Shopping_mall>;
    ];
    ao:activity <http://dbpedia.org/resource/Shopping>.
```

Fig. 2. A Fresnel Lens associated to a mobile context using PRISSMA

User. Represents the target mobile user associated to a `fresnel:Purpose` and consists in a `foaf:Person` sub-class. To provide more flexibility, the class can be used to model both user stereotypes and specific users.

Device. Represents the mobile device on which Web of Data resource consumption takes place, enabling device-specific data representation. The class inherits from W3C Delivery Context Ontology [7] `dcn:Device` that provides an extensible and fine-grained model for mobile device features.

Environment. Models the physical context in which the Web of Data resource consumption takes place, therefore enabling customized resource representation according to specific situations. Different dimensions are involved in modelling the surrounding environment. As anticipated in Section 3, PRISSMA does not provide a comprehensive, fine grained model. The approach is to delegate refinements and extensions to domain specialists.

Location is modelled with the notion of Point of Interest (POI). The `prissma:POI` class consists in a simplified, RDFized version of W3C Point of Interest Core specifications [15]. POIs are defined as entities that "describe information about locations such as name, category, unique identifier, or civic address". Each `prissma:POI` consists in a `geo:SpatialThing` and can be associated to a given `geo:Point` coupled with a physical radius via the `prissma:extension` property. The properties `prissma:poiCategory` and `prissma:poiLabel` are used to assign a category (e.g. `http://dbpedia.org/resource/Shopping_mall`) and a label (e.g. `http://dbpedia.org/page/Harrods`). Address details can be provided as well, using the vCard ontology class `v:Address` [9].

Temporal dimension is modelled extending the `time:TemporalEntity` class
[11]. The `prissma:descriptivePeriod` property associates a description to each
Time entity (e.g. `http://dbpedia.org/resource/Afternoon`).

Other dimensions are considered: the `motion` property associates any given
high-level representation of motion to an `Environment`. The environmental proximity of a generic object can trigger different resource representations: nearby objects are associated to the Environment with the `prissma:nearbyEntity` property. The `prissma:Activity` class consists in a placemark aimed at connecting
third-party solutions focused on inferring high-level representations of user actions
(e.g.'running', 'driving', 'working', 'shopping', etc.). The Association Ontology's
`ao:activity` [5] is used to associate this class to the rest of the model.

Figure 2 provides an example related to the scenario presented in Section 1:
`personInterestLens` is a Fresnel lens that shows the interests of a `foaf:Person`.
The lens is activated by the purpose `shoppingPurpose`, which is defined by the
`prissma:User` "compulsive buyer" and by the `prissma:Environment` "shopping
in a mall".

The original Fresnel proposition does not specify the representation paradigm
that the user agent should adopt (e.g. web page-based box model, voice output,
etc). Fresnel implementations output a paradigm-independent *abstract box model*
that is rendered appropriately by the application. Being able to address this aspect in a declarative way is important in a context-aware environment, where the
representation paradigm might vary dynamically. The class `prissma:Paradigm`
models therefore a Fresnel box model customized for a given representation
paradigm. The property `prissma:usedParadigm` allows the association to a
`fresnel:Purpose`.

5   Discussion and Perspectives

This paper presents an approach towards contextual adaptation of Web of Data
resources on mobile devices and describes the PRISSMA vocabulary used as
foundation.

Relying on a presentation-level approach guarantees independence from the
chosen Web of Data access strategy: PRISSMA operates after this step, and is
therefore compatible with the follow-your-nose approach, SPARQL endpoints
querying and indexing services APIs. PRISSMA relies on Fresnel, and com-
plies therefore to the well-known presentation vocabulary, thus adding context
adaptability as a new scenario of use. A Fresnel-based approach favours the
sharing and reuse of contextual presentation knowledge across mobile applications consuming the Web of Data. The adoption of RDF formalism guarantees
homogeneity within presentation knowledge, i.e. context-related information are
expressed in RDF and no other formalism is needed. An extensible vocabulary
provides support to further contextual information.

On the other hand, it must not be underestimated that a context-savvy mobile
presentation layer based on an RDF vocabulary introduces intrinsic complexity
that has to be taken into account while developing context-adapting, RDF-based
applications.

The PRISSMA vocabulary and the algorithms solving the research steps described in Section 3 need to undergo a proper evaluation. A mobile Web of Data browser implementing the proposal described in this paper is to be provided as a test bench. The prototype is meant to contextually adapt the representation of requested instances, enabling therefore a testing campaign to compare the resulting adaptation of Web of Data resources to user expectations.

References

1. Auer, S., Doehring, R., Dietzold, S.: LESS - Template-Based Syndication and Presentation of Linked Data. In: Aroyo, L., Antoniou, G., Hyvönen, E., ten Teije, A., Stuckenschmidt, H., Cabral, L., Tudorache, T. (eds.) ESWC 2010. LNCS, vol. 6089, pp. 211–224. Springer, Heidelberg (2010)
2. Dadzie, A.-S., Rowe, M., Petrelli, D.: Hide the Stack: Toward Usable Linked Data. In: Antoniou, G., Grobelnik, M., Simperl, E., Parsia, B., Plexousakis, D., De Leenheer, P., Pan, J. (eds.) ESWC 2011, Part I. LNCS, vol. 6643, pp. 93–107. Springer, Heidelberg (2011)
3. Dey, A.K.: Understanding and using context. Personal Ubiquitous Computing 5, 4–7 (2001)
4. Eissele, M., Weiskopf, D., Ertl, T.: Interactive context-aware visualization for mobile devices. In: Butz, A., Fisher, B., Christie, M., Krüger, A., Olivier, P., Therón, R. (eds.) SG 2009. LNCS, vol. 5531, pp. 167–178. Springer, Heidelberg (2009)
5. Ferris, B., Inkster, T.: The Association Ontology 0.4 (2010), http://purl.org/ontology/ao/core#
6. Fonseca, J.M.C., González Calleros, J.M., Meixner, G., Paternò, F., Pullmann, J., Raggett, D., Schwabe, D., Vanderdonckt, J.: Model-Based UI XG Final Report (2010), http://www.w3.org/2005/Incubator/model-based-ui/XGR-mbui/
7. Fonseca, J.M.C., Lewis, R.: Delivery Context Ontology, W3C Working Draft (2009), http://www.w3.org/TR/2009/WD-dcontology-20090616/
8. Gasimov, A., Magagna, F., Sutanto, J.: CAMB: context-aware mobile browser. In: Proceedings of the 9th International Conference on Mobile and Ubiquitous Multimedia, pp. 22:1–22:5. ACM (2010)
9. Halpin, H., Suda, B., Walsh, N.: An Ontology for vCards (2006), http://www.w3.org/2006/vcard/ns-2006.html
10. Heath, T., Bizer, C.: Linked Data: Evolving the Web into a Global Data Space, 1st edn. Morgan & Claypool (2011)
11. Hobbs, J.R., Pan, F.: Time Ontology in OWL (2006), http://www.w3.org/TR/owl-time/
12. Lemlouma, T., Layaida, N.: Context-Aware Adaptation for Mobile Devices. In: IEEE International Conference on Mobile Data Management (2004)
13. Nathanail, S., Tsetsos, V., Hadjiefthymiades, S.: Sensor-Driven Adaptation of Web Document Presentation. In: Stephanidis, C. (ed.) HCI 2007. LNCS, vol. 4556, pp. 406–415. Springer, Heidelberg (2007)
14. Pietriga, E., Bizer, C., Karger, D., Lee, R.: Fresnel: A Browser-Independent Presentation Vocabulary for RDF. In: Cruz, I., Decker, S., Allemang, D., Preist, C., Schwabe, D., Mika, P., Uschold, M., Aroyo, L.M. (eds.) ISWC 2006. LNCS, vol. 4273, pp. 158–171. Springer, Heidelberg (2006)
15. Womer, M.: Points of Interest Core, W3C Working Draft (2011), http://www.w3.org/TR/poi-core/

DC Proposal: Towards an ODP Quality Model

Karl Hammar

Jönköping University, Jönköping, Sweden
karl.hammar@jth.hj.se

Abstract. The study of ontology design patterns (ODPs) is a fairly recent development. Such patterns simplify ontology development by codifying and reusing known best practices, thus lowering the barrier to entry of ontology engineering. However, while ODPs appear to be a promising addition to research and while such patterns are being presented and used, work on patterns as artifacts of their own, i.e. methods of developing, identifying and evaluating them, is still uncommon. Consequently, little is known about what ODP features or characteristics are beneficial or harmful in different ontology engineering situations. The presented PhD project aims to remedy this by studying ODP quality characteristics and what impact these characteristics have on the usability of ODPs themselves and on the suitability of the resulting ontologies.

1 Research Problem

In spite of the development of several well-defined ontology construction methods, efficient ontology development continues to be a challenge. One reason for this is that such work requires both extensive knowledge of the domain being modeled, and a sufficient understanding of ontology languages and logical theory. Ontology Design Patterns (ODPs) are considered a promising contribution to this challenge [2]. Such patterns encode best practices, which helps reduce the need for extensive experience when developing ontologies. Additionally, by promoting a limited set of best practice outlooks on common problems, patterns help harmonize how these problems are viewed, supporting interoperability of ontologies developed using them.

Previous study into the state of ODP research indicates that while patterns are being used as tools in the ontology research community, and while there are patterns being developed and presented, patterns are rarely studied as artifacts of their own. The amount of work done on evaluation of patterns and pattern development and usage methods is limited [10]. There exists no established theory or model of ODP quality, and consequently there is no way of telling with certainty what is a good pattern for solving a particular type of problem based on pattern features. We will within the presented PhD project attempt to remedy this situation, by studying what quality characteristics or features of Ontology Design Patterns that are beneficial in different types of ontology engineering situations, and inversely, what such features or characteristics that could be considered harmful.

L. Aroyo et al. (Eds.): ISWC 2011, Part II, LNCS 7032, pp. 277–284, 2011.
© Springer-Verlag Berlin Heidelberg 2011

To this end, three research questions have been established:

1. Which quality characteristics of ODPs can be differentiated, and how can these be measured?
2. Which quality characteristics of ODPs affect the suitability of the resulting ontologies for different uses?
3. How do the quality characteristics of ODPs affect one another?

The remainder of this paper is structured as follows: Section 2 covers some other work that has been published on the topic of ontology evaluation, conceptual model quality, and design patterns. Section 3 presents what we hope this PhD project will contribute to the research community. The state of the work so far is presented in Sect. 4, hypotheses based on this tentative research is presented in Sect. 5, and the road ahead is mapped out in Sect. 6.

2   Related Work

2.1   Ontology Design Patterns

The use and understanding of ontology design patterns has been heavily influenced by the work taking place in the FP6 NeOn Project, the results of which include a pattern typology [14] and the eXtreme Design collaborative ontology development methods, based on pattern use [3]. While this view is influential and the NeOn typology is referenced frequently, it is not the only perspective on patterns - for instance, Blomqvist [1] presents a different typology based on the level of abstraction and granularity of the reusable solution.

eXtreme Design (XD) is defined as *"a family of methods and associated tools, based on the application, exploitation, and definition of ontology design patterns (ODPs) for solving ontology development issues"* [13]. The XD approach to selecting patterns is based on the pattern containing a written description of the type of problem for which the original pattern developer considers it appropriate (the *Generic Use Case*), that the ontology developer can match against his/her modelling problem (in XD parlance the *Local Use Case*). Additionally, to find candidate patterns for a given problem, search against pattern keywords can be performed.

2.2   Ontology Evaluation

While the amount of work on ODP evaluation is limited, there are quite a few methods and frameworks proposed for the evaluation of ontologies that may be relevant for ODPs also. The semiotic metaontology O^2 is used as a basis for instantiating *oQual*, a proposed ontology for the evaluation and selection of ontologies for a given task [6]. These two ontologies are complemented by and used in the *QoodGrid* methodology of ontology selection [7]. The *QoodGrid* framework is detailed and proposes some useful measures and formalizations of properties, but it has not been tested extensively. Also, it is unknown how well suited this complex framework would be for small reusable solutions.

One of the most cited works in ontology evaluation is [9], introducing *Onto-Clean*. This method tests whether an ontology is consistent with the real world it is supposed to model by applying a number of formal metaproperties (*essence, identity, unity*, etc.) to the concepts in the ontology and then checking for any inconsistencies in how these metaproperties are instantiated.

2.3   Conceptual Model Quality

It seems reasonable that we in this PhD project also consider established knowledge regarding the quality of other types of conceptual models, such as UML or ER models. While the such models may differ from ODPs in terms of computational functionality and logic formalisms used, research performed on them may still give guidance on how humans interpret and understand graphical and text representation of ODPs, i.e. the semiotics of conceptual computer models.

Genero et al. [8] present a simple experiment on the effect of structural complexity in ER models to the understandability and modifiability of such models. While the complexity measures used are not directly transferrable, the experimental method and test measures used are very applicable to our case. The difference between objective/quantitative metrics and subjective/qualitative ones are emphasized by Moody & Shanks [12], who also work with ER models. In [12] we also find an important discussion of the effects of various metrics/qualities on one another (for instance, the negative correlation between completeness of a model and implementability of that same model).

Lindblad et al. [11] present a discussion on the quality of more logically abstract models. They emphasize that models should be minimal in order not to overconstrain systems and users, while at the same time being *feasible complete*, i.e. that they should contain all the statements describing the domain such that the cost of adding the statement does not outweigh the benefit it brings to the model (a kind of scoping of the problem). It should be interesting to consider how this view of appropriate model size is impacted by the open world assumption of the semantic web. Lindblad et al. also stress the importance of comprehensibility of models requiring support for visualization and filtering.

3   Contribution

The ODP selection method proposed in XD [13] is appropriate for finding patterns that has been described as satisfactorily solving a particular problem from a larger set of patterns. It may also be possible to automate, provided that an appropriate logical vocabulary for describing *Local Use Cases* and *Global Use Cases* is developed. It does however not guide the user in selecting, from a given set of functionally appropriate patterns, the one that is best suited for use in their situation. The right choice then could depend on non-functional requirements on the ontology as a whole (expandability, performance, testability, etc), or it could depend on quality attributes of the pattern itself (how easy is it to apply, how it is documented, is there an example ontology using it, etc). An ODP quality model should in this scenario guide the developer in selecting patterns to use that, apart from solving the functional requirements of their modeling problem, also has features and qualities that are appropriate and helpful to them.

We also anticipate that the quality model resulting from this PhD work will be helpful in creating or extracting new high quality ODPs for a variety of purposes. While there exists already some work on these topics (primarily in [14]), this work is described in terms of specific technical issues, dealing with how to go about specializing patterns, generalizing existing models, and reengineering from other types of logic models. The evaluation which features that ought to be present in an ODP for a particular purpose is left with the ontology engineer. Providing this ontology engineer with a defined quality model may help them in developing better ODPs.

It can be argued that ontologies are similar enough to OOP inheritance hierarchies that an understanding of object-oriented design pattern quality should be sufficient also when constructing ontologies, and that the specific study of ODPs is therefore not required. We find this comparison and argument to be flawed, for two reasons. To begin with, OOP design patterns are used in the modeling and production of information systems only, whereas ontologies have a much wider applicability as shared conceptualizations for various purposes, and therefore are often more general in nature. Secondly, OOP languages/designs are not based in description logic and do not not make use of classification and inferencing as ontologies do. It is our opinion that such capability is of key importance in many ontology usage scenarios, such as document classification, situation recognition, constraint modeling, etc. Consequently, ontologies are different enough from object-oriented designs that developing an understanding of quality as it applies to ontology design patterns specifically is a worthy contribution.

4   Current State

At the time of writing, the PhD project is just over one year into its four year runtime. The time elapsed so far has been spent on familiarizing the author with semantic technologies and their applications, establishing that there is a need for this research (see [10]), initiating some small-scale initial experiments, and developing a metamodel for representing ODP qualities.

4.1   Quality Metamodel

The ODP quality metamodel is illustrated in Fig. 1. The purpose of this metamodel is to systematize how we discuss and think about ODP quality, to establish a vocabulary and theory that experiments can build upon and observations relate to. It is obviously subject to change as the PhD project proceeds.

The topmost half of the figure displays the relation R(D,ODPU,OU) where D denotes a domain, ODPU denotes an ODP Use (*ontology engineering, ontology matching, transformation*, etc), and OU denotes a use to which the deliverable result of the ODP usage is put.

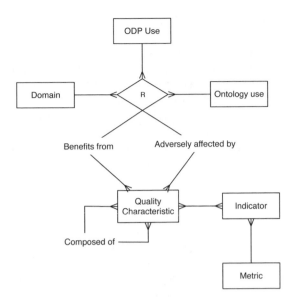

Fig. 1. Proposed Quality Metamodel

An instantiation of this relation R could be said to form a usage example of ODPs. For each such instantiation R a number of quality characteristics affect the result, for better or for worse. Some quality characteristics are abstract, consisting of compositions of other quality characteristics. Quality characteristics are mapped to concrete indicators that are measurable using some metric.

4.2   Initial Data Gathering

For the purpose of eliciting hypotheses, some small scale experiments have been performed. The XD methodology has been used in the scope of an internal project dealing with expert finding in the academic domain. The problem was decomposed according to XD methodology and a number of patterns from the ontologydesignpatterns.org repository were used to iteratively develop solutions for the problems, integrating the solutions at the end of each iteration. The end result was a rather complex ontology, highly heterogeneous in terms of annotation quality and density.

Several interviews have been staged with master student groups using ODPs in the development of ontologies used within master thesis projects. From these interviews it was learned that when selecting patterns students prioritized high quality documentation, both in terms of annotations on the example pattern instantiation (if one is provided), and in terms of written usage examples and instructions. They also preferred smaller patterns as they were easier to "wrap one's head around", however, the patterns should not be as small as to be trivial.

5   Hypotheses

Based on the initial data gathering efforts discussed in Sect. 4.2, a number of hypotheses regarding the influence of certain ODP characteristics on a resulting ontology have been developed, as detailed in Table 1. For formal definitions of the structural quality characteristics mentioned, see [5]. These attributes are not yet aligned to the previously presented quality metamodel. The belief is that it would be wiser to collect data proving or disproving their usability before instantiating a quality model based on them, since this data collection process may result in a changed understanding of ODP quality that necessitates changing the underlying metamodel also.

Table 1. ODP quality characteristics influences on resulting ontology

ODP quality characteristic	Resulting ontology feature/characteristic
Small size	Low average depth, high breadth
Small size	High MO count & high cycle count
Small size	Variable density
Small size	Varying annotation quality
Small size	Complex result
Small size	Poor computational performance

All of these first hypotheses deal with the issue of size, i.e. the number of classes and properties in an ODP. Essentially, the intuition and experience from the initial small experiments is that if one uses smaller patterns, one has to use more of them to cover a particular problem or domain. By using more patterns, the total complexity of the end result will increase, as each pattern-based module will be integrated via subsumption links to many of the other small pattern-based modules that make up the ontology. If one instead used larger patterns, each such module would in itself hold a certain set of functionality, and fewer links would need be added to integrate the modules. This complexity in integration is also believed to lead to a higher subsumption hierarchy cycle count, which is detrimental to computational performance.

Some further hypotheses presented in Table 2 deal with the impact of particular features on the comprehensibility and reusability of the ODPs themselves. As indicated by the performed interviews with students, smaller patterns are easier to understand. We also believe that using a simpler logical language with less advanced constructs (i.e. RDFS rather than OWL) will help in understanding the ODP. Further, we say that a highly abstract pattern likely will be more general in nature and thus more reusable in many different scenarios. At the same time, it will be more difficult to understand and apply than a more concrete and well defined pattern. Finally, minimalism (defined as fulfilling the requirements of one problem and containing no extraneous statements) is believed to make a pattern easier to understand and applicable in a greater set of scenarios.

Table 2. ODP quality characteristics impact on ODP comprehensibility/reusability

ODP quality characteristic	Impact on ODP
Small size	High comprehensibility
Simple language	High comprehensibility
High abstraction level	High reusability
High abstraction level	Low comprehensibility
Minimalism	High comprehensibility
Minimalism	High reusability

6   The Road Ahead

Two more experiments using bachelor and master students have been initiated and will be completed during the fall. In each of these experiments, two groups of students have been provided with two sets of patterns, with which to solve a set of ontology engineering tasks. The first set of patterns are representative of the patterns available in the ontologydesignpatterns.org repository. The second set have been constructed based on solutions in [4]. The latter are generally larger and more complex, but also more thoroughly described. An analysis of the resultant ontologies will be performed to study the consequences of these differences. Also, interviews with the student groups will be performed to gauge the perceived comprehensibility of the two sets of patterns.

Later in the fall we will work on a project that deals with situation recognition using input from a number of sensor subsystems. We have been granted access to a reference implementation of a particular system for this purpose (not using ontologies), as well as the design documentation and specifications for the system. The goal is build an ontology-based equivalent using ODPs in the construction of the system ontology. Key considerations in the ontology development process are computational performance of the system, high documentation quality, and the ability to reconfigure the ontology depending on available sensor subsystems or situation recognition contexts.

Testing how well OntoClean works on ODPs looks to be an interesting experiment - on the one hand, ODPs are intended to be general solutions, and as such, OntoClean metaproperties might not apply cleanly to the more general concepts they contain. On the other hand, ODPs are small enough that it in terms of effort expended would be quite simple to apply the OntoClean metaproperties to them (applying OntoClean to a full ontology is generally considered to be a rather time-consuming process). We hope to be able to test such use of Onto-Clean on ODPs during the spring and if the results of this work are positive, will attempt to integrate OntoClean metaproperties and constraints into our quality model.

Finally, discussions are underway with an industry partner where we hope to be able to test the usability of ODPs in an information logistics context. The scenario involves a large set of heterogeneous production data that needs to be integrated and made easily searchable. This is a type of scenario where it is common to see ontologies used as shared vocabularies. Such ontologies need not be very complex, as they are not subject of much reasoning and classification. However, they need to be flexible and general, in order to accommodate the rapid growth of input data. For these reasons, this project should be an interesting contrast to the previously mentioned situation recognition project.

References

1. Blomqvist, E.: Semi-automatic ontology construction based on patterns. Ph.D. thesis, Department of of Computer and Information Science, Linköpings universitet (2009)
2. Blomqvist, E., Gangemi, A., Presutti, V.: Experiments on pattern-based ontology design. In: Proceedings of the fifth International Conference on Knowledge Capture, pp. 41–48. ACM (2009)
3. Daga, E., Blomqvist, E., Gangemi, A., Montiel, E., Nikitina, N., Presutti, V., Villazon-Terrazas, B.: NeOn Deliverable D2.5.2 Pattern based ontology design: methodology and software support. NeOn Project (2008), http://www.neon-project.org
4. Fowler, M.: Analysis patterns: reusable object models. Addison-Wesley (1997)
5. Gangemi, A., Catenacci, C., Ciaramita, M., Lehmann, J.: Ontology evaluation and validation: an integrated formal model for the quality diagnostic task. Tech. rep., Laboratory for Applied Ontology, ISTC-CNR, Roma/Trento (Italy) (2005)
6. Gangemi, A., Catenacci, C., Ciaramita, M., Lehmann, J.: Modelling Ontology Evaluation and Validation. In: Sure, Y., Domingue, J. (eds.) ESWC 2006. LNCS, vol. 4011, pp. 140–154. Springer, Heidelberg (2006)
7. Gangemi, A., Catenacci, C., Ciaramita, M., Lehmann, J.: Qood grid: A metaontology-based framework for ontology evaluation and selection. In: Proceedings of the EON 2006 Workshop (2006)
8. Genero, M., Poels, G., Piattini, M.: Defining and Validating Measures for Conceptual Data Model Quality. In: Pidduck, A.B., Mylopoulos, J., Woo, C.C., Ozsu, M.T. (eds.) CAiSE 2002. LNCS, vol. 2348, pp. 724–727. Springer, Heidelberg (2002)
9. Guarino, N., Welty, C.: Evaluating ontological decisions with OntoClean. Communications of the ACM 45(2), 61–65 (2002)
10. Hammar, K., Sandkuhl, K.: The State of Ontology Pattern Research: A Systematic Review of ISWC, ESWC and ASWC 2005–2009. In: Workshop on Ontology Patterns: Papers and Patterns from the ISWC Workshop (2010)
11. Lindland, O., Sindre, G., Sølvberg, A.: Understanding quality in conceptual modeling. IEEE software, 42–49 (1994)
12. Moody, D., Shanks, G.: What makes a good data model? Evaluating the quality of entity relationship models. In: Loucopoulos, P. (ed.) ER 1994. LNCS, vol. 881, pp. 94–111. Springer, Heidelberg (1994)
13. Presutti, V., Daga, E., Gangemi, A., Blomqvist, E.: eXtreme Design with Content Ontology Design Patterns. In: Proceedings of the Workshop on Ontology Patterns (WOP 2009), collocated with ISWC (2009)
14. Presutti, V., Gangemi, A., David, S., de Cea, G., Surez-Figueroa, M., Montiel-Ponsoda, E., Poveda, M.: NeOn Deliverable D2. 5.1. A Library of Ontology Design Patterns: reusable solutions for collaborative design of networked ontologies. NeOn Project (2008), http://www.neon-project.org

DC Proposal: Automation of Service Lifecycle on the Cloud by Using Semantic Technologies

Karuna P. Joshi*

Computer Science and Electrical Engineering
University of Maryland, Baltimore County,
Baltimore, MD 21250, USA
kjoshi1@umbc.edu

Abstract. Managing virtualized services efficiently over the cloud is an open challenge. We propose a semantically rich, policy-based framework to automate the lifecycle of cloud services. We have divided the IT service lifecycle into the five phases of requirements, discovery, negotiation, composition, and consumption. We detail each phase and describe the high level ontologies that we have developed to describe them. Our research complements previous work on ontologies for service descriptions in that it goes beyond simple matchmaking and is focused on supporting negotiation for the particulars of IT services.

Keywords: Cloud Computing, services lifecycle, automation.

1 Research Motivation

Information Technology (IT) products and services which were previously either in-house or outsourced are now being replaced by a new delivery model where businesses purchase IT components like software, hardware or network bandwidth or human agents as services from providers who can be based anywhere in the world. Such services will increasingly be acquired "on demand", and composed on the fly by combining pre-existing components. In such scenario, multiple providers will collaborate to create a single service and each component service will be virtualized and might participate in many composite service orchestrations. The service, in effect, will be virtualized on the cloud. In any organization service acquisition will be driven by enterprise specific processes and policies that will constraint this acquisition.

The academic community has recently become interested in automating steps needed to acquire services from the cloud. However, researchers have concentrated on issues like service discovery or service composition. For our research, we have worked on automating the entire lifecycle of services on the cloud from discovery, negotiation to composition and performance monitoring. We outline our proposed methodology in section 3.

Motivation: Currently, providers decide how the services are bundled together and delivered to the consumers on the cloud. This is typically done statically and as a manual process. There is a need to develop reusable, user-centric mechanisms that will

---

* Advisor: Yelena Yesha and Tim Finin.

L. Aroyo et al. (Eds.): ISWC 2011, Part II, LNCS 7032, pp. 285–292, 2011.
© Springer-Verlag Berlin Heidelberg 2011

allow the service consumer to specify their desired quality constraints, and have automatic systems at the providers end control the selection, configuration and composition of services. This should be without requiring the consumer to understand the technical aspects of services and service composition. The U.S. National Institute of Standards and Technology (NIST) has identified [18] on-demand self-service as an essential characteristic of the cloud model where a consumer can unilaterally provision computing capabilities as needed automatically, without requiring human interaction with each service's provider.

Our Approach: We believe that a semantically rich, policy-based framework will facilitate the automation of the lifecycle of virtualized services. We have developed an integrated methodology that encompasses the entire data and process flow of services on the cloud from inception, creation/composition to consumption/ monitoring and termination. We have identified processes and policies to automate this lifecycle. We have used semantically rich descriptions of the requirements, constraints, and capabilities that are needed by each phase of the lifecycle. This methodology is complementary to, and leverages, previous work on ontologies, like OWL-S [14], for service descriptions in that it is focused on automating processes needed to procure services on the cloud. We have concentrated on enabling multiple iterations of service negotiation with constraints being relaxed during each iteration till a service match is obtained. Using Semantic web technologies like OWL [8] and Jena[15], we have also created high level ontologies for the various phases. These can be reasoned over to automate the phases guided by high level policy constraints provided by consumers, service customers, or service providers. Semantic web technologies also address the interoperability and portability issue of cloud computing.

Evaluation: As this is a new line of research, there is currently no success criteria based on comparisons with previous work. We are developing a system as a proof of concept and will validate it against example enterprise policies obtained from various organizations. Towards that we are collaborating with NIST, a federal government organization. In collaboration with NIST, we are designing a prototype of our proposed methodology which will work as a "pilot" to demonstrate automatic acquisition of services on the cloud. This pilot is being developed by using Semantic Web technologies like OWL [8], RDF [10], and SPARQL [11]. We are also collaborating with an international financial organization and the UMBC IT department to understand the complete process and policies that are applied towards procuring IT services. We will calibrate our system to demonstrate how distinct processes of service acquisition can be captured and how enterprise policies can be expressed using our ontology and other policy languages to show that the service acquisition process can be automated.

One of our research goals is to develop appropriate ways to evaluate our system. One measure is existential -- can we create a system that will automate the service acquisition process via our lifecycle's realization. As that our approach relies on semantically rich policies, another measure is how well our ontologies and policy mechanisms handle a given real-world policy. Given the large and diverse organization from which we are seeking policies, we hope to get a good representative sample to determine which enterprise policies can be automated for service acquisition and which cannot be. So our evaluation might also enable us to determine how useful such an automated service procurement system will be in the real world. These are our preliminary ideas on evaluating our research which we plan to refine further.

2 Related Work

Most approaches to automating the acquisition or use of online services have been limited to exploring a single aspect of the lifecycle like service discovery, service composition or service quality. There is no integrated methodology for the entire service lifecycle covering service planning, development and deployment in the cloud. In addition, most of the work is limited to the software component of the service and does not cover the service processes or human agents which are a critical component of IT Services.

Papazoglou and Heuvel [1] have proposed a methodology for developing and deploying web services using service oriented architectures. Their approach, however, is limited to the creation and deployment of web services and does not account for virtualized environment where services are composed on demand. Providers may need to combine their services with other resources or providers' services to meet consumer needs. Other methodologies, like that proposed by Bianchini et al. [2], do not provide this flexibility and are limited to cases where a single service provider provides one service. Zeng et al. [3] address the quality based selection of composite services via a global planning approach but do not cover the human factors in quality metrics used for selecting the components. Maximilien and Singh [4] propose an ontology to capture quality of a web service so that quality attributes can be used while selecting a service. While their ontology can serve as a key building block in our system, it is limited by the fact that it considers single web services, rather than service compositions.

Black et al. [5] proposed an integrated model for IT service management, but it is limited to managing the service from only the service provider's perspective. Paurobally et al. [12] have described a framework for web service negotiation using the iterated Contract Net Protocol [13]. However their implementation is limited to pre-existing web services and does not extend to virtualized services that are composed on demand. Our negotiation protocol, as detailed in [21], accounts for the fact that the service will be composed only after the contract/SLA listing the constraints is finalized. GoodRelations [17] is an ontology developed for E-commerce to describe products. While this ontology is useful for describing service components that already exist on the cloud, it is difficult to describe composite virtualized services being provided by multiple vendors using this ontology. Researchers like Sbodio et. al [16] have proposed algorithms for service discovery using SPARQL language. We are using SPARQL and other associated Semantic Web technologies to allow complex negotiation process between service providers and service consumers.

The Information Technology Infrastructure Library (ITIL) is a set of concepts and policies for managing IT infrastructure, development and operations that has wide acceptance in the industry. The latest version of ITIL lists policies for managing IT services [7] that cover aspects of service strategy, service design, service transition, service operation and continual service improvement. However, it is limited to interpreting "IT services" as products and applications that are offered by in-house IT department or IT consulting companies to an organization. This framework in its present form does not extend to the service cloud or a virtualized environment that consists of one or more composite services generated on demand.

3   Research Contribution: Service Lifecycle Ontology

We have developed a new methodology which integrates all the processes and data flows that are needed to automatically acquire, consume and manage services on the cloud. We divide this IT service lifecycle on a cloud into five phases. In sequential order of execution, they are requirements, discovery, negotiation, composition, and consumption. We have described these phases in detail along with the associated metrics in [19]. Figure 1 is a pictorial representation detailing the processes and data flow of the five phases. We have developed the ontology for the entire lifecycle in OWL 2 DL profile, available at [22].

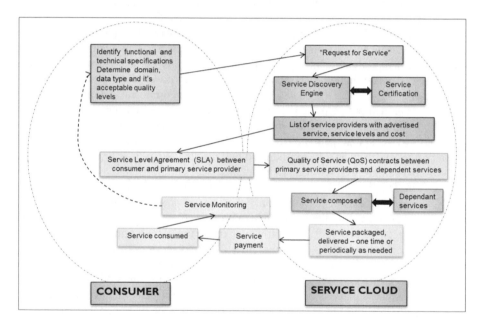

Fig. 1. The IT service lifecycle on a virtualized cloud comprises five main phases: requirements, discovery, negotiation, composition and consumption

3.1   Service Requirements Phase

In this phase, the consumer details the technical and functional specifications that a service needs to fulfill and also non-functional attributes such as characteristics of the providing agent, constraints and preferences on data quality and required security policies for the service. Service compliance details like required certifications, standards to be adhered to etc. are also identified. Depending on the service cost and availability, a consumer may be amenable to compromise on the service quality. Functional specification describe in detail what functions/tasks should a service help automate The technical specifications lay down the hardware, software, application standards and language support policies to which a service should adhere. Once the consumers have identified and classified their service needs, they issue a Request for Service (RFS). This request could be made by directly contacting a few service providers for their quotes. Alternatively, consumers can use a service discovery engine or service broker on the cloud to procure the service.

The two top classes of this phase are the Specification class and the "Request For Service" class. The Specification class consists of six main classes that define the functional specifications, technical specifications, Human agent specifications, security policies, service compliance policies and data quality policies. The technical specs contain information about the Hardware, Operating System and other compatible services/applications that the desired service should conform to. Human Agent specifications also list the technical and domain expertise that the service providing agent should have. Part of our ongoing work is to use existing ontologies that have been developed for classes like standard hardware, operating systems and computer applications. Semantic Web policy languages can be used to describe service specifications and constraints in machine-processable format. For instance, a consumer may opt for a service with poor data quality to take advantage of low cost of the service.

3.2 Service Discovery Phase

Service providers are discovered by comparing the specifications listed in the RFS. The discovery is constrained by functional and technical attributes defined, and also by the budgetary, security, compliance, data quality and agent policies of the consumer. While searching the cloud, service search engines or service brokers can be employed. This engine runs a query against the services registered with a central registry or governing body and matches the domain, data type, compliance needs, functional and technical specifications and returns the result with the service providers matching the maximum number of requirements listed at the top.

One critical part of this phase is service certification, in which the consumers will contact a central registry, like UDDI [6], to get references for providers that they narrow down to.

This phase uses the RFS class from the requirements phase to search for service providers and generate a list of providers with which to begin negotiations. The class Service certification validates the provider's credentials.

If the consumers find the exact service within their budgets, they can begin consuming the service immediately upon payment. However, often the consumers will get a list of providers who will need to compose a service to meet the consumer's specifications. The consumer will then have to begin negotiations with the service providers which is the next phase of the lifecycle. Each search result will return the primary provider who will be negotiating with the consumer.

3.3 Service Negotiation Phase

The service negotiation phase covers the discussion and agreement that the service provider and consumer have regarding the service delivered and its acceptance criteria. The service delivered is determined by the specifications laid down in the RFS. Service acceptance is usually guided by the Service Level Agreements (SLA) [9] that the service provider and consumer agree upon. SLAs define the service data, delivery mode, agent details, quality metrics and cost of the service. While negotiating the service levels with potential service providers, consumers can explicitly specify service quality constraints (data quality, cost, security, response time, etc.) that they require.

At times, the service provider will need to combine a set of services or compose a service from various components delivered by distinct service providers in order to meet the consumer's requirements. The negotiation phase also includes the discussions that the main service provider has with the other component providers. When the services are provided by multiple providers (composite service), the primary provider interfacing with the consumer is responsible for composition of the service. The primary provider will also have to negotiate the Quality of Service (QoS) with the secondary service providers to ensure that SLA metrics are met. The negotiation steps are detailed in [21].

This phase uses the RFS class from the requirements phase and the provider's list class from the discovery phase to negotiate the contracts between consumer and primary provider and between the various component providers themselves. The key deliverable of this phase is the service contract between the service consumer and service provider. The SLA is a key part of this service contract and will be used in the subsequent phases to compose and monitor the service. Another deliverable of this phase are the service sub contracts between the service provider and component (or dependent services) providers. The QoS are the essential part of the service sub-contracts and are used in the consumption phase to monitor service performance.

3.4 Service Composition Phase

In this phase one or more services provided by one or more providers are combined and delivered as a single service. Service orchestration determines the sequence of the service components.

The main class of this phase is the Service class that combines the various components into a single service. We include the OWL-S [14] Composite Process class ontology. The Service class takes inputs from the Specification, Service Contracts and Service Level Agreement classes defined in the earlier phases to help determine the orchestration of the various components.

3.5 Service Consumption/Monitoring Phase

The service is delivered to the consumer based on the delivery mode (synchronous/asynchronous, real-time, batch mode etc.) agreed upon in the negotiation phase. After the service is delivered to the consumer, payment is made for the same. The consumer then begins consuming the service. In a cloud environment, the service usually resides on remote machines managed by the service providers. Hence the onus for administrating, managing and monitoring the service lies with the provider. In this phase, consumer will require tools that enable service quality monitoring and service termination if needed. This will involve alerts to humans or automatic termination based on policies defined using the quality related ontologies. The Service Monitor measures the service quality and compares it with the quality levels defined in the

SLA. This phase spans both the consumer and cloud areas as performance monitoring is a joint responsibility. If the consumer is not satisfied with the service quality, s/he should have the option to terminate the service and stop service payment.

The composite service is composed of human agents providing the service, the service software and dependent service components. All the three elements, agents, software and dependent services, must be monitored to manage the overall service quality. For the service software providers have to track its performance, reliability, assurance and presentation as they will influence customer's satisfaction rating (CSATs). Since the dependent services/components will be at the backend and will not interface directly with the consumers, the service provider only needs to monitor their performance. We have proposed a framework to manage quality based on fuzzy-logic for such composed services delivered on the cloud in [20].

4 Summary

In this paper we have defined the integrated lifecycle for IT services on the cloud. To the best of our knowledge, this is the first such effort, and it is critical as it provides a "big" picture of what steps are involved in deploying IT services. This methodology can be referenced by organizations to determine what key deliverables they can expect at any stage of the process.

We are currently refining the ontology described in [21] to capture the steps and metrics we have identified in the lifecycle using semantic web languages. We are also developing a prototype in collaboration with NIST to demonstrate the service lifecycle for a storage service on the cloud. We are creating the prototype by using Semantic web technologies like SPARQL, OWL and RDF.

References

1. Papazoglou, M., Van Den Heuvel, W.: Service-oriented design and development methodology. International Journal of Web Engineering and Technology 2(4), 412–442 (2006)
2. Bianchini, D., De Antonellis, V., Pernici, B., Plebani, P.: Ontology-based methodology for e-service discovery. International Journal of Information Systems, The Semantic Web and Web Services 31(4-5), 361–380 (2006)
3. Zeng, L., Benatallah, B., Dumas, M., Kalagnanam, J., Sheng, Q.: Quality driven web services composition. In: Proceedings of the 12th International Conference on World Wide Web, pp. 411–421 (2003)
4. Maximilien, E.M., Singh, M.: A Framework and Ontology for Dynamic Web Services Selection. IEEE Internet Computing 8(5), 84–93 (2004)
5. Black, J., et al.: An integration model for organizing IT service Management. IBM Systems Journal 46(3) (2007)
6. Ran, S.: A model for web services discovery with QoS. ACM SIGecom Exchanges 4(1), 1–10 (2003)
7. Van Bon, J., et al.: Foundations of IT service management based on ITIL V3. Van Haten Publishing (2008)
8. McGuinness, D., Van Harmelen, F., et al.: OWL web ontology language overview, W3C recommendation, World Wide Web Consortium (2004)

9. Whats in a Service Level Agreement?, SLA@SOI,
   `http://sla-at-soi.eu/?p=356` (retrieved on July 30, 2009)
10. Lassila, O., Swick, R., et al.: Resource Description Framework (RDF) Model and Syntax Specification. World Wide Web Consortium (1999)
11. Prud'hommeaux, E., Seaborne, A.: SPARQL Query Language for RDF. W3C recommendation, `http://www.w3.org/TR/rdf-sparql-query/` (retrieved on April 27, 2011)
12. Paurobally, S., Tamma, V., Wooldrdige, M.: A Framework for Web Service Negotiation. ACM Transactions on Autonomous and Adaptive Systems 2(4) Article 14 (November 2007)
13. Smith, R.: The Contract Net Protocol: High-Level Communication and Control in a Distributed Problem Solver. IEEE Transactions on computers C-29(12), 1104–1113 (1980)
14. Martin, D., et al.: Bringing Semantics to Web Services: The OWL-S Approach. In: Cardoso, J., Sheth, A.P. (eds.) SWSWPC 2004. LNCS, vol. 3387, pp. 26–42. Springer, Heidelberg (2005)
15. Jena–A Semantic Web Framework for Java, `http://jena.sourceforge.net/` (retrieved on May 10, 2011)
16. Sbodio, M.L., Martin, D., Moulin, C.: Discovering Semantic Web services using SPARQL and intelligent agents. Journal of Web Semant. 8(4), 310–328 (2010)
17. Hepp, M.: GoodRelations: An Ontology for Describing Products and Services Offers on the Web. In: Gangemi, A., Euzenat, J. (eds.) EKAW 2008. LNCS (LNAI), vol. 5268, pp. 329–346. Springer, Heidelberg (2008)
18. NIST initiative to develop standards for Cloud Computing,
   `http://www.nist.gov/itl/cloud/` (last retrieved July 2011)
19. Joshi, K., Finin, T., Yesha, Y.: Integrated Lifecycle of IT Services in a Cloud Environment. In: Proceedings of The Third International Conference on the Virtual Computing Initiative (ICVCI 2009), Research Triangle Park, NC (October 2009)
20. Joshi, K., Joshi, A., Yesha, Y.: Managing the Quality of Virtualized Services. In: Proceedings of the SRII Global Conference, San Jose (March 2011)
21. Joshi, K., Finin, T., Yesha, Y.: A Semantic Approach to Automate Service Management in the Cloud, UMBC Technical Report, TR-CS-11-02 (June 2011)
22. Joshi, K.: OWL Ontology for Lifecycle of IT Services on the Cloud,
   `http://ebiquity.umbc.edu/ontologies/itso/1.0/itso.owl`

DC Proposal: Knowledge Based Access Control Policy Specification and Enforcement

Sabrina Kirrane

Digital Enterprise Research Institute,
National University of Ireland, Galway
sabrina.kirrane@deri.org
http://www.deri.ie

Abstract. The explosion of digital content and the heterogeneity of enterprise content sources have resulted in a pressing need for advanced tools and technologies, to support enterprise content search and analysis. Semantic technology and linked data may be the long term solution to this growing problem. Our research explores the application of access control to a knowledge discovery platform. In order to ensure integrated information is only accessible to authorised individuals, existing access control policies need to be associated with the data. Through in-depth analysis we aim to propose an access control model and enforcement framework which can be used to represent and enforce various access models both inside and outside the enterprise. Furthermore, through experimentation we plan to develop a methodology which can be used as a guideline for the lifting of distributed access control policies from the individual data sources to a linked data network.

Keywords: Policy, Access Control, Reasoning, Information Analysis, Knowledge Discovery, Linked Data.

1 Introduction

The Internet is growing exponentially, fuelling research into the next generation of Internet technologies. Over the past two decades much research has gone into the use of semantic technology and linked data for data integration, search and information analysis. This exponential expansion of data and many of the challenges that come with it are mirrored within the enterprise. Our use case examines how such techniques can be used to build an integrated data network for enterprise content analysis and knowledge discovery. Furthermore, we investigate how subsets of interconnected enterprise data can be shared with partner companies or exposed on the Internet.

When we integrate data from multiple line of business (LOB) applications and document repositories and share a subset of this information externally we need to ensure that we don't expose sensitive data to unauthorised individuals. For example, an application could be developed to extract and link data from both a document management system (DMS) and a human resource (HR) application. If several documents were written by the management team over the past

L. Aroyo et al. (Eds.): ISWC 2011, Part II, LNCS 7032, pp. 293–300, 2011.
© Springer-Verlag Berlin Heidelberg 2011

month, discussing future plans for the company, terms such as "takeover", "acquisition", "redundancy", "office closure", "pay cuts" and "lay-offs" may feature prominently in a list of last month's frequent phrases. Worse still, by integrating the data from both systems it might be possible to determine the offices that are scheduled to be closed and the individual employees that will be laid off. Much of the information within the enterprise is highly sensitive and as such the access policies of the underlying data source applications need to be extracted, represented and enforced in the linked data network.

In recent years, there has been a great deal of research into the use of policies for access control [1], [2] and [3]. The term "policy", in this context, is used to refer to the access control model which describes the blueprint, the policy language which defines both the syntax and the semantics of the rules and the framework which is a combination of the access control model, the language and the enforcement mechanism. Natural language, programming languages, XML based languages and ontologies can all be used to express policies. XML and ontologies are two popular choices for representing policy languages as they benefit from flexibility, extensibility and runtime adaptability. However, ontologies are better suited to modelling the semantic relationship between entities. In addition, the common framework and vocabulary used by ontologies, to represent data structures and schemas, provides greater interpretability and interoperability. Regardless of the language chosen, a logic based underlying formalisation is crucial for automatic reasoning in a linked data network. To date, very little research has been done on the elevation and the representation of policies in the linked data network. Many of the proposed access control models lack a formal enforcement framework.

In this paper, we identify the need for a policy language which can represent multiple access control models and a framework which will allow enterprises to abstract information from existing content sources and publish subsets of this data to the linked open data (LOD) cloud. With a view to addressing the aforementioned issues our research focuses on the:

- Examination of existing enterprise content and access control models, with a view to proposing an integrated model
- Analysis of the expressivity of existing policy languages and their ability to represent multiple access control models
- Identification of a reasoning and enforcement framework taking into consideration flexibility, interoperability, scalability and usability
- Development of a methodology for the automatic lifting of policies from the individual data source to the linked data network
- Implementation of a prototype in order to evaluate the overall effectiveness of the access control framework

The remainder of the paper is structured as follows: In section 2, we detail our approach, the chosen research methodology and our evaluation strategy. In section 3, we describe the shape of the literature with respect to related access control models, languages and enforcement frameworks. Finally, we conclude and give the direction of our future work in section 4.

2   Research Objectives and Plan

2.1   Approach

In this section we detail our strategy for the conceptualisation and realisation of a linked data access control model, policy language and enforcement framework.

Access Control Model. Our initial objective is to examine existing access control models and to generate an integrated model which can be used to represent the access control policies of a linked data network either inside or outside the enterprise. Our analysis will consider heterogeneous access policies from existing enterprise content sources. It will be designed to be flexible enough to cater for both authenticated and non authenticated users. In addition, the model will take into account existing standards such as the initiative by the OASIS Integrated Collaboration Object Model (ICOM)[1] and the Distributed Management Task Force (DMTF) Common Information Model (CIM)[2].

Policy Language. Once we have produced the integrated access control model we will analyse existing policy languages to determine their suitability for representing the authorisations, obligations and rules in a linked data network and their degree of support for our integrated access control features. In order to facilitate reasoning over policies in the linked data network, the policy language will need to have underlying formal semantics. The language must also provide support for propagation based on the semantic relationship between all entities and both positive and negative authorisations and obligations. Our analysis will also take into account non functional requirements such as flexibility, extensibility, runtime adaptability and interoperability. We will examine the level of support without need for change, the adaptability of the policy language and the scope of the changes required. Depending on the results of our analysis we will choose the policy language that most closely meets our needs or, if appropriate, we will integrate and extend components from one or more languages.

Enforcement Framework. We will examine each of the reasoning and enforcement frameworks based on functional requirements identified in the literature as important for the representation and enforcement of access control policies. These are namely: propagation and inference [1] [2] [3] [4], conflict management [1] [2] [3] [4], distributivity [2] [3], exception handling [1] [2] and support for standards [5]. Given our use case investigates the lifting and representation of data from existing distributed content sources we will add provenance, policy lifting and policy merging to the list of criteria. In addition, we will consider non functional requirements such as flexibility, interoperability, scalability and usability. Our objective is to propose a new framework which builds on existing work and addresses any shortcomings.

---

[1] http://www.oasis-open.org/committees/icom/
[2] http://www.dmtf.org/standards/cim

In order to examine the effectiveness of the model, language, lifting methodology and framework we plan to implement a prototype. Figure 1, provides a high level overview of the components that make up the proposed information analysis and knowledge discovery platform. Domain data and relationships will be modelled as ontologies in the abstraction layer. The LOD cloud will be used to enrich the semantics of both the enterprise and access control models. Instance data will be abstracted from enterprise content repositories and LOB applications, stored in the enterprise data integration layer and appropriate indexes will be generated. The enforcement framework in turn will be composed of several interfaces, a rules engine, a query engine and local storage for both ontologies and access control rules. The policy integration interface will facilitate the retrieval of existing data access policies. Data from existing content sources will be obtained through the data integration interface. The policy specification interface will provide authenticated access to the policies and the data governed by the relevant policies. All data access queries will be intercepted by the query reasoning engine which will grant or deny access to data based on a combination of rules and inference.

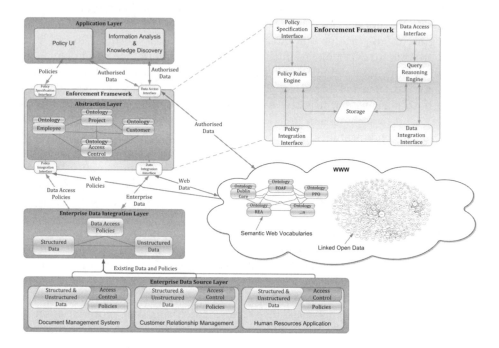

Fig. 1. Information analysis and knowledge discovery platform

We will investigate the applicability of existing standard technologies for conceptualization as they offer many well known benefits such as interoperability, extensibility, scalability and re-usability. We will utilise a number of methods in order to determine the best way to extract access constraints from the LOB applications. Through experimentation we aim to development a methodology which can be used to as a guideline for the lifting of access control policies from the individual data sources to the abstraction layer. Many researchers are currently investigating the representation and enforcement of privacy on the web in general and in semantically aware environments in particular. Although we do not plan to specifically address this problem our prototype will take into consideration current work in this area.

2.2   Methodology

We have chosen two complimentary methodologies, the Design Science Research (DSR) methodology proposed by Peffers *et al* [6] and Action Research (AR) as defined in [7], to guide our research. DSR involves the identification of the research problem and the motivation behind the research, the specification of the objectives and the development of a research artefact to meet these objectives. The artefact is subsequently evaluated and the problem, process, outcomes etc are communicated to academics and practitioners alike. AR is an iterative process which involves the identification of a problem, the taking of action and the examination of the outcomes of the action. If the examination determines that the problem has not been solved, the researcher takes what they have learned and further refines the problem and takes further action in an attempt to resolve the problem. This process continues until the researcher is satisfied that they have solved the problem or concludes that the problem can't be solved. The relationship between AR and DSR is multifaceted e.g. AR can be wrapped around DSR, DSR can be wrapped around AR or a number of AR iterations could form part of one or more individual DSR steps. In our research project we have adopted the DSR methodology to guide the overall research from conception through to conclusion. The AR methodology will in turn be used in the design and development step of the DSR methodology to enable the refinement of both the access control framework and the lifting methodology through iterative reflection.

2.3   Evaluation

A number of evaluation methods will be used throughout the research project, both artificial and naturalistic. The access control model, language, enforcement framework and the lifting methodology will be constructed and evaluated in an iterative manner as part of the AR methodology. The model and the policy language will be evaluated using qualitative means, such as interviews and desk studies, to confirm that they are flexible enough to support both authenticated and non authenticated users and a variety of access control models. The lifting methodology will be evaluated using precision and recall metrics to determine the effectiveness of the methodology. Finally the enforcement framework will be evaluated in our real world use case scenario through the implementation of a prototype to ensure that it addresses both the functional and non functional requirements highlighted in our approach as important for a knowledge discovery platform. A combination of experiments and interviews will be used to evaluate the expressivity of the policy language and scalability, performance and usability of our prototype.

3   Related Work

Access Control Models. Role-based access control (RBAC), task-based authorisation control (TBAC) and attribute-based access control (ABAC) are three of the most common approaches proposed for restricting access to information systems. In RBAC a number of roles are defined, users are assigned to appropriate roles and access to resources is granted to one or more roles [8]. RBAC policies have been the de facto authorisation mechanism used in enterprises for many years. The TBAC model proposed by [9] facilitates task oriented access controls required for workflow and agent based distributed systems. Essentially, TBAC models access controls from a task-oriented as opposed to a subject-orientated perspective. ABAC [10], which was also designed for distributed systems, provides an alternative means of access control where the requester is unknown prior to the submission of the request. ABAC grants or denies access to resources, based on attributes of the requester and/or the resource. Early work on semantic access control policies by [11] was based on RBAC. However, most of the recent work in this area is based on ABAC [4], [1], [2]. We propose an integrated approach which will include a number of access control models and will be flexible enough to represent access control both inside the enterprise and on the web.

Policy Language. Policy languages can be categorised as general or specific. In the former the syntax caters for a diverse range of functional requirements (e.g. access control, query answering, service discovery, negotiation etc...), whereas in contrast the latter focuses on just one functional requirement. Two of the most well-known access control languages, KAos [12] and Rei [13], are in fact general policy languages. A major drawback of both KAos [12] and Rei [13] is they have little or no support for policy propagation. Although Kolovski *et al* [14] propose a description logics formalisation for XACML it also does not consider policy propagation based on the semantic relationship between entities. Concept-level access control (CLAC) [4], semantic-based access control (SBAC) [1] and the semantic network access control model proposed by [2] all focus on the policy specification and reasoning in the conceptual layer of the semantic aware environment. Qin *et al* [4] allow propagation of access controls based on the semantic relationships among concepts. [1], [2] and [3] enhance the semantic relations by allowing propagation based on the semantic relationships between the subjects, objects, and permissions. The framework proposed in Amini *et al* [3] enables the specification of policy rules in both conceptual and ground levels. However [3] specifically states that access controls should be based on attributes as opposed to identity. In open distributed environments such as the web ABAC is the ideal choice however, in an enterprise setting we can't ignore the RBAC that already exist in the underlying data sources. None of the authors consider either TBAC or a combination of access control models.

Enforcement Framework. A framework is made up of the syntax of the policy language, the semantics of the language and the execution model. In addition to providing a mechanism to represent and enforce the criteria specified by the policy constructs, features such as distributivity, propagation and inference, policy merging, conflict management, exception handling and provenance need to be catered for by the framework. At first glance it may appear as if well known access control frameworks such as KAos [12], Rein [13] and PROTUNE [5] support many of the aforementioned features. However, as their underlying policy languages do not consider semantic relations between all of the policy entities the frameworks themselves can not fully exploit propagation and inference based on subject, objects and actions. The primary issues with (CLAC) [4] are the lack of both an underlying formalism and an enforcement mechanism. The policy language proposed by [1] is not based on well defined semantics and their solution does not scale well due to their centralised approach. Although [3] caters for many of the features listed above it does not support exception policies and more importantly their policy language is not based on standards and as such raises questions with respect to usability and interoperability. The framework detailed in [2] is the closest match for our requirements. However, they do not take existing policy representation and provenance into consideration and it is not clear the extent of their use of existing standard based technologies. [1] and [3] propose access control models and enforcement frameworks for linked data networks. However they both dismissed RBAC stating that is it not suitable for semantically aware environments. We argue that linked data network access control framework must cater for various access models e.g. RBAC, TBAC and ABAC. In addition, in an enterprise setting it makes sense to populate the linked data network based on data stored in content repositories and LOB applications. None of the authors have provided any guidance with respect to the automatic population of the linked data network in general and the access controls in particular, both of which we plan to address in our research.

4 Conclusions and Future Work

Our research explores the application of access control to a linked data network. We identify the need for a policy language which can represent multiple access control models and highlight the importance of representing existing access control policies in the linked data network. The research problem, motivation, objectives and the approach have been described in this paper. The next step is to analyse existing access control models, languages, frameworks and standards in detail and to generate a detailed design based on both the requirements and the results of our in depth analysis. The prototype will subsequently be developed based on the design. Once we have completed the development of the prototype we will investigate if the artefact meets the objectives or if further iterations of design, development and reflection are required.

Acknowledgements. This work is supported in part by the Science Foundation Ireland under Grant No. SFI/08/CE/I1380 (Lion-2), the Irish Research Council for Science, Engineering and Technology Enterprise Partnership Scheme and Storm Technology Ltd.

References

1. Javanmardi, S., Amini, M., Jalili, R., GanjiSaffar, Y.: SBAC: A Semantic Based Access Control Model. In: 11th Nordic Workshop on Secure IT-systems (NordSec 2006), Linkping, Sweden (2006)
2. Ryutov, T., Kichkaylo, T., Neches, R.: Access Control Policies for Semantic Networks. In: 2009 IEEE International Symposium on Policies for Distributed Systems and Networks, pp. 150–157. IEEE (July 2009)
3. Amini, M., Jalili, R.: Multi-level authorisation model and framework for distributed semantic-aware environments. IET Information Security 4(4), 301 (2010)
4. Qin, L., Atluri, V.: Concept-level access control for the Semantic Web. In: Proceedings of the 2003 ACM Workshop on XML Security - XMLSEC 2003, Number Cimic, p. 94. ACM Press (2003)
5. Bonatti, P.A., De Coi, J.L., Olmedilla, D., Sauro, L.: Rule-Based Policy Representations and Reasoning. In: Bry, F., Małuszyński, J. (eds.) Semantic Techniques for the Web. LNCS, vol. 5500, pp. 201–232. Springer, Heidelberg (2009)
6. Peffers, K., Tuunanen, T., Rothenberger, M.: A design science research methodology for information systems research. Management Information Systems 24, 45–77 (2007)
7. Checkland, P., Holwell, S.: Action Research: Its Nature and Validity. Systemic Practice and Action Research 11(1), 9–21 (1998)
8. Sandhu, R., Coyne, E., Feinstein, H., Youman, C.: Role-based access control: a multi-dimensional view. In: Tenth Annual Computer Security Applications Conference, pp. 54–62 (1994)
9. Thomas, R., Sandhu, R.: Task-based authorization controls (TBAC): A family of models for active and enterprise-oriented authorization management. Database Security, 166–181 (1998)
10. McCollum, C., Messing, J., Notargiacomo, L.: Beyond the pale of MAC and DAC-defining new forms of access control. In: Proceedings of 1990 IEEE Computer Society Symposium on Research in Security and Privacy, 1990, pp. 190–200. IEEE (1990)
11. Yague, M., Maña, A., López, J., Troya, J.: Applying the semantic web layers to access control. In: Proceedings of 14th International Workshop on Database and Expert Systems Applications, 2003, pp. 622–626. IEEE (2003)
12. Bradshaw, J., Dutfield, S., Benoit, P., Woolley, J.: KAoS: Toward an industrial-strength open agent architecture. In: Software Agents, pp. 375–418 (1997)
13. Kagal, L., Finin, T.: A policy language for a pervasive computing environment. In: Proceedings POLICY 2003. IEEE 4th International Workshop on Policies for Distributed Systems and Networks, pp. 63–74. IEEE Comput. Soc. (2003)
14. Kolovski, V., Hendler, J., Parsia, B.: Analyzing web access control policies. In: Proceedings of the 16th International Conference on World Wide Web WWW 2007, p. 677 (2007)

DC Proposal: Online Analytical Processing of Statistical Linked Data

Benedikt Kämpgen

Institute AIFB, Karlsruhe Institute of Technology, 76128 Karlsruhe, Germany
benedikt.kaempgen@kit.edu

Abstract. The amount of Linked Data containing statistics is increasing; and so is the need for concepts of analysing these statistics. Yet, there are challenges, e.g., discovering datasets, integrating data of different granularities, or selecting mathematical functions. To automatically, flexibly, and scalable integrate statistical Linked Data for expressive and reliable analysis, we propose to use expressive Semantic Web ontologies to build and evolve a well-interlinked conceptual model of statistical data for Online Analytical Processing.

1 Introduction

An important part of the Semantic Web comprises statistical Linked Data (SLD). Typically, SLD contain dimensions of metadata some of which hold temporal properties, e.g., for time-series. Also, most SLD contain numerical values that may represent aggregations from raw operational data and that often are further aggregated for analysis. According to the Linked Data principles[1] SLD should use unambiguous URIs for all relevant entities, e.g., datasets; entity URIs should be resolvable using the HTTP protocol to offer useful metadata, e.g., the location where data points of a dataset can be found; metadata should use Semantic Web standards such as RDF and SPARQL to be understandable to machines; and data should be reusing URIs from other datasources, e.g., so that relationships between datasets can be discovered.

Using Linked Data principles for publishing and consuming statistical data for decision support bears advantages such as easier integration and enrichment with other datasources. Efforts such as the Linking Open Data project, data.gov, and data.gov.uk have resulted in the release of useful SLD. Such open data is not restricted to any specific usage, its full values can be unlocked, driving innovation. First projects have demonstrated useful consumption of SLD, e.g. the Open Data Challenge.

Thus, we assume that the amount of Linked Data containing statistical information will be increasing; and so will the need for concepts of consuming SLD. There are still open questions of how to publish SLD. For instance, the W3C Working Group on Government Linked Data[2] is working on best practices and

---

[1] http://www.w3.org/DesignIssues/LinkedData
[2] http://www.w3.org/2011/gld/charter.html

L. Aroyo et al. (Eds.): ISWC 2011, Part II, LNCS 7032, pp. 301–308, 2011.
© Springer-Verlag Berlin Heidelberg 2011

standard vocabularies to publish SLD from governmental institutions. Although publication issues will be inherently important for our work, we want to focus on concepts of consumption, namely the analysis of SLD.

Consider the task of comparing metrics that quantify a country's well-being with numbers that describe employees' perceived satisfaction at work. Here, we want to integrate, for instance, the European Commission's publication of the Gross Domestic Product growth of all European countries per year as provided by Eurostat, and a dataset with survey data about employees' fear of unemployment in the last few years, also published as Linked Data.

Research Question. When we try to fulfil similar scenarios of analysing SLD, we encounter specific challenges:

Distributed Datasources. Single information pieces about datasets may be distributed over servers and files and published by different parties and in different formats. Permanent availability and performance is not guaranteed.

Heterogeneous Datasets. Several heterogeneous ontologies for describing SLD are in use. There is no common agreement on how to make important aspects of statistical data self-descriptive, e.g., hierarchies of categorisations and conversion or aggregation functions.

Varying Data Quality. SLD may be incomplete, inaccurate, sparse, imprecise and uncertain; best-effort answers may be required. Generally, data may not be self-descriptive enough to aid machines in interpretation and analysis. Again, there is no common agreement on how to attach sufficient provenance information to data. Aggregating data with varying quality should be transparent for users. Also, data may imply access restrictions that need to be considered.

Scale of Linked Data. Datasets may be large and need to be explored iteratively and interactively; direct querying and analysis of SLD using ad-hoc queries on the SLD will not scale. Data warehouses are needed to (temporarily) store, pre-process, and analyse the data. Also, values may need to be pre-computed for fast look-ups of calculations. However, SLD are dynamic and may be updated or refined continuously; this requires an automatic approach to building and evolving data warehouses.

Based on these challenges, we want to approach the following research question: How to automatically (with few manual effort), flexibly (integrating many heterogeneous datasources), and fast (in comparison to other integration systems) integrate distributed SLD to one conceptual model for expressive (e.g., aggregations by hierarchies, conversions of metrics, complex calculations) and reliable (e.g., transparent best-effort answers) analysis.

Approach. To sufficiently fulfil our research question, we propose to exploit expressive Semantic Web ontologies to automatically map SLD to a well-interlinked conceptual model, so that the data can be analysed using Online Analytical Processing (OLAP).

OLAP is a commonly used decision support analysis method characterized by a multidimensional view of data and interactive exploration of data using simple to understand but data-complex queries, e.g., selection, drill-down/roll-up, and slice/dice[17]. The necessary Multidimensional Model (MDM) describes statistical data by data points, Facts, in a coordination system forming a Hypercube (Cube) of n axes, or Dimensions. Dimension Values can be grouped along Hierarchies of one or more Levels. Dimensions also can be Measures. If subsuming sets of Facts, Measures are aggregated using aggregation functions, e.g. *sum*. Cubes that share Dimensions and Values are put together into Multicubes.

Figure 1 shows an MDM that fulfils our scenario. Here, the average GDP and the cumulated number of answers given in the employment survey are made comparable in a Multicube with shared geographic and temporal Dimensions.

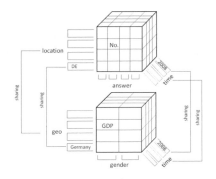

Fig. 1. Example of a Multicube

For OLAP, typically, data is extracted from heterogeneous data sources, transformed into a well-interlinked MDM, and more or less temporarily loaded into a data warehouse. Common problems[14] include: Manual effort needed for developing and maintaining such ETL pipelines; semantic gap between conceptual model and logical implementation; and inflexibility to change. Semantic Web ontologies allow to make data self-descriptive; to represent consensus about the meaning of data; to find implicit knowledge and inconsistencies; and to ease the integration effort. Although usage of Semantic Web ontologies is not explicitly required by the Linked Data principles, we assume that expressive ontological structures will make it possible to overcome the challenges of SLD analysis[10].

2   Related Work

Our research question has been addressed by roughly two kinds of work: approaches to integrate and analyse statistical data from the Web; and approaches to apply Semantic Web concepts to data warehousing and OLAP.

Publishing and consuming statistical data over the Web often is based on XML[12]. There are XML standards to transfer statistical information, e.g.,

XBRL, SDMX, and DDI. However, these approaches have problems with integrating heterogeneous datasources. They lack the concept of semantically describing statistical data. Efforts to apply Semantic Web concepts to such standards are at an early stage (e.g., SDMX[1]). Other related approaches retrieve statistical information from the Web, automatically integrate the data and let the user analyse it: Google Squared, Google Refine, and Needlebase use keyword searches and structured background information to structure data from the Web in tables. They rely more on concepts and techniques from Information Retrieval, Machine Learning, NLP and Pattern Matching, and less on ontologies and Linked Data. Google Public Data Explorer allows expressive analyses. There is work on analysing Linked Data about sensors, however, it does not allow expressive queries[13].

Niinimäki and Niemi[9] describe an ETL approach to first transform data into an ontology for multidimensional models and then serialise the ontology for use with an Online Analytical Processing server for expressive analysis. They put much focus on their specific ontology, which directly models a multidimensional model and which needs to be deployed manually for the statistical data at hand. With SLD, manually mapping the data to a conceptual model is not an option. We intend to use SLD that is sufficiently semantically described to be automatically mapped to a meaningful conceptual model. There is recent work on creating data warehouses using general ontologies[15,7]; however, they do not deal with the problem of integrating datasets described by heterogeneous ontologies. Nebot et al.[6] do so, however, they limit their work to static datasources, which is not realistic with Linked Data. Also, they require the user to manually control the building of a conceptual model; our work focuses on automatically retrieving a valid conceptual model from SLD. Much work regarding consumption of SLD has been done for Semantic Sensor data. There is work on OLAP for Semantic Web ontologies describing sensor data, however, it is not dealing with challenges of Linked Data[16].

3   Research Plan

In this section, we describe in more detail our approach of analysing SLD, and also, how we intend to measure our success. There are several Multidimensional Models (MDM)[11] with different expressivity and focus. So far, no MDM has been adopted as a standard[14]. Similarly, there are several ontologies but no commonly agreed standard to describe SLD[18].

In previous work[2], we have developed a proof-of-concept mapping between a basic MDM and the RDF Data Cube vocabulary, an ontology that is already used by some publishers of SLD. We have implemented this mapping and used the prototype in experiments with real world data for a preliminary evaluation. This approach, however, does not cope with our mentioned challenges of SLD. In the following, we describe our plan to extend our approach towards this. Mostly, this will require to automatically build and evolve more expressive MDMs from SLD.

Distributed Datasources. require us to find, select and retrieve datasets. In our current system, an analysis is started from URIs of datasets to be integrated and analysed. An analysis could also start from a business question, e.g., a multidimensional query. The system then would automatically look for suitable datasets that can answer the query. Datasets and ontologies can be found in repositories and catalogs such as CKAN, or by Semantic Search engines such as Sindice and be automacally matched to users' information needs.

For instance, if we want to compare the GDP with survey results measuring the people's fear of becoming unemployed, datasets containing such metrics could be automatically added to the MDM. Also, if one dataset only contains the relevant measures for one country, additional datasets covering other countries could be recommended.

URIs of datasets are resolved to retrieve information about the datasets. This may provide new URIs, which are resolved, iteratively. At the moment, URIs are not distinguished; if once collected, they are every time used for querying the datasets, resulting in longer query times as actually needed. Also, at the moment, we do not consider data that is available in a form other than plain RDF.

Heterogeneous Datasets. require us to integrate SLD using various ontologies. At the moment, our mapping only supports a certain ontology. Datasets may even be described without any specific ontology for statistical data but still bear interesting statistics. For instance, datasets describing large numbers of people or institutions – such as the Billion Triple Challenge dataset – contain useful statistics, e.g., for each pair of institutions the number of people that know each other. Similarly of interest may be Linked Data and ontologies for geo-spatial, sensor and social-network data. In order to integrate information from different datasources a more complex MDM may be needed. E.g., many MDMs only support many-to-one relationships between Facts and Values of one specific Dimension. In real world scenarios, many-to-many relationships are possible, e.g., a patient having several diagnoses at the same time.

There are many possible heterogeneity issues when integrating SLD, e.g., how to handle time aspects such as the notion of "now". Or how to handle special-purpose values such as "unknown", "explicitly not inserted", and "not applicable". Also, different levels of granularity regarding hierarchies and calculations may need to be aligned for integration:

Hierarchies aid the user to retrieve correct and useful information. At the moment, most Dimensions only have one Hierarchy and Level, represented by *rdfs:label* from the Dimension Value. Only for time dimensions, we consider the natural time hierarchy of year, month and day. More complex Hierarchies may be useful[4]. For instance a Cube of people having their Hierarchy of supervisors as one Dimension. The supervisor Dimension is asymmetric as it may contain varying numbers of Levels, depending on the person. Linked Data provides ontologies to explicitly describe hierarchies, however, current datasets do not make use of this. Hierarchies may be contained implicitly and retrieved and enriched automatically using other sources[5].

Calculations are implicitly contained in Measures queried from an MDM. It is still an open problem how to automatically retrieve useful aggregation functions. At the moment, we use a simple heuristic to determine aggregation functions for a Measure: For each possible aggregation function we create a Measure; e.g., sum, avg, min, max, count, count, and distinct count for numerical values. However, not all aggregation functions make sense, e.g., to use as aggregation function the sum operator for a Measure giving the current stock of a product in a certain period of time. This is known as summarizability problem. Ontological structures can be of use, here[7]. Similarly, there is no common agreement on how to represent and convert between heterogeneous representations of mathematical information[18]. For instance, this would require to state how Measures were created and to uniquely represent Measure attributes such as units. Another open issue is to represent and share complex Measures over heterogeneous datasets, e.g., *precision* and *recall* in one dataset to analyse another.

Varying Data Quality. requires us to integrate incomplete, inaccurate, sparse, imprecise and uncertain information and to give best-effort answers to business questions. For instance, missing values could be filled with most probable values or values from another, less trusted source. Yet, the process of automatically selecting values to be integrated and aggregated should be transparent and comprehensible. Also, data may imply access restrictions that need to be considered. For that, the MDM could be enriched with information describing users, privileges and policies that serve to articulate an access control and audit (ACA) policy[10].

Scale of Linked Data. requires us to incrementally build and update data warehouses. In our current system no versioning of the MDM is done. However, SLD and ontologies may be continuously changing, e.g., new Facts added, Measures corrected, and Dimensions modified. Queries over such changes may be interesting, e.g., whether a Fact Dimension Value has been modified several times (known as "slowly changing dimensions"[11]). User queries may allow to restrict the search space of possibly useful MDMs[8].

For evaluation, we will implement an information system and analyse its capability to automatically, flexibly, and scalable allow expressive and reliable analysis of SLD. See Figure 2 for our planned integration system architecture which we have generalized from an early prototype[2].

The user queries for answers from SLD using an OLAP client (1). To answer this query, an RDF/SPARQL engine finds, selects, and retrieves SLD (2). In an integration engine an MDM is built or updated as a common conceptual model for the retrieved data (3). Our work will be centered around this integration engine. The MDM may be serialized in an OLAP server (4.1). Then, either this OLAP server (4.2) or the MDM directly gives the answer (5).

To overcome the challenges in analysing SLD, we intend to represent the MDM as an expressive Semantic Web ontology and to make use of concepts and techniques from fields such as Ontology Engineering and Matching, as well as Reasoning.

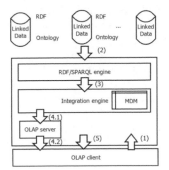

Fig. 2. Integration System Architecture

We do not intend to evaluate our concepts using qualitative usage analyses. Still, we plan to use the system in real-world use cases and compare its suitability with other systems. More concretely, for an analysis task, we plan to compare the amount of manual effort, the amount of data available, the expressivity of possible business questions, the performance of the system, and the quality of given answers. Also, we consider benchmarks such as the Business Intelligence Use Case from the Berlin SPARQL Benchmark and general quality criteria for MDMs [3].

4   Conclusion

We have proposed to exploit expressive Semantic Web ontologies to automatically create a well-interlinked conceptual model from various sources of statistical Linked Data that can be interactively and reliably analysed using Online Analytical Processing (OLAP). We have described challenges that we plan to work on, e.g., discovering datasets, integrating data of different granularities, or selecting mathematical functions. For evaluation, we intend to implement an information system suitable for real-world scenarios.

Acknowledgements. This work was supported by the German Ministry of Education and Research (BMBF) within the SMART project (Ref. 02WM0800) and the European Community's Seventh Framework Programme FP7/2007-2013 (PlanetData, Grant 257641). I thank Andreas Harth, Elena Simperl, and Denny Vrandečić for guidance and insights.

References

1. Cyganiak, R., Field, S., Gregory, A., Halb, W., Tennison, J.: Semantic Statistics: Bringing Together SDMX and SCOVO. In: Proceedings of the WWW 2010 Workshop on Linked Data on the Web, pp. 2–6 (2010)

2. Kämpgen, B., Harth, A.: Transforming Statistical Linked Data for Use in OLAP Systems. In: Proceedings of the 7th International Conference on Semantic Systems. I-SEMANTICS 2011. ACM (2011)
3. Lechtenbörger, J., Vossen, G.: Multidimensional normal forms for data warehouse design. Information Systems Journal 28(5), 415–434 (2003)
4. Malinowski, E., Zimányi, E.: Hierarchies in a multidimensional model: From conceptual modeling to logical representation. Data Knowl. Eng. 59, 348–377 (2006)
5. Mazón, J.N., Trujillo, J., Serrano, M., Piattini, M.: Improving the Development of Data Warehouses by Enriching Dimension Hierarchies with WordNet. In: Collard, M. (ed.) ODBIS 2005/2006. LNCS, vol. 4623, pp. 85–101. Springer, Heidelberg (2007)
6. Nebot, V., Berlanga, R., Pérez, J.M., Aramburu, M.J., Pedersen, T.B.: Multidimensional Integrated Ontologies: A Framework for Designing Semantic Data Warehouses. In: Spaccapietra, S., Zimányi, E., Song, I.-Y. (eds.) Journal on Data Semantics XIII. LNCS, vol. 5530, pp. 1–36. Springer, Heidelberg (2009)
7. Niemi, T., Niinimäki, M.: Ontologies and summarizability in OLAP. In: Proceedings of the 2010 ACM Symposium on Applied Computing SAC 2010, p. 1349 (2010)
8. Niemi, T., Nummenmaa, J., Thanisch, P.: Constructing OLAP cubes based on queries. In: Proceedings of the 4th ACM international workshop on Data warehousing and OLAP. DOLAP 2001. ACM (2001)
9. Niinimäki, M., Niemi, T.: An ETL Process for OLAP Using RDF/OWL Ontologies. In: Spaccapietra, S., Zimányi, E., Song, I.-Y. (eds.) Journal on Data Semantics XIII. LNCS, vol. 5530, pp. 97–119. Springer, Heidelberg (2009)
10. Pardillo, J., Mazón, J.N.: Using Ontologies for the Design of Data Warehouses. Journal of Database Management 3(2), 73–87 (2011)
11. Pedersen, T.B., Jensen, C., Dyreson, C.E.: A foundation for capturing and querying complex multidimensional data. Information Systems Journal 26, 383–423 (2001)
12. Perez, J.M., Berlanga, R., Aramburu, M.J., Pedersen, T.B.: Integrating Data Warehouses with Web Data: A Survey. IEEE Transactions on Knowledge and Data Engineering 20, 940–955 (2008)
13. Phuoc, D.L., Hauswirth, M.: Linked Open Data in Sensor Data Mashups. In: Proceedings of the 2nd International Workshop on Semantic Sensor Networks (SSN 2009) in conjunction with ISWC 2009 (2009)
14. Rizzi, S., Abelló, A., Lechtenbörger, J., Trujillo, J.: Research in data warehouse modeling and design: dead or alive? In: Proceedings of the 9th ACM International Workshop on Data Warehousing and OLAP, pp. 3–10 (2006)
15. Romero, O., Abelló, A.: Automating Multidimensional Design from Ontologies. In: Proceedings of the ACM Tenth International Workshop on Data Warehousing and OLAP DOLAP 2007 (2007)
16. Shah, N., Tsai, C.F., Marinov, M., Cooper, J., Vitliemov, P., Chao, K.M.: Ontological On-line Analytical Processing for Integrating Energy Sensor Data. Iete Technical Review 26, 375 (2009)
17. Vassiliadis, P.: Modeling Multidimensional Databases, Cubes and Cube Operations. In: Proc. of the 10th SSDBM Conference. pp. 53–62 (1998)
18. Vrandečić, D., Lange, C., Hausenblas, M., Bao, J., Ding, L.: Semantics of Governmental Statistics Data. In: Proceedings of the WebSci 2010 (2010)

DC Proposal: Model for News Filtering with Named Entities

Ivo Lašek

Czech Technical University in Prague, Faculty of Information Technology,
Prague, Czech Republic
lasekivo@fit.cvut.cz

Abstract. In this paper we introduce the project of our PhD thesis. The subject is a model for news articles filtering. We propose a framework combining information about named entities extracted from news articles with article texts. Named entities are enriched with additional attributes crawled from semantic web resources. These properties are then used to enhance the filtering results. We described various ways of a user profile creation, using our model. This should enable news filtering covering any specific user needs. We report on some preliminary experiments and propose a complex experimental environment and different measures.

Keywords: Information Filtering, User Modelling, Evaluation Methods.

1 Introduction

We are flooded with information nowadays. No one is able to keep track of all news articles in the world. Tools enabling us to filter the every day information stream are required.

This paper addresses the problem of news information filtering. We propose a framework, which maintains a complex user profile in order to perform content based news filtering. The user profile is put together not only based on traditional information retrieval techniques. We count also with semantic information hidden behind named entities that can be extracted from the article text. The additional semantic information and traditional information retrieval techniques are put together to form a unified news filtering framework.

2 Main Contributions

- *More detailed articles filtering*, not only based on predefined categories. The categorization is determined based on a concrete user feedback.
- *Model combining textual (unstructured) and semantic information.* Rather than representing an article as a bag of words, we represent it as a bag of contained named entities and their properties.
- We implemented *a prototype of the framework for news filtering based on extracted named entities.*
- We propose *evaluation methods* to test our model and scenarios for user experiments.

L. Aroyo et al. (Eds.): ISWC 2011, Part II, LNCS 7032, pp. 309–316, 2011.
© Springer-Verlag Berlin Heidelberg 2011

3   Related Work

There are two main approaches to news information filtering. To some extent domain independent collaborative filtering was used in [2] and earlier in [3]. However in the case of news filtering, the collaborative filtering has significant drawbacks as described in [1]. This approach tends to recommend generally popular topics at the expense of the more specific ones. Also, it takes some time, before the system learns the preferences on a new article.

Contrary, the other approach - content based filtering - is able to recommend a new article practically immediately. Good comparison of content based filtering approaches is given in [4]. Inspirational examples of news filtering using Bayesian classifier are described in [5,6,7]. In [6] and [7], authors distinguish long-term and short-term interest. To learn the long-term interest a Bayesian classifier is used. For the short-term, the nearest neighbour algorithm is used. The articles are transformed to tf-idf vectors and then compared based on a cosine similarity measure.

Another related problem is a recommendation of related articles mentioned in [18]. Articles are recommended not only based on their similarity, but also according to their novelty and coherence. A comparison of various information retrieval techniques for such recommendation is provided in [8]. Used metrics for modelling the relatedness of articles include apart from traditional cosine similarity of term vectors also BM25 [19] and Language Modelling [20]. Here, often one document is used as a query to find related documents. We use different approach in this sense. In our case, the user profile serves as a query to filter relevant incoming articles.

An interesting example of exploitation of semantic information in connection with unstructured text is presented by BBC [9]. In this case, semantic data obtained from DBpedia serve to interlink related articles, through identifying similar topics. The topics are determined based on extracted named entities, mapped to DBpedia ontology. We extend this idea and build a user profile, using the data obtained from semantic web resources.

4   Proposed Methodology

4.1   Articles Processing Pipeline

The articles processing pipeline is shown in Fig. 1. First, we collect news articles using RSS[1] feeds. Sometimes RSS feeds contain only a fragment of the whole article. In this case, the rest of the article is automatically downloaded from its original web page. We analyse the DOM tree of the web page and locate non-repeating blocks, containing bigger amount of text. We build on heuristics introduced in the work of our colleagues [10].

---

[1] Really Simple Syndication - a family of web feed formats used to publish frequently updated works.

Fig. 1. The articles processing pipeline

In downloaded articles, named entities are identified. A ready made tool is used. Currently, we delegate this task to OpenCalais[2]. Apart from named entities themselves, OpenCalais provides often their basic attributes and links to other Linked Data[3] resources containing additional information too.

Currently, we evaluate the tool made by our colleagues for annotation of named entities in news articles [22]. Thus users get the possibility to mark additional entities, the system was not able to identify. The annotation tool is currently implemented as a plug-in to a web browser.

Extracted entities and their properties (if available) are used to query additional Linked Data resources (e.g. DBpedia or Freebase). If there are some relevant resources, additional information about extracted entities is crawled.

Articles modelling and user profile creation is cowered in detail in the following Section 5.

4.2 User Feedback Collection

In order to build a user profile, we need to collect the user feedback about filtered articles. In the initial learning phase, the user may provide general information about her interests. This is done by providing RSS feeds of news portals, he usually reads. As if she was using an ordinary RSS reader.

This initial setup partially overcomes the cold start problem.

When we talk about user feedback, we mean user rating of the article at the scale from 1 to 5, assuming an explicit user feedback.

5 Modelling News Articles

To model the content of an article, we distinguish three types of features: Subject-Verb-Object triples (SVO), terms (like in the information retrieval vector model) and named entities together with their properties.

For weighting of *terms*, we use ordinary tf-idf metric [12]. The results are normalized to range from 0 to 1. In case of *SVO triples*, we use only the binary measure. Either the triple is present in an article (1) or it is not (0).

In our previous work, we used an adaptation of tf-idf for entities too. Entity identifiers were used instead of terms. The *entity frequency* (we denote it as *ef*) was then counted based on the number of occurrences of the entity in the article. However, during the course of our experiments, we observed that the idf part disqualifies some popular entities, because they are often mentioned. But the fact, that an entity is often mentioned does not mean it is less important for the user.

---

[2] OpenCalais. http://www.opencalais.com/
[3] Linked Data. http://linkeddata.org/

Often the opposite is true. This problem is the subject of our future experiments. Currently, we tend to omit the idf part and count only with entity frequencies as the weight. The frequency of an entity i in an article j is counted as follows:

$$ef_{i,j} = \frac{e_{i,j}}{\sum_k e_{k,j}} \tag{1}$$

In equation 1 $e_{i,j}$ is the number of occurrences of the considered entity in a particular article and the denominator is the sum of the number of occurrences of all entities identified in the document.

We use entities identified during the named entity extraction phase. Additionally, we gather their properties in the crawling phase. The properties of an entity are presented in the form of a predicate object pair. Each such a pair has its own identifier. Weights of properties correspond to frequencies of entities. The weight of property k in the context of an article j is computed as follows:

$$pf_{j,k} = \sum_{ef_{i,j} \in E_{j,k}} \alpha \times ef_{i,j} \tag{2}$$

Where $E_{j,k}$ is the set of entities contained in an article j, having property k. And α is the proportion of the importance of entity properties to the importance of the entity. In following examples, we count with $\alpha = 1$. Thus properties are equally important as entities.

Additionally, *an important feature of an article is its rating*, given by each user. Consider for example following two sentences representing two articles:

A1: Google launches a new social site.

A2: Microsoft recommends reinstalling Windows.

A possible representation of these two articles is denoted in Table 1, 2 and 3.

Table 1. Normalized term weights in an article

Article	google	launch	new	social	site	microsoft	recommend	reinstall	windows
A1	0.8	0.8	0.8	0.8	0.8	0	0	0	0
A2	0	0	0	0	0	1	1	1	1

Table 2. Subject-Verb-Object representation of an article

Article	Google-to-launch-site	Microsoft-to-recommend-reinstalling
A1	1	0
A2	0	1

Table 3. Entities and their properties representing an article

Article	Google	Microsoft	Windows	Google, Microsoft type:Company	locatedIn:USA	Windows type:Product
A1	$ef_{1,1} = 1$	$ef_{2,1} = 0$	$ef_{3,1} = 0$	$pf_{1,1} = 1$	$pf_{1,2} = 1$	$pf_{1,3} = 0$
A2	$ef_{1,2} = 0$	$ef_{2,2} = 0.5$	$ef_{3,2} = 0.5$	$pf_{2,1} = 0.5$	$pf_{2,2} = 0.5$	$pf_{2,3} = 0.5$

6   User Profile Creation

With this representation of articles, we may use various machine learning approaches to identify user needs. For some of the algorithms, the described representation using numeric feature weights is fine. Some of the algorithms (e.g. apriori) require features (or attributes) to be nominal.

To transform the model to suitable representation, we can use binary representation - simply register the presence or absence of a given feature. The other option preserving the semantic of various weights is to discretize weights, using predefined bins (e.g. low, medium, high).

Naive Approach. First approach, we were evaluating in our previous work [13], counted with only two types of user feedback (positive and negative). The user profile was composed of features extracted from articles rated by the user and is divided in two parts: P^+ (set of features extracted from positively rated articles) and P^- (set of features extracted from negatively rated articles).

So far, we counted only with entities and omitted SVO triples and terms. The rank of a new article j is then computed as follows:

$$rank_j = \sum_{entity_i \in P^+} ef_{i,j} - \sum_{entity_i \in P^-} ef_{i,j} \qquad (3)$$

If the rank is higher than a certain threshold, the article is considered as interesting for the user. This approach worked good for simple profiles. But with more complex user needs, only summing the weights is not flexible.

Apriori Algorithm. Apriori algorithm [14] may be applied to articles rated by a user. Having the representation described in Section 5, we can try to find association rules having the user rating of an article on its right side. A user profile is then composed of these rules. Sample rules may look like this:

```
type:Company ^ locatedIn:USA => rating4 (confidence 0.20)
subject:Google ^ Google type:Company => rating4 (confidence 0.80)
```

The first rule gives us the information that an article containing information about entity of type `Company` with the property `locatedIn` set to `USA` would the user rate with rating 4 with the confidence of 20%. The second rule reflects SVO triples.

Clustering. Interesting results achieved by using centroid based approach to document classification [15] inspired us to consider clustering as another method of creation of a user profile. Clustering may help to identify rather abstract concepts than concrete entities, the user is interested in. Given the model introduced in Section 5, clustering of articles rated by a particular user is performed. K-Means algorithm [16] is used for clustering. The cosine function is used to measure the similarity of vectors (SVO, terms and entities vectors) representing a particular articles.

For each cluster the average rating of articles contained in this cluster is computed. Any new article gets the average rating of the cluster it belongs to. We assign a new article to appropriate cluster separately for each type of features and then combine computed ratings:

$$rating_j = \alpha * rating_j^{SVO} + \beta * rating_j^{entities} + \gamma * rating_j^{terms} \qquad (4)$$

The coefficients α, β and γ sum to one.

Formal Concept Analysis. Application of formal concept analysis [17] to articles rated by a particular user to a common grade may bring interesting results. We propose to analyse properties of named entities contained in articles in order to identify common concepts. The user profile creation and evaluation composes of following steps:

- Collect articles rated by the user as interesting.
- Identify properties and entities contained in all the collected articles.
- Use the identified concepts to identify new articles, that would be rated on the same grade.

An opened question remains, if the formal concept analysis is not too restrictive in this scenario. In this context application of fuzzy concept lattices [21] may bring interesting results.

7   Test Data Collection

We intend to test the whole system in three different ways:

- *Golden standard* - We collect data that represent our golden standard using web browser plug-in to save user ratings of arbitrary articles. Precision and recall metrics as well as F-measure metric and Kendall's tau coefficient are used to evaluate results.
- *Explicit feedback collection* - While using the system and rating recommended articles, users provide an important feedback. It can be used to evaluate results of the system. Same metrics as for golden standard apply.
- *Implicit feedback collection* - Finally, the system may collect the implicit feedback too. One of possible metrics is the really opened (user have clicked on them) to total recommended articles ratio:

$$succ = \frac{\#opened\_articles}{\#recommended\_articles} \qquad (5)$$

8   Research Progress

We developed and tested a simple form of the proposed model consisting of named entities [13]. The user profile was constructed using the naive approach described in Section 6. Each entity was represented by its ef-idef (entity frequency - inverse document entity frequency) weights. We evaluated its ability to recommend one particular topic. The results were promising. However, the use case was constrained to one particular topic. We used only the extracted entities, without employing their properties. Such a system filters the news accurately, but it is too constrained. We believe, the use of entity properties may add the necessary generalization of extracted concepts. We implemented a framework to collect news articles, to identify named entities using OpenCalais and to crawl additional information about identified entities from semantic web resources. For crawling of semantic web resources, we use LDSpider [11].

9   Conclusion

In this paper, we introduced the idea of news information filtering, using not only the text of news articles, but also information hidden behind named entities. We introduced the unified model of articles for news filtering. We believe, our approach can be combined with current information retrieval methods and improve their results. In Section 6 we described our plan of future work and named various approaches to user profile creation, using the proposed model. Several evaluation methods were described.

Acknowledgements. This work has been partially supported by the grant of The Czech Science Foundation (GAČR) P202/10/0761 and by the grant of Czech Technical University in Prague registration number SGS11/085/OHK3/1T/18.

References

1. Liu, J., Dolan, P., Pedersen, E.R.: Personalized News Recommendation Based on Click Behavior. In: IUI 2010: Proceedings of the 2010 International Conference on Intelligent User Interfaces, pp. 31–40 (2010)
2. Das, A.S., Datar, M., Garg, A., Rajaram, S.: Google news personalization: scalable online collaborative filtering. In: Proceedings of the 16th International Conference on World Wide Web (WWW 2007), pp. 271–280. ACM, New York (2007)
3. Resnick, P., Iacovou, N., Suchak, M., Bergstrom, P., Riedl, J.: GroupLens: an open architecture for collaborative filtering of netnews. In: Proceedings of the 1994 ACM Conference on Computer Supported Cooperative Work (CSCW 1994), pp. 175–186. ACM, New York (1994)
4. Pazzani, M., Billsus, D.: Content-Based Recommendation Systems. In: Brusilovsky, P., Kobsa, A., Nejdl, W. (eds.) Adaptive Web 2007. LNCS, vol. 4321, pp. 325–341. Springer, Heidelberg (2007)
5. Carreira, R., Crato, J.M., Gonalves, D., Jorge, J.A.: Evaluating adaptive user profiles for news classification. In: Proceedings of the 9th International Conference on Intelligent User Interfaces, pp. 206–212. ACM (2004)
6. Billsus, D., Pazzani, M.: A Hybrid User Model for News Story Classification (1999)
7. Billsus, D., Pazzani, M.J.: User Modeling for Adaptive News Access. In: User Modeling and User-Adapted Interaction, vol. 10, pp. 147–180. Springer, Netherlands (2000)

8. Bogers, T., Bosch, A.: Comparing and evaluating information retrieval algorithms for news recommendation. In: Proceedings of the ACM Conference on Recommender Systems (2007), pp. 141–144 (2007)

9. Kobilarov, G., Scott, T., Raimond, Y., Oliver, S., Sizemore, C., Smethurst, M., Bizer, C., Lee, R.: Media Meets Semantic Web – How The BBC Uses DBpedia and Linked Data to Make Connections. In: Aroyo, L., Traverso, P., Ciravegna, F., Cimiano, P., Heath, T., Hyvönen, E., Mizoguchi, R., Oren, E., Sabou, M., Simperl, E. (eds.) ESWC 2009. LNCS, vol. 5554, pp. 723–737. Springer, Heidelberg (2009)

10. Maruščák, D., Novotný, R., Vojtáš, P.: Unsupervised Structured Web Data and Attribute Value Extraction. In: Proceedings of 8th Annual Conference Znalosti 2009, Brno (2009)

11. Robert, I., Jurgen, U., Christian, B., Andreas, H.: LDSpider: An open-source crawling framework for the Web of Linked Data. In: Proceedings of 9th International Semantic Web Conference (ISWC 2010). Springer, Heidelberg (2010)

12. Robertson, S.E., Jones, K.S.: Relevance weighting of search terms. Journal of the American Society for Information Science, 129–146 (1976)

13. Lašek, I., Vojtáš, P.: Semantic Information Filtering - Beyond Collaborative Filtering. In: 4th International Semantic Search Workshop (2011), http://km.aifb.kit.edu/ws/semsearch11/11.pdf (accessed June 13, 2011)

14. Agrawal, R., Imielienski, T., Swami, A.: Mining Association Rules between Sets of Items in Large Databases. In: Proceedings of Conference on Management of Data, pp. 207–216. ACM Press, New York (1993)

15. Han, E.-H., Karypis, G.: Centroid-Based Document Classification: Analysis and Experimental Results. In: Zighed, D.A., Komorowski, J., Żytkow, J.M. (eds.) PKDD 2000. LNCS (LNAI), vol. 1910, pp. 424–431. Springer, Heidelberg (2000)

16. MacQueen, J.B.: Some methods for classification and analysis of multivariate observations. In: Proceedings of the Fifth Symposium on Math, Statistics, and Probability, pp. 281–297. University of California Press, Berkeley (1967)

17. Ganter, B., Wille, R.: Formal Concept Analysis: Mathematical Foundations. Springer, Heidelberg (1999)

18. Lv, Y., Moon, T., Kolari, P., Zheng, Z., Wang, X., Chang, Y.: Learning to model relatedness for news recommendation. In: Proceedings of the 20th International Conference on World Wide Web, WWW 2011, pp. 57–66. ACM Press (2011)

19. Robertson, S., Walker, S.: Some simple effective approximations to the 2-poisson model for probabilistic weighted retrieval. In: Proceedings of SIGIR 1994, pp. 232–241. ACM Press, New York (1994)

20. Ponte, J.M., Croft, W.B.: A language modeling approach to information retrieval. In: Proceedings of SIGIR 1999, pp. 275–281. ACM Press, New York (1998)

21. Krajci, S., Krajciova, J.: Social Network and One-sided Fuzzy Concept Lattices. In: Proceedings of FUZZ-IEEE 2007, IEEE International Conference on Fuzzy Systems, pp. 1–6. Imperial College, London (2007)

22. Fišer, D.: Sémantická anotace doménově závislých dat. Katedra softwarového inženýrství, MFF UK (2011)

DC Proposal: Graphical Models and Probabilistic Reasoning for Generating Linked Data from Tables*

Varish Mulwad

Computer Science and Electrical Engineering
University of Maryland, Baltimore County
varish1@cs.umbc.edu

Abstract. Vast amounts of information is encoded in tables found in documents, on the Web, and in spreadsheets or databases. Integrating or searching over this information benefits from understanding its intended meaning and making it explicit in a semantic representation language like RDF. Most current approaches to generating Semantic Web representations from tables requires human input to create schemas and often results in graphs that do not follow best practices for linked data. Evidence for a table's meaning can be found in its column headers, cell values, implicit relations between columns, caption and surrounding text but also requires general and domain-specific background knowledge. We describe techniques grounded in graphical models and probabilistic reasoning to infer meaning associated with a table. Using background knowledge from the Linked Open Data cloud, we jointly infer the semantics of column headers, table cell values (e.g., strings and numbers) and relations between columns and represent the inferred meaning as graph of RDF triples. A table's meaning is thus captured by mapping columns to classes in an appropriate ontology, linking cell values to literal constants, implied measurements, or entities in the linked data cloud (existing or new) and discovering or and identifying relations between columns.

Keywords: Linked Data, Tables, Entity Linking, Machine Learning, Graphical Models.

1 Introduction

Most of the information found on the Web consists of text written in a conventional style, e.g., as news stories, blogs, reports, letters, advertisements, etc. There is also a significant amount of information encoded in structured forms like tables and spreadsheets, including stand-alone spreadsheets or table as well as tables embedded Web pages or other documents. Cafarella et al. [2] estimated that the Web contains over 150 million high quality relational tables. In some

---

* Advisor: Tim Finin, University of Maryland, Baltimore County. This research was supported in part by NSF awards 0326460 and 0910838, MURI award FA9550-08-1-0265 from AFOSR, and a gift from Microsoft Research.

L. Aroyo et al. (Eds.): ISWC 2011, Part II, LNCS 7032, pp. 317–324, 2011.
© Springer-Verlag Berlin Heidelberg 2011

ways, this information is easier to understand because of its structure but in other ways it is more difficult because it lacks the normal organization and context of narrative text. Both integrating or searching over this information will benefit from a better understanding of its intended meaning.

A wide variety of domains that are interesting both technically and from a business perspective have tabular data. These include medicine, healthcare, finance, e-science (e.g., biotechnology), and public policy. Key information in the literature of these domains, which can be very useful for informing public policy, is often encoded in tables. As a part of Open Data and transparency initiative, fourteen nations including the United States of America share data and information on websites like www.data.gov in structured format like CSV, XML. As of May 2011, there are nearly 390,000 raw datasets available. This represents a large source of knowledge, yet we do not have systems that can understand and exploit this knowledge.

In this research, we will present techniques to evaluate our claim that "It is possible to generate high quality linked data from tables by jointly inferring the semantics of column headers, values (string and literal) in table cells, relations between columns, augmented with background knowledge from open data sources such as the Linked Open Data cloud."

2 Motivation

Ever since its inception in 2001, the Semantic Web has laid strong foundations for representing and storing knowledge in machine understandable formats such as RDF and OWL. The principles of Linked Data further strengthens the move from a web of documents to a web of data.

While the Semantic Web was able to lay strong foundations, its growth remains slow because of the lack of quality of data available on the Semantic Web. Even though data.gov has more than 390,000 datasets, only 0.071% of those are available as RDF datasets. Existing technology on the Semantic Web either rely on user's knowledge or generate "low quality" data which in some cases is as useless as raw data. Our proposed framework can easily be used to convert legacy datasets stored in tabular formats to RDF and publish it on the Semantic Web and the Linked Data Cloud. Not only does our framework generate RDF from tables, but it also produces high quality linked RDF.

Many real world problems and applications can benefit from exploiting information stored in tables including evidence based medical research [14]. Its goal is to judge the efficacy of drug dosages and treatments by performing meta-analyses (i.e systematic reviews) over published literature and clinical trials. The process involves finding appropriate studies, extracting useful data from them and performing statistical analysis over the data to produce a evidence report.

Key information required to produce evidence reports include data such as patient demographics, drug dosage information, different types of drugs used, brands of the drugs used, number of patients cured with a particular dosage. Most of this information is encoded in tables, which are currently beyond the scope of regular text processing systems and search engines. By adding semantics to such tables, we can develop systems that can easily correlate, integrate and search over different tables from different studies to be combined for a single meta-analysis.

3   Related Work

Several systems have been implemented to generate Semantic Web data from databases [15,19,12], spreadsheets [5,7] and csv [3] . Virtually all are manual or semi-automated and none have focused on automatically generating *linked* RDF data. Current systems on the Semantic Web either require users to specify the mapping to translate relational data to RDF or systems that do it automatically focus only on a part of the table (like column header strings). These systems have mainly focused on relational databases or simple spreadsheets.

The key shortcoming in such systems is that they rely heavily on users and their knowledge of the Semantic Web. Most systems on the Semantic Web also do not automatically link classes and entities generated from their mapping to existing resources on the Semantic Web. The output of such systems turns out to be just "raw string data" represented as RDF, instead of generating high quality linked RDF.

In the web tables domain, Wang et al. [21] present a table understanding system which identifies a concept to be associated with the table based on the evidence provided by the column header and strings in the "entity column" of the table. The concepts come from their knowledge base Probase created from the text on the World Wide Web which can be noisy and "semantically poor" as compared to concepts from the Linked Open Data cloud.

Ventis et al. [20] identify concepts to be associated with the column headers in a table based on the evidence provided by strings in a given column. They also identify relations between the "subject column" and other columns in the table. However they also rely on a isA database they create from the text on the Web which can be noisy as well as "semantically poor".

Limaye et al. [8] present a probabilistic graphical model based framework that identifies concepts to be associated with the column headers, links table cell values to entities and identifies relations between columns with Yago as a background knowledge base.

None of the current table understanding systems propose or generate any form of linked data from the inferred meaning. A key missing component in current systems is tackling literal constants. The work mentioned above will work well with string based tables. To the best of our knowledge, no work has tackled the problem on interpreting literals in tables and using them as evidence in the table interpretation framework. The framework we present is complete automated interpretation of a table that focuses on all aspects of a table - column headers, row values, relations between columns. Our framework will tackle strings as well as literals.

City	State	Mayor	Population
Baltimore	MD	S.Rawlings	640,000
Philadelphia	PA	M.Nutter	1,500,000
New York	NY	M.Bloomberg	8,400,000
Boston	MA	T.Menino	610,000

(a)

```
@prefix rdfs: <http://www.w3.org/2000/01/rdf-schema#>.
@prefix dbpedia: <http://dbpedia.org/resource/>.
@prefix dbpedia-owl: <http://dbpedia.org/ontology/>.
@prefix dbpprop: <http://dbpedia.org/property/>.

"City"@en is rdfs:label of dbpedia-owl:City.
"State"@en is rdfs:label of dbpedia-owl:AdminstrativeRegion.
"Baltimore"@en is rdfs:label of dbpedia:Baltimore.
dbpedia:Baltimore a dbpedia-owl:City.
"MD"@en is rdfs:label of dbpedia:Maryland.
dbpedia:Maryland a dbpedia-owl:AdministrativeRegion.
```

(b)

Fig. 1. This example shows a simple table about cities in the United States and some output of the prototype system that represents the extracted information as linked data annotated with additional metadata

4   Interpreting a Table

Generating high quality linked data from a table requires understanding its intended meaning. The meaning of a table is often encoded in column headers, table cells and implicitly conveyed via structure and relations between columns in a table.

Consider the table shown in Figure 1(a). The column headers suggest the type of information in the columns: *city* and *state* might match classes in a target ontology such as DBpedia [1]; *mayor* and *population* could match properties in the same or related ontologies. Examining the data values, which are initially just strings, provides additional information that can confirm some possibilities and disambiguate between possibilities for others. For example, the strings in column one can be recognized as entity mentions that are instances of the *dbpedia-owl:Place* class. Additional analysis can automatically generate a narrower description such as major cities located in the United States.

Consider the strings in column three. The string by themselves suggest that they are politicians. The column header provides additional evidence and better interpretation that the strings in column three are actually mayors. Discovering relations between columns is important as well. By identifying relation between column one and column three, we can infer that the strings in column three are mayors of cities presented in column one. Linking the table cell values to known entities enriches the table further. Linking S.Rawlings to *dbpedia:Stephanie_C._Rawlings-Blake*, T.Menino to *dbpedia:Thomas_Menino*, M.Nutter to *dbpedia:Michael_Nutter* we can automatically infer additional information that all three belong to the Democratic party, since the information will be associated with the linked entities.

Column four in this table presents literal values. The numbers in the column are values of the property *dbpedia-owl:populationTotal* and this property can be associated with the cities in column one. All the values in the column are in the range of 100,000. They provide evidence that the column may be representing the property population. Once relation between column one and column four is discovered, we can also look up on DBpedia, where the linked cities in column one will further confirm that the numbers represent population of the respective cities.

Producing an overall interpretation of a table is a complex task that requires developing an overall understanding of the intended meaning of the table as well as attention to the details of choosing the right URIs to represent both the schema as well as instances. We break down the process into following tasks: (a) assign every column (or row header) a class label from an appropriate ontology; (b) link table cell values to appropriate linked data entities, if possible; (c) discover relationships between the table columns and link them to linked data properties; and (d) generate a linked data representation of the inferred data.

5   Approach

We first developed a baseline system to evaluate the feasibility of tackling the problem. The baseline system is a sequential multi-step framework which first maps every column header to a class from an appropriate ontology. Using the predicted class as additional evidence, the frameworks then links table cell values to entities from the Linked Data Cloud. The final step in the framework is discovering relations between table columns. Once this information is inferred, the framework generates a linked data representation of the interpretation (see figure 1(b)). The details of the baseline system and its evaluation is described in [11]. Based on the evaluation of our baseline system, we present a framework grounded in the principles of graphical models and probabilistic reasoning.

The baseline system makes local decision at each step of the framework. The disadvantage of such a system is that error can percolate from the previous phase to the next phase in the system thus leading to an overall poor interpretation of a table. To overcome this problem, we need to develop a framework that performs joint inference over the evidence available in the table and jointly assign values to the column headers, table cells and relations between columns. Probabilistic graphical models [6] provide a convenient framework for expressing a joint probability over a set of variables and perform inferencing over them.

Constructing a graphical model involves the following steps: a) Identifying variables in the system b) Identifying interactions between variables and representing it as a graph c) Parametrizing the graphical structure d) Selecting an appropriate algorithm for inferencing. In the following sections we first present our work on the first three tasks in constructing a graphical model.

Variables in the System. The column headers, the table cells and the relation between columns in a table represent the set of variables in an interpretation framework.

Graphical Representation. We choose a Markov network based graphical representation,since the interaction between the column headers, table cell values and relation between table columns are symmetrical. The interaction between a column header and cell values in the column is captured by inserting an edge between the column header and each of the values in the column in the graph. To correctly disambiguate what a table cell value is, evidence from the rest of the values in the same row can be used. This is captured by inserting edges between every pair of cell values in a given row. Similar interaction exists between the column headers and is captured by the edges between every pair of table column headers.

Parametrizing the Network. To represent the distribution associated with the graph structure, we need to parametrize the structure. One way to parametrize a Markov network is representing the graph as a factor graph. A factor graph is an undirected graph containing two types of nodes : variable nodes and factor nodes. A factor node captures and computes the afinity between the variables interacting at that factor node. Variable nodes can also have associated "node potentials". Our parametrized model has two node potentials - ψ_1 and ψ_2 and three factor nodes - ψ_3, ψ_4 and ψ_5.

The node potential ψ_1 captures the affinity between the column header string in the table and the class to which the column header is being mapped, i.e., the affinity between *State* and *dbpedia:AdminstrativeRegion*. We define ψ_1 as the exponential of the product of a weight vector and a feature vector computed for column header. Thus, $\psi_1 = exp(w_1^T.f_1(C_i, L_{C_i}))$, where w_1 is the weight vector, L_{C_i} is the class label associated with column header C_i. The feature vector f_1 is composed of the following features : the Levenshtein distance, Dice score [16], semantic similarity between the column header string and the class label and the information content of the class label. We use the semantic similarity measure defined in [4].

We also compute the Information content for a given class in an ontology. Based on the semantic similarity and information content measures defined in [13], the information content for a class in given ontology is defined as follows: $I.C(L_C) = -log_2[p(L_C)]$ where $p(L_C)$ is the probability of the class L_C. We compute the probability by counting the number of instances that belong to the class L_C and divide it by the total number of instances. More specific classes in an ontology present more information as compared to more general classes. For example it is better to infer that a column header is of type of *dbpedia-owl:City* as compared to inferring that as *dbpedia-owl:Place* or *owl:Thing*. The information content measure precisely captures that.

ψ_2 is the node potential that captures affinity between the string in the table cell and the entity that the cell is being mapped to, for example, affinity between *Baltimore* and *dbpedia:Baltimore_Maryland*. We define ψ_2 as the exponential of the product of a weight vector and a feature vector computed for a cell value. Thus, $\psi_2 = exp(w_2^T.f_2(R_{i,j}, E_{i,j}))$, where w_2 is the weight vector, $E_{i,j}$ is the entity associated with the value in table cell $R_{i,j}$. The feature vector f_2 is composed of the following features : Levenshtein distance, Dice score, Page Rank, Page Length and Wikitology[17] index score. Along with a set of similarity metrics, we choose a set of popularity metrics, since when it is difficult to disambiguate the more popular entity is more likely the correct answer. Presently, the popularity metrics are Wikipedia based metrics, but these can be easily changed and adapted to a more general sense of popularity. The weight vectors w_1, w_2 can be learned using standard machine learning procedures.

ψ_3 is a factor node that captures the affinity between the class label assigned to a column header and the entities linked to the cell values in the same column. The behavior (function) of the factor node is still to be defined. ψ_4 is a factor node that captures the affinity between the entities linked to the values in the table cells in a given row in the table, i.e., the affinity between *dbpedia:Baltimore_Maryland, dbpedia:Maryland dbpedia:Stephanie_Rawlings-Blake*. We define ψ_4 as the product of Point wise mutual information between each pair of entities in the given row. Thus, $\psi_4 = \Pi_{i,k,i\neq k}PMI(E_{i,j}, E_{k,j})$, where $PMI(E_{i,j}, E_{k,j})$ computes the point wise mutual information between the entity $E_{i,j}$ in column i, row j and $E_{k,j}$ in column k, row j. Point wise mutual information between any two entities will provide us the association between the two – in some sense it is the probability of their co-occurrence. If the entities are associated (which will be the case in the context of table rows), PMI will be high. If the entities are not associated, PMI will be low.

ψ_5 is a factor node that captures the affinity between classes that have been assigned to all the column headers in the table, i.e., the affinity between *dbpedia-owl:City, dbpedia-owl:AdministrativeRegion dbpedia-owl:Mayor*. We again rely on point wise mutual information to capture the association between the class labels assigned to column headers. We define ψ_5 as the product of Point wise mutual information between each pair of column headers in the table.

6 Evaluation Plan

We will use two different types of evaluation to measure the effectiveness of our proposed techniques. In the first evaluation, we will compute accuracy for correctly predicted class labels for column headers, entities linked to table cells and relation between columns using the ground truth from the dataset of over 6000 web tables from [8]. While we have proposed a automatic framework for interpreting and representing tabular data as linked data, it may be helpful to develop a framework with human in the loop to make the linked data more useful and customized for certain applications. For such cases, our algorithms can generate a ranked list of candidates for each of the column headers, table cells and relation between columns. We will use Mean Average Precision [9] to compare the ranked list generated by our algorithms against a ranked list produced by human evaluators.

References

1. Bizer, C., Lehmann, J., Kobilarov, G., Auer, S., Becker, C., Cyganiak, R., Hellmann, S.: Dbpedia - a crystallization point for the web of data. Journal of Web Semantics 7(3), 154–165 (2009)
2. Cafarella, M.J., Halevy, A.Y., Wang, Z.D., Wu, E., Zhang, Y.: Webtables: exploring the power of tables on the web. PVLDB 1(1), 538–549 (2008)
3. Ding, L., DiFranzo, D., Graves, A., Michaelis, J.R., Li, X., McGuinness, D.L., Hendler, J.A.: Twc data-gov corpus: incrementally generating linked government data from data.gov. In: Proc 19th Int. Conf. on the World Wide Web, pp. 1383–1386. ACM, New York (2010)

4. Han, L., Finin, T., McNamee, P., Joshi, A., Yesha, Y.: Improved pmi utility on word similarity using estimates of word polysemy. TKDE (2011) (under review)
5. Han, L., Finin, T., Parr, C., Sachs, J., Joshi, A.: RDF123: From Spreadsheets to RDF. In: Sheth, A.P., Staab, S., Dean, M., Paolucci, M., Maynard, D., Finin, T., Thirunarayan, K. (eds.) ISWC 2008. LNCS, vol. 5318, pp. 451–466. Springer, Heidelberg (2008)
6. Koller, D., Friedman, N.: Probabilistic Graphical Models: Principles and Techniques. MIT Press (2009)
7. Langegger, A., Wöß, W.: XLWrap – Querying and Integrating Arbitrary Spreadsheets with SPARQL. In: Bernstein, A., Karger, D.R., Heath, T., Feigenbaum, L., Maynard, D., Motta, E., Thirunarayan, K. (eds.) ISWC 2009. LNCS, vol. 5823, pp. 359–374. Springer, Heidelberg (2009)
8. Limaye, G., Sarawagi, S., Chakrabarti, S.: Annotating and searching web tables using entities, types and relationships. In: Proc. 36th Int. Conf. on Very Large Databases (2010)
9. Manning, C.D., Raghavan, P., Schütze, H.: Introduction to Information Retrieval, 1st edn. Cambridge University Press (July 2008)
10. Mulwad, V., Finin, T., Syed, Z., Joshi, A.: T2LD: Interpreting and Representing Tables as Linked Data. In: Proc. Poster and Demonstration Session at the 9th Int. Semantic Web Conf. (November 2010)
11. Mulwad, V., Finin, T., Syed, Z., Joshi, A.: Using linked data to interpret tables. In: Proc. 1st Int. Workshop on Consuming Linked Data, Shanghai (2010)
12. Polfliet, S., Ichise, R.: Automated mapping generation for converting databases into linked data. In: Proc. 9th Int. Semantic Web Conf. (November 2010)
13. Resnik, P.: Semantic similarity in a taxonomy: An information-based measure and its application to problems of ambiguity in natural language. Journal of Artificial Intelligence Research 11(1), 95–130 (1999)
14. Sackett, D., Rosenberg, W., Gray, J., Haynes, R., Richardson, W.: Evidence based medicine: what it is and what it isn't. Bmj 312(7023), 71 (1996)
15. Sahoo, S.S., Halb, W., Hellmann, S., Idehen, K., Thibodeau Jr., T., Auer, S., Sequeda, J., Ezzat, A.: A survey of current approaches for mapping of relational databases to rdf. Tech. rep., W3C (2009)
16. Salton, G., Mcgill, M.J.: Introduction to Modern Information Retrieval. McGraw-Hill, Inc., New York (1986)
17. Syed, Z., Finin, T.: Creating and Exploiting a Hybrid Knowledge Base for Linked Data. Springer, Heidelberg (April 2011)
18. Syed, Z., Finin, T., Mulwad, V., Joshi, A.: Exploiting a Web of Semantic Data for Interpreting Tables. In: Proc. 2nd Web Science Conf. (April 2010)
19. Vavliakis, K.N., Grollios, T.K., Mitkas, P.A.: Rdote - transforming relational databases into semantic web data. In: Proc. 9th Int. Semantic Web Conf. (2010)
20. Venetis, P., Halevy, A., Madhavan, J., Pasca, M., Shen, W., Wu, F., Miao, G., Wu, C.: Recovering semantics of tables on the web. In: Proc. 37th Int. Conf. on Very Large Databases (2011)
21. Wang, J., Shao, B., Wang, H., Zhu, K.Q.: Understanding tables on the web. Tech. rep., Microsoft Research Asia (2011)

DC Proposal: Evaluating Trustworthiness of Web Content Using Semantic Web Technologies

Jarutas Pattanaphanchai*

Electronics and Computer Science, Faculty of Physical and Applied Science,
University of Southampton,
Southampton, SO17 1BJ, United Kingdom
{jp11g09,kmo,wh}@ecs.soton.ac.uk

Abstract. Trust plays an important part in people's decision processes for using information. This is especially true on the Web, which has less quality control for publishing information. Untrustworthy data may lead users to make wrong decisions or result in the misunderstanding of concepts. Therefore, it is important for users to have a mechanism for assessing the trustworthiness of the information they consume. Prior research focuses on policy-based and reputation-based trust. It does not take the information itself into account. In this PhD research, we focus on evaluating the trustworthiness of Web content based on available and inferred metadata that can be obtained using Semantic Web technologies. This paper discusses the vision of our PhD work and presents an approach to solve that problem.

1 Introduction

Trust plays an important role in the process of consuming data in many different circumstances such as communication between humans or data exchange between a human and a computer. Untrustworthy data may lead to wrong decisions or may make users misunderstand the concept or story, especially on the Web which has an abundance of information, but there is a lack of any control over the quality of its publications. With the incredible increase in the amount of Web content, it is becoming more necessary for users to be able to evaluate the trustworthiness of the information they use in order to judge whether to trust and use it or not. Therefore, having a trust mechanism for users to assess whether information is trustworthy is important and useful for the proper consumption of such information. However, trust is a subjective issue and dealing with it is a complicated task because it depends on the context in which the information is being considered. For example, people would trust a doctor to provide complete, accurate, and correct information about their health, but not about their finances.

The Semantic Web [1] is a technology that has been designed to make computers more intelligent by allowing them to understand the semantics of information

---

* Advisors: Dame Wendy Hall and Kieron O'Hara.

L. Aroyo et al. (Eds.): ISWC 2011, Part II, LNCS 7032, pp. 325–332, 2011.
© Springer-Verlag Berlin Heidelberg 2011

and process it properly. RDF (Resource Description Framework) is used to describe things in a way that computers can understand and also to portray the relationships between those things. Therefore, RDF can be used to represent metadata about Web resources. It also is used by machines to process this metadata in order to provide trust information to users. With this feature, less effort is needed on the side of humans than if they were to manually assess the trustworthiness of the information themselves [12,4]. RDF provides an opportunity to produce an effective trust model which uses metadata that is available in the Semantic Web for evaluating the trustworthiness or credibility of Web content. It also helps the Semantic Web at its current stage to be more successful because end users can have greater confidence in it.

Recent work on trust has focused on authentication-, reputation-, and policy-based trust [6,8,2], but it does not consider the content itself. The concept of content trust was first proposed by Gil and Artz [4]. They defined content trust as *"a trust judgement on a particular piece of information or some specific content provided by an entity in a given context"*. They also described the factors which influence a user's decision on whether to trust content on the Web. Moreover, they introduced the content trust model, which solves the problem of assessing the reliability of Web resources, by inferring the trustworthiness of the content of these Web resources [4]. Their work proposed the preliminary concept of content trust which can be explored more to produce a reasonable model. Similarly, our work considers the trustworthiness of the information on the Web based on the sources of that information. Credibility is another concept which is similar to evaluating the trustworthiness of the information on the Web. It focuses on studying and analysing factors that influence a user's decision on whether to trust the information on the Web. Several works have studied and proposed criteria for use in evaluating the credibility of Web sites or Web information. For example, the authority of the source that creates the information, the accuracy of the information, the appearance/presentation (such as the user interface, graphic design, and navigation) and the speed of loading the document [3,11,15]. However, each unique set of criteria presented in the different pieces of research has its limitations (e.g. it is hard to collect the information based on that criterion directly from the Web or it only slightly reflects the credibility of the information content itself). Therefore, we have to select the criteria that can be used in practice and that have a significant impact on the evaluation of the trustworthiness of Web content.

In this paper, we propose a model to evaluate the trustworthiness of information on the Web. This model uses Semantic Web technologies to gather metadata, which is collected based on our credibility evaluation criteria. The main contribution of our work is integrating metadata to build a data model that can be used to evaluate the trustworthiness of the information on the Web. We present these integrated metadata in an easily understandable form to the users who will, in turn, use this information to support their decisions of whether or not to trust the information provided on this Web. The rest of this paper is structured as follows. We review related work in Section 2. In Section 3, we introduce our proposed trust model, describe the concept of the model, and discuss the criteria that we use in our work. In Section 4, we propose our methodology to address the problem of evaluating the trustworthiness of content and present future work. Finally, in Section 5, we conclude our work.

2   Related Work

In the early stages of trust research, researchers focused on policy-based and reputation-based methods. The policy-based method assesses trust based on a set of rules. Bizer and Cyganiak [2] proposed a framework called *WIQA (Information Quality Assessment Framework)*, which is a set of software components that can be employed by applications in order to filter information based on quality policies. The reputation-based methods estimate trustworthiness by using other users' opinions or recommendations. Several trust metrics have been studied, and algorithms to compute trust across trust networks or social networks which have been presented [7,9,14]. The work presented so far focuses on evaluating the trustworthiness of the entities, which are judged based on their identities and their behaviours; for example using digital signatures or rating from recommendation system. However, this work did not take the information provided by such entities into account. More recent research has proposed evaluating trust based on content. Content trust is a concept which judges the trustworthiness of data based on features of that content or information resources. Some research uses RDF or annotations to present information about the source and content of desired information, which can then be used to determine that information's trustworthiness [5]. Other approaches discuss the factors which influence users' decisions on whether to trust the content and use these factors as criteria to evaluate the trustworthiness or credibility of information [4,13,3,15,11]. Instead of considering only the content, the entity that publishes that content should also be considered– we should assess whether or not authors who provide the information can be considered trustworthy. Therefore, we should consider both dimensions (entity and content), and this will help produce a more reasonable approach to evaluating the trustworthiness of information, and to provide support for making decisions.

3   The Proposed Model

3.1   Basic Concept

The Semantic Web is an extension of the existing Web, designed with the goal of letting computers deal with data rather than just documents. It describes facts about things and how they are related using RDF (Resource Description Framework) in the form of subject-predicate-object expressions called triples. RDF allows both structured and semi-structured data on the Web to be combined, exposed and shared across different sources or applications. In addition, it allows both users and software to follow links to discover more information related to these data [10]. Accordingly, the Semantic Web provides a way to gather metadata that is useful for evaluating the trustworthiness of information and also provides an opportunity to adopt trust into the Semantic Web itself.

In our work, we define an entity as a source which provides or publishes information. Therefore, we will evaluate an entity based on its credentials or its identity. In addition, we also analyse the available metadata to estimate the trustworthiness of the information on the Web. Therefore, in this paper, we consider the trustworthiness of the information on the Web as an evaluation of the metadata based on a set of trustworthiness criteria. Our proposed model deals with integrated metadata which is provided alongside the content of the Web (explicit metadata) and metadata inferred from a Semantic Web data resource (implicit metadata). The following sections describe the criteria that we use in our model and the architecture of the model.

3.2 Evaluation Criteria

Previous research proposed several factors that affect users' decisions on trusting content provided by an information resource on the Web [4,13,3,15,11]. These factors range from the source of that information to the information which is provided. An example of these factors are the author or the organization which publishes the information, the graphic design of the Web page, bias, and likelihood (the probability of the content being correct). We investigated the criteria from those pieces of research. We found that some of these criteria can be adopted for implementation in practice, such as the currency criterion that can be assessed based on the time stamp from the system and the authority of the information. However, some of these proposed criteria required data that is difficult to gather or do not have significant impact. For example, the time a document needs to be loaded may indicate the performance of the system and may influence trust but it does not reflect the information on the Web itself and thus it has less impact on the trustworthiness of information.

Figure 1 shows the criteria for evaluating the credibility or the trustworthiness of the information on the Web which is defined in studies that were discussed above. It shows that a number of characteristics, such as authority (source), accuracy, currency, and relevance, appear three times across the four studies which studied different domains and participants. This indicates that these criteria are the common criteria which can be used across the different circumstances to evaluate the trustworthiness of the information on the Web and therefore, we also use them in our approach. In more detail, we consider the following factors:

In conclusion, we selected the following four factors to use in our model since they have been common factors in several domains and can be adopted for automatic analysis:

- *Authority*: The reputation of the source that produced the content. We consider this criterion on two levels, the institutional and the individual level.
- *Currency*: Whether the content of the document is up-to-date or is regularly updated.

Gil and Aetz (2007)	Wathen and Burkell (2002)	Fogg et.al. (2000)	Rieh and Belkin (1998)

• Topic
• Context and criticality
• Popularity
• Authority
• Direct experience
• Recommendation
• Related resources
• Provenance
• User expertise
• Bias
• Incentive
• Limited resources
• Agreement
• Specificity
• Likelihood
• Age
• Appearance
• Deception
• Recency
	• Surface credibility
	  - Appearance/
	    presentation e.g.
	    colour, graphics, font
	    size
	  - Usability/interface
	    design e.g. menus,
	    navigability
	  - Organization of
	    information e.g. layers,
	    choice of detail level
	• Message credibility
	  - Source e.g. expertise,
	    credential.
	  - Message e.g. relevance,
	    currency, accuracy
		• Real world presence e.g.
		  an organization's physical
		  address, a phone number,
		  or photographs of the
		  members.
		• Error e.g. typographical
		  error, grammatical error
		• User interface/navigation
		• advertisements
			• Source
			  - Institutional level e.g.
			    URL, name of institution.
			  - Individual level e.g.
			    identification of author/
			    creator, affiliation,
			    qualification of creator.
			• Content e.g. theories from
			  education sites,
			  bibliographies, contact
			  information, get
			  information that meet
			  users' need.
			• Format e.g. graphical
			  images, information
			  structure.
			• Presentation e.g. writing
			  style, references, the size
			  of document.
			• Currency
			• Accuracy
			• Speed of loading

Fig. 1. Comparison of Credibility/Trustworthiness Influence Factors in Four Studies

– *Accuracy*: Whether the information in a document is accurately expressed i.e. it is grammatically correct and lacks spelling mistakes.
– *Relevance*: Whether the content meets the users' needs, which means it is useful for them.

We use these criteria to evaluate the trustworthiness of content. Our assessment approach is inspired by Wang et al. [13]. We use a range of metadata, in order to evaluate the trustworthiness of the information based on each criterion as follows:

– *Authority* is determined by the expertise of the author, the author's credentials and their institutional affiliations (the name of a public organization).
– *Currency* is calculated as the difference between the time stamp of the creation of the document or the time stamp of submission of the document and the current time. In the case that the system provides the time stamp of the last modification of the document, we will consider this metadata in our model.
– *Accuracy* is estimated according to the number of errors that appear in the document. Specifically, we measure the percentage of words which are spelled incorrectly or are part of a grammatically incorrect statement.
– *Relevance* is determined by the proportion of the users' search terms which are present in the key areas namely the title and the abstract of the publication. In addition, we will adopt the ontology concept for finding related terms and then use these terms to match in the key area in the future work. This allows the model to match the relevant data more efficiently than matching only exact keywords in the key areas of the document.

3.3   Architecture

The content trust model consists of three main modules. Specifically, the input, trustworthiness criteria and metacollection, and output modules, as shown in

Figure 2. Firstly, the input module gathers information from the Web page according to the user's keywords. In addition, our model considers the context and time frame as input from the environment to the model because evaluating trustworthiness according to a specific context. Also, the time frame has an affect on the judgement of whether or not to trust information (information that was trusted in the past is not guaranteed to be trusted in the future). Our model takes the time frame into account, since it evaluates the trustworthiness of the information every time the user interacts with the system. This means the system obtains the most recent information at the time at which the evaluation is performed. In other words, the time frame is automatically supplied to system. Then, the metacollection function will extract metadata provided alongside the content and aggregate the relevant information of this metadata from Semantic Web data sources based on the factors in each criterion from the trustworthiness criteria and the context from the input module. The trustworthiness criteria define the metadata that the system should collect for each criterion. Finally, the output module displays the result by presenting those collected metadata in simple sentences. The aim of this module is to provide the needed information to support users' judgement of the trustworthiness of the information on the Web. As trust is a very subjective notion, it is important to present information about trustworthiness in meaningful ways to the user. This is a complex problem which will be an important part of our future work.

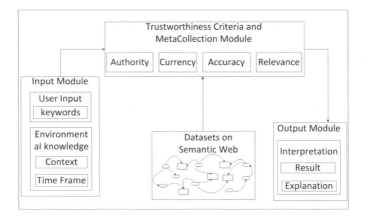

Fig. 2. The Content Trust Model Architecture

4   Methodology

Our methodology for evaluating the trustworthiness/credibility of Web content using Semantic Web technologies has three main phases.

In the first phase, we will build our data model to be used in the system. We will use an RDF graph to represent the data model, called a metadata graph. This metadata is gathered from information provided alongside the content (explicit metadata) along with information from a Semantic Web data store (implicit metadata). In more detail, the explicit metadata is the data that is published as RDF, alongside the Web page itself. It can be parsed to obtain metadata about the information on the page. In contrast, the implicit metadata is obtained by a query to the Semantic Web data store. We submit a query to the data store to search for more information based on the explicit metadata.

In the second phase, based on the metadata graph, we provide an explanation of this metadata, in order to present it in an easily understandable format. We give the information corresponding to each criterion we presented in Section 3.2. The explanation process will query the data from the metadata graph based on the criteria. Then, the system will interpret these metadata as simple sentences which are then presented to the users as an explanation. As a result, each criterion produces an explanation which is combined with the explanations from other criteria. In the initial implementation, we limit the Web content and Web information to academic publications from the University of Southampton to show that we can collect explicit and implicit data to evaluate the trustworthiness of Web content. We will develop a browser plug-in to present the explanation to users when they browse the Web. We also consider the case in which metadata required for a criterion is not available, in which case we will use other data in the content to help in assessing trustworthiness.

In the third phase, we will evaluate our approach by conducting user based test cases to test our model and conduct a survey to evaluate our system. We will use the publications of the Web Science Conference to be a test case because it is a new conference which has only been established in recent years. Therefore, there might be less information for evaluating the trustworthiness of the publications since we have little background knowledge of the authors who are new researchers in this research area. In addition, the Web Science Conference is a conference for a new, challenging research area which integrates several domains. For this reason, it covers a wide area of research which provides a large range of information. We will evaluate our model by using the questionnaire methodology from the expertise and the general user. For the expert evaluation, we will set up the experiment by choosing the information resources to be evaluated. We will ask the experts to evaluate the trustworthiness of the information. Then, we will compare the result from the experts and the result from our system. For a general user, we will build a questionnaire system to ask the user about their satisfaction with the system when taking part in the group of experiments.

5   Conclusion

In this paper, we proposed a trust model to solve the problem of evaluating the trustworthiness of the information on the Web from available metadata using Semantic Web technologies. We proposed a method to collect the explicit and implicit metadata to build a metadata graph and present this metadata, which will enable the users to judge whether this information is trustworthy. The results that have been obtained so far show that, by using Semantic Web technologies, we can retrieve relevant data and evaluate the trustworthiness of information based on its information credibility. For the first phase of the PhD work, we can build a metadata graph focusing on the authority and currency criteria and present these metadata to users. In the next step, we will work on dealing with the missing metadata and evaluating our approach.

References

1. Berners-Lee, T., Hendler, J., Lassila, O.: The semantic web. Scientific American 284(5), 34–43 (2001)
2. Bizer, C., Cyganiak, R.: Quality-driven information filtering using the WIQA policy framework. Web Semantics: Science, Services and Agents on the World Wide Web 7(1), 1–10 (2009)
3. Fogg, B.J., Marshall, J., Osipovich, A., Varma, C., Laraki, O., Fang, N., Paul, J., Rangnekar, A., Shon, J., Swani, P., Treinen, M.: Elements that Affect Web Credibility: Early Results from a Self-Report Study. In: Human Factors in Computing Systems, pp. 287–288 (2000)
4. Gil, Y., Artz, D.: Towards content trust of web resources. Web Semantics: Science, Services and Agents on the World Wide Web 5(4), 227–239 (2007)
5. Gil, Y., Ratnakar, V.: Trusting Information Sources One Citizen at a Time. In: Horrocks, I., Hendler, J. (eds.) ISWC 2002. LNCS, vol. 2342, pp. 162–176. Springer, Heidelberg (2002)
6. Golbeck, J., Hendler, J.: Accuracy of Metrics for Inferring Trust and Reputation in Semantic Web-Based Social Networks. In: Motta, E., Shadbolt, N.R., Stutt, A., Gibbins, N. (eds.) EKAW 2004. LNCS (LNAI), vol. 3257, pp. 116–131. Springer, Heidelberg (2004)
7. Golbeck, J., Parsia, B.: Trust network-based filtering of aggregated claims. International Journal of Metadata, Semantics and Ontologies 1(1), 58–65 (2006)
8. Heath, T., Motta, E.: Reviews and ratings on the semantic web. In: The 5th International Semantic Web Conference (ISWC 2006), pp. 7–8 (2006)
9. Heath, T., Motta, E., Petre, M.: Computing word-of-mouth trust relationships in social networks from semantic web and web2. 0 data sources. In: Proceedings of the Workshop on Bridging the Gap between Semantic Web and Web, Citeseer, vol. 2, pp. 44–56 (2007)
10. RDF Working Group: Resource Description Framework (RDF) (2004), http://www.w3.org/RDF/
11. Rieh, S., Belkin, N.: Understanding judgment of information quality and cognitive authority in the WWW. In: Proceedings of the 61st Annual Meeting of the American Society for Information Science, Citeseer, vol. 35, pp. 279–289 (1998)
12. W3C: RDF Primer (2004), http://www.w3.org/TR/2004/REC-rdf-primer-20040210/
13. Wang, W., Zeng, G., Tang, D.: Using evidence based content trust model for spam detection. Expert Systems with Applications 37(8), 5599–5606 (2010)
14. Wang, X., Zhang, F.: A New Trust Model Based on Social Characteristic and Reputation Mechanism for the Semantic Web. In: First International Workshop on Knowledge Discovery and Data Mining (WKDD 2008), pp. 414–417 (January 2008)
15. Wathen, C.N., Burkell, J.: Believe it or not: Factors influencing credibility on the Web. Journal of the American Society for Information Science and Technology 53(2), 134–144 (2002)

DC Proposal: Decision Support Methods in Community-Driven Knowledge Curation Platforms

Razan Paul

School of ITEE, The University of Queensland,
St. Lucia, QLD 4072, Australia
razan.paul@uq.edu.au

Abstract. Skeletal dysplasias comprise a group of genetic diseases characterized by highly complex, heterogeneous and sparse data. Performing efficient and automated knowledge discovery in this domain poses serious challenges, one of the main issues being the lack of a proper formalization. Semantic Web technologies can, however, provide the appropriate means for encoding the knowledge and hence enabling complex forms of reasoning. We aim to develop decision support methods in the skeletal dysplasia domain by applying uncertainty reasoning over Semantic Web data. More specifically, we devise techniques for semi-automated diagnosis and key disease feature inferencing from an existing pool of patient cases – that are shared and discussed in the SKELETOME community-driven knowledge curation platform. The outcome of our research will enable clinicians and researchers to acquire a critical mass of structured knowledge that will sustain a better understanding of these genetic diseases and foster advances in the field.

Keywords: Decision Support Methods, Semantic Web, Skeletal Dysplasias.

1 Background and Problem Statement

Skeletal dysplasias are a heterogeneous group of genetic disorders affecting skeletal development. Currently, there are over 450 recognized bone dysplasias, structured in 40 groups. Patients with skeletal dysplasias have complex medical issues including short stature, bowed legs, a larger than average head and neurological complications. However, since most skeletal dysplasias are very rare (<1:10,000 births), data on clinical presentation, natural history and best management practices is sparse. Another reason for data sparseness is the small number of phenotypic characteristics typically exhibited by patients from the large range of possible phenotypic and radiographic characteristics usually associated with these diseases. Due to the rarity of these conditions and the lack of mature domain knowledge, correct diagnosis is often very difficult. In addition, only a few centers worldwide have expertise in the diagnosis and management of these disorders. As there are no defined guidelines, the diagnosis of new cases relies strictly on parallels to past case studies.

Medical decision support systems can assist clinicians and researchers both in the research of skeletal dysplasias, as well as in the decision making process. However, the absence of mature domain knowledge and a lack of well documented, well

L. Aroyo et al. (Eds.): ISWC 2011, Part II, LNCS 7032, pp. 333–340, 2011.
© Springer-Verlag Berlin Heidelberg 2011

structured past cases has hindered the development of decision support methods. Additionally, the general sparseness and disperse nature of skeletal dysplasia data has limited the development and availability of authoritative databases via the leading clinical and research centres. To make diagnoses, improve understanding and identify best treatments, clinicians need to analyze historical dysplasia patient data, verify known facts and relationships and discover new and previously unknown facts and relationships among the phenotypic, radiographic and genetic attributes associated with existing and new cases. In order to do this, they currently need to query many heterogeneous data sources and to effectively aggregate diverse types of data relating to phenotypic, radiographic and genetic observations. This integration step represents a significant challenge due to the extreme heterogeneity of the data models, metadata schemas and vocabularies, data formats and inconsistencies in naming and identification conventions.

The above-mentioned issues also limit the potential of successfully applying existing or traditional knowledge representation and decision support methods, such as Rule Based Systems [1], Neural networks [2], Fuzzy cognitive maps (FCMs) [3], Fuzzy Rule based classification [4] or Clustering algorithms [4], to the bone dysplasia domain. Creating a decision support model (e.g., a rule base) requires a set of well-established data models, data acquisition guidelines and mature domain knowledge. In this domain, clinicians have to diagnose patients with little or no similarity to past cases – this requires the generation of new evidence by combining existing evidence. This scenario combined with the general data scarcity issue has determined that distributed ontology reasoning [5] is a necessity within the skeletal dysplasia domain. Finally, because the study and understanding of skeletal dysplasias is still relatively immature, the justifications underpinning the decisions and the decision support methods, also need to be documented. All these elements make the knowledge representation and decision support methods in the bone dysplasia domain a very exciting and potentially productive area of research.

2   Aim and Objectives

Our hypothesis is that representing both the knowledge and the data via Semantic Web formalisms, together with the application of inductive and statistical reasoning on the resulting knowledge base, can support the development of efficient decision support methods in the skeletal dysplasia domain. More specifically, such approaches will enable the inferencing of key disease features from an existing pool of patients and the semi-automated diagnosis of the specific disease affecting new patients. This hypothesis can be further decomposed into the following research questions:

1. How can the generalized evidential statements (evidences), including their probabilistic uncertainties, be induced from existing patient cases stored in a Semantic Web knowledge base?
2. How can we efficiently build a comprehensive ontology that is capable of capturing the induced generalized evidences?
3. How can the probabilistic uncertainty of an evidential statement be incorporated to improve the precision of both the evidence learning (inductive reasoning) and the statistical reasoning?

4. How can the diagnosis of an undiagnosed skeletal dysplasia patient and the key features of a particular dysplasia be determined using the induced generalized evidential statements (including the associated justification)?
5. What is the optimum approach for combining existing skeletal dysplasia evidential statements to form new evidence?

In order to answer these research questions, this thesis will aim to achieve the following objectives:

a) The development of an ontology to store generalized evidential statements (evidences) that lay the foundation for further reasoning tasks.
b) Inducing the generalized knowledge (evidences) from the existing patient cases and encoding this knowledge in an interoperable manner.
c) Reasoning with this induced generalized knowledge to develop decision support methods. Instead of relying on a mature domain knowledge, or approach is to determine solutions directly from similar past examples. Our reasoning approach will utilize the generalized and inductive knowledge of past cases, concrete problem situations (new cases) and combine the learned knowledge from past cases to form new knowledge that will assist in solving specific problems.

3   Related Work

Existing online knowledge bases, such as the European Skeletal Dysplasia Network (ESDN) (www.esdn.org) and the Queensland Bone Dysplasia Registry (QBDR) (http://qbdc.org/bone-dysplasia-registry/) are ideal approaches for encouraging community-driven content exchange and curation. However, the underlying content is static, lacks formally defined semantics, and lacks decision support methods, thus making it difficult for the content to be reused, reasoned across and recombined for different purposes.

Most prior work in representing generalized knowledge for medical decision support methods [1-4, 6] use some non-standard formalisms or proprietary formats which hinder integration, interoperability and efficient knowledge reasoning. They also lead to unjustified results by fusing all generalized knowledge into a black box system or assume a mature established domain knowledge. Moreover, some of these previous methods cannot evolve over time, due to their shallow knowledge representation formalisms. Case-based reasoning [6], on the other hand, cannot combine past evidences to form a new evidence for a given problem where no past similar evidence exists. This scenario is typical for rare diseases like skeletal dysplasias. It also uses non-generalized evidences, which does not guarantee correctness.

Rule based systems [1] and fuzzy rule-based classification [4] use exact matching on rules built on mature and established domain knowledge - which is inapplicable in a domain that suffers from data sparseness. The neural network approach [2], cannot provide justification for the resulting knowledge because it fuses all the evidence into the internal weights, whereas in the skeletal dysplasia domain, justification is very important to both clinicians and researchers in order to understand the underlying causal elements.

It is widely accepted that uncertainty is an indispensable aspect of medical data. Bayesian reasoning [7], a widely used probability formalism, presents issues when applied in this domain due to the estimation of the prior and conditional probabilities. Fuzzy sets are commonly used models to manage vagueness and imprecision in the medical domain [8]. Dempster-Shafer theory [9] is an alternative to representing probabilistic uncertainty mathematically. This is a potentially valuable tool to be used in the decision making process when precise knowledge is missing [9]. An important aspect of this theory is the combination of evidence obtained from multiple sources with the computation of a degree of belief that takes into account all the available evidence. Finally, as opposed to Bayesian reasoning, Dempster-Shafer theory does not require an estimation of the prior and conditional probabilities of the individual constituents of the set.

Today's decision support systems require the automatic integration of knowledge from multiple sources. However, the lack of interoperability and standard formalisms impede these systems to take advantage of the connectivity provided by the Web. Decision support systems [10, 11]   using Semantic Web standards are being developed to overcome the above challenges. Semantic Web rule-based reasoning has been used for domain specific decision support methods, for example, in the Ambient Intelligence domain [12]. However, such approaches cannot make use of underlying trends in instance data that have not been encoded as ontological background knowledge and cannot handle probabilistic uncertainties within the knowledge. Moreover, they cannot form new evidence by combining existing evidence via reasoning, where there exist no prior examples.

A recent related effort [11] presents a novel fuzzy expert system for a diabetes decision support application using a 5-layer fuzzy ontology and a semantic decision support agent. However, as with its predecessors, this system also depends on mature and established domain knowledge, and uses fuzzy rule-based reasoning [13], which follows an exact matching approach.

Medical decision support systems have emerged from the co-evolution of research in decision support systems and medical informatics. In [14], a Semantic Web based Clinical Decision Support System is presented to provide evidence-guided recommendations for follow-up after treatment for Breast Cancer. ControlSem [15], a medical decision support system using Semantic Web technologies, was developed with the goal of controlling medical procedures. Similarly, in [16], the authors present a medical expert system for heart failure. These expert systems use general purpose rule base reasoning (deductive reasoning) [13] because the underlying domain has well-defined rules and a mature background knowledge.

4   Research Plan

We aim to develop decision support methods via an ontology-based interoperable framework tailored towards the skeletal dysplasia domain. This research combines ontological techniques with inductive and statistical reasoning techniques, and will be integrated within the SKELETOME[1] community-driven knowledge curation platform. Figure 1 presents the high level building blocks of the framework. The SKELETOME ontology[2], developed to capture the essential knowledge of skeletal dysplasia domain,

---

[1] http://itee.uq.edu.au/~eresearch/projects/skeletome/
[2] http://purl.org/skeletome/bonedysplasia

is a foundational prerequisite of our framework. While the SKELETOME ontology stores past and newly emerging patient cases, the Evidence ontology stores generalized evidences. Taking into account the lack of mature domain knowledge, the evidence extraction process induces generalized evidences from the existing patient cases and encodes the resulting knowledge in the Evidence ontology. Reasoning over the induced generalized evidences enables the development of the targeted decision support methods, i.e., automated diagnosis and identification of key disease features.

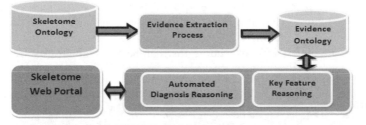

Fig. 1. Research methodology – building blocks

SKELETOME Ontology: The main role of the SKELETOME Ontology is to improve the highly static and rigid format of the ISDS Nosology [17], by enabling a more flexible classification of the disorders and integration with existing Web resources, such as the Gene Ontology, the Human Phenotype Ontology and the NCI Thesaurus. This ontology captures the complex relations between the phenotypic, radiographic and genetic elements that characterize all skeletal dysplasias.

Evidence Ontology: The Evidence Ontology models uncertainty (both fuzzy / vagueness and probabilistic uncertainty) by re-using concepts from Fuzzy Theory, such as *fuzzy value, fuzzy variable, fuzzy set, membership value, fuzzy term* and *probabilistic uncertainty*. It enables the representation of uncertain generalized evidences and helps to simplify uncertain knowledge representation in OWL. OWL cannot encode Fuzzy and probabilistic uncertainty semantics. The crisp syntax of OWL DL will be used with the Evidence ontology to enable the encoding of Fuzzy and probabilistic uncertainty semantics.

Evidence Extraction Process: Generalized evidence extraction from past patient cases stored in the SKELETOME ontology is a crucial prerequisite for the implementation of the automated diagnosis and key feature inference capabilities. Without the extracted evidence, uncertainty reasoning cannot be performed. The actual extraction process will use Machine Learning techniques, and more specifically, a level wise search algorithm [18] that is able to infer evidences from the instances of the SKELETOME ontology concepts, made available by domain experts. The effectiveness of the method will be evaluated empirically. An *updating* module will be developed to ensure the continuous synchronization of the Evidence ontology instance base with the current patient repository.

Automated Diagnosis Reasoning: Clinicians can determine, to a level of approximation, possible diseases based on the medical symptoms that the patient presents. The vagueness (fuzziness) of medical symptoms is modelled by the *fuzzy set* concept in the Evidence ontology. The probabilistic uncertainty of the medical fact is modelled by attaching a confidence/conditional probability value to each evidence. The automated diagnosis will infer a possible skeletal dysplasia based on a set of symptoms using Dempster–Shafer theory and fuzzy set theory (see Fig. 2). Generalized evidences are represented using the Evidence ontology, stored in the SKELETOME knowledge base and include probabilistic uncertainty values based on existing phenotype, radiographic and genetic information. The candidate hypothesis for an undiagnosed patient will be computed via reasoning. Dempster–Shafer theory will then be applied to the set of patient symptoms to determine the diagnosis based on the evidence stored in Evidence. In our case, the Dempster–Shafer calculation will only consider fuzzy terms, linguistic variables and probabilistic uncertainty, excluding the membership value of each fuzzy term.

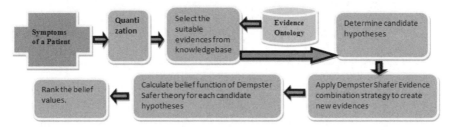

Fig. 2. Automated diagnosis reasoning based on a set of symptoms

Key Feature Reasoning: A key feature is a feature that is deemed highly characteristic of a specific skeletal dysplasia (for example, short fingers are a characteristic of Platyspondylic Lethal skeletal dysplasia). Inferring the key feature of a skeletal dysplasia from diverse phenotypic or genetic information is critical for the diagnosis of new patients. To determine such key features, our algorithm (depicted in Fig. 3) firstly determines the candidate hypotheses from the evidence stored in the Evidence Ontology. Then it applies domain-oriented ranking functions on the candidate hypotheses and presents the top K to the clinician using the system.

Fig. 3. Finding the Key Feature of a Dysplasia

5   Evaluation

The results of our research will be evaluated empirically. The patient cases used for evaluation will be collected from the ESDN and the QBDR - 90% of the cases will be used to construct our knowledge base, while the rest will act as test cases.

Evaluating the Decision Support Methods: The decision support methods will be evaluated on the basis of following criteria: (1) Decision-Making Effectiveness; (2)

Performance; (3) Ease of use and understating; (4) Scalability. Effectiveness will be measured using experimental methods combined with a usability study. Three metrics will be used to assess effectiveness: *Accuracy*, *Precision* and *Recall*. For the automated diagnosis aspect: *Accuracy* is the overall percentage of correctly diagnosed cases; *Precision* is the percentage of the correct diagnoses of a particular dysplasia; and *Recall* is the percentage of cases of a particular dysplasia that were correctly identified. For the key feature reasoning aspect: *Accuracy* is the overall percentage of correctly extracted features; *Precision* is the percentage of correctly selected key features and *Recall* is the percentage of dysplasias associated with a particular key feature that were correctly identified. To measure the performance of the decision methods, the following metrics will be used: Run time, Load time, Memory usage (main memory), and Memory usage (disk storage). The ease of use and understanding will be determined via a questionnaire with Likert-scale answers. We will define the testing environment and set of tasks to be performed by the participants. Observation data will also be collected from the usability study and used to complement to questionnaire. Scalability will measure how the methods scale as the number of patients in the KB increases. We will test our decision support methods against a number of different sized instance datasets and observe the changes in response times.

Evaluating the Evidence Ontology: Task-based evaluations [19] will be used to measure the generalized uncertain evidence representation capability of the Evidence ontology. A set of use-cases, formulated as parameterized test questions and answer keys will be leveraged to characterize the ontology in terms of accuracy, insertion errors, deletion errors and substitution errors.

Evaluation of the Evidence Extraction Process: To quantitatively assess the quality of the evidence extraction process, we will measure the evidence retrievability (recall) [20] and the evidence spuriousness (precision) [20]. Evidence retrievability measures how well the underlying trends in past data have been discovered. Although retrievability provides a good estimate of the fraction of detected patterns in the data, it does not provide an estimate of the quality of the found patterns. The quality of a pattern will be measured using spuriousness, which quantifies the number of items in the pattern that are not associated with the matching base pattern.

6 Conclusions

No prior research has investigated knowledge integration and decision support methods for the skeletal dysplasia domain - although it suffers from two important problems: (1) existing data is represented in a heterogeneous and non-interoperable manner, and (2) there are no mechanisms for building a consolidated and evolving knowledge base to support the decision making process. Our proposed research aims to address these problems by (1) using Semantic Web standards to formalize both the knowledge and data in the domain; (2) developing decision support methods based on past patient case studies, combined evidence and aggregated knowledge discovery. We believe that this research will advance the knowledge of the skeletal dysplasia community and expedite their understanding and diagnosis of skeletal dysplasias.

Acknowledgments. The work presented in this paper is supported by the Australian Research Council (ARC) under the Linkage grant SKELETOME - LP100100156.

References

1. Hudson, D.L.: Medical Expert Systems. In: Encyclopedia of Biomedical Engineering. John Wiley and Sons (2006)
2. Chan, K., et al.: Diagnosis of hypoglycemic episodes using a neural network based rule discovery system. Expert Systems with Applications (2011)
3. Papageorgiou, E.I., et al.: Fuzzy Cognitive Map Based Approach for Assessing Pulmonary Infections. In: Rauch, J., Raś, Z.W., Berka, P., Elomaa, T. (eds.) ISMIS 2009. LNCS, vol. 5722, pp. 109–118. Springer, Heidelberg (2009)
4. Gadaras, I., Mikhailov, L.: An interpretable fuzzy rule-based classification methodology for medical diagnosis. Artificial Intelligence in Medicine 47(1), 25–41 (2009)
5. Schlicht, A., Stuckenschmidt, H.: Towards distributed ontology reasoning for the web. IEEE (2008)
6. Begum, S., et al.: Case-Based Reasoning Systems in the Health Sciences: A Survey of Recent Trends and Developments. IEEE Transactions on Systems, Man, and Cybernetics–Part C: Applications and Reviews (2010)
7. Wang, H.Q., Dash, D., Druzdzel, M.J.: A method for evaluating elicitation schemes for probabilistic models. IEEE Transactions on Systems Man and Cybernetics Part B-Cybernetics 32(1), 38–43 (2002)
8. Lekkas, S., Mikhailov, L.: Evolving fuzzy medical diagnosis of Pima Indians diabetes and of dermatological diseases. Artificial Intelligence in Medicine 50(2), 117–126 (2010)
9. Dymova, L., Sevastjanov, P.: An interpretation of intuitionistic fuzzy sets in the framework of the dempster-shafer theory. Springer, Heidelberg (2010)
10. Goossen, F., et al.: News personalization using the CF-IDF semantic recommender. ACM (2011)
11. Lee, C.S., Wang, M.H.: A Fuzzy Expert System for Diabetes Decision Support Application. IEEE Transactions on Systems Man and Cybernetics Part B-Cybernetics 41(1), 139–153 (2011)
12. Patkos, T., Chrysakis, I., Bikakis, A., Plexousakis, D., Antoniou, G.: A Reasoning Framework for Ambient Intelligence. In: Konstantopoulos, S., Perantonis, S., Karkaletsis, V., Spyropoulos, C.D., Vouros, G. (eds.) SETN 2010. LNCS, vol. 6040, pp. 213–222. Springer, Heidelberg (2010)
13. Straccia, U.: Managing Uncertainty and Vagueness in Description Logics, Logic Programs and Description Logic Programs. In: Baroglio, C., Bonatti, P.A., Małuszyński, J., Marchiori, M., Polleres, A., Schaffert, S. (eds.) Reasoning Web. LNCS, vol. 5224, pp. 54–103. Springer, Heidelberg (2008)
14. Hussain, S., Raza Abidi, S., Raza Abidi, S.S.: Semantic Web Framework for Knowledge-Centric Clinical Decision Support Systems. In: Bellazzi, R., Abu-Hanna, A., Hunter, J. (eds.) AIME 2007. LNCS (LNAI), vol. 4594, pp. 451–455. Springer, Heidelberg (2007)
15. Andreasik, J., Ciebiera, A., Umpirowicz, S.: ControlSem–distributed decision support system based on semantic web technologies for the analysis of the medical procedures. In: 3rd Conference on Human System Interactions (HSI). IEEE (2010)
16. Prcela, M., Gamberger, D., Jovic, A.: Semantic web ontology utilization for heart failure expert system design. Studies in health technology and informatics 136, 851 (2008)
17. Warman, M.L., et al.: Nosology and classification of genetic skeletal disorders: 2010 revision. American Journal of Medical Genetics Part A (2010)
18. Paul, R., Hoque, A.S.M.: Mining irregular association rules based on action & non-action type data. In: Fifth International Conference on Digital Information Management (ICDIM). IEEE, Thunder Bay (2010)
19. Porzel, R., Malaka, R.: A task-based approach for ontology evaluation, Citeseer (2004)
20. Gupta, R., et al.: Quantitative evaluation of approximate frequent pattern mining algorithms. ACM (2008)

DC Proposal: Towards Linked Data Assessment and Linking Temporal Facts

Anisa Rula

University of Milano-Bicocca,
Department of Computer Science, Systems and Communication (DISCo),
Innovative Techonologies for Interaction and Services (Lab),
Viale Sarca 336, Milan, Italy
anisa.rula@disco.unimib.it

Abstract. Since the Linked Data is continuously growing on the Web, the quality of overall data can rapidly degrade over time. The research proposed here deals with the quality assessment in the Linked Data and the temporal linking techniques. First, we conduct an in-depth study of appropriate dimensions and their respectively metrics by defining a data quality framework that evaluates, along these dimensions, linked published data on the Web. Second, since the assessment and improvement of the Linked Data quality such as accuracy or the resolution of heterogeneities is performed through record linkage techniques, we propose an extended technique that apply time in similarity computation which can improve over traditional linkage techniques. This paper describes the core problem, presents the proposed approach, reports on initial results, and lists planned future tasks.

Keywords: Linked Data Quality, Quality Assessment, Temporal Linking.

1 Problem Definition

Data quality is an important issue for data driven applications which should be deeply investigated and understood. As a consequence of non controlled quality of the data that flows across information systems, the overall data can rapidly degrade over time. The literatures provides a wide range of techniques used to assess and improve the quality of data, such as record linkage, business rules, and similarity measures [2]. However, the quality becomes more complex and controversial as a consequence of networked-based structure (such as the web), where the amount of data evolve and it becomes more complex to be controlled.

Our focus is based on the assessment of data sets represented by structured data published on the web, known as Linked Data [4]. The aspect of quality in Linked Data is considered as an important task to consumers for a number of obvious reasons: they need data to be correct, thus, they need to have the ability to select and compare data from different sources to detect and correct errors in the data sets. Missing values or duplication can lead to applications not realizing the full potential of exchanging data.

L. Aroyo et al. (Eds.): ISWC 2011, Part II, LNCS 7032, pp. 341–348, 2011.
© Springer-Verlag Berlin Heidelberg 2011

Techniques as record linkage are mostly used to assess and improve the data quality of the information systems. Although these techniques have been adapted in the Linked Data context [11], they ignore as well as the traditional one that real-world entity can evolve over time and can fall short for temporal data. For example, a person can change her phone number and address and so facts that describe the same real-world entity at different times can contain different values. Identifying entities that refer to the same concept enables interesting longitudinal data analysis over such data. Thus, the representation of temporal entities within the Linked Data cloud is an essential step of the linking which provide linking in a temporal context and likewise evaluate the quality of the Linked Data.

As a possible scenario, we can consider the Digital Bibliography & Library Project (DBLP)[1], published as Linked Data. It is one of the largest collection of bibliographic metadata about computer science publications over many decades. Although in general the DBLP data is of a very high quality, we have noticed some quality problems by querying that data. We wish to identify individual authors such that we can list all publications by each author. This data set contains temporal entities over a long period of time; each entity is associated with a time stamp and describes some aspects of a real-world entity at that particular time. A necessary extension to the traditional linking should be approached by incorporating the time concept to the entities.

In particular, this PhD work will concentrate on assessing the quality of Linked Data which is divided into two parts. First, to solve the above mentioned problem we aim at providing data quality dimensions as well as related metrics and validation tools which are mandatory for the assessment of quality of the published data (Linked Data). Second, as a continuously work of quality assessment we aim at providing an in-depth study about the matching heuristics defined in the context of Linked Data and the application of the approach proposed in [16] for linking temporal facts that describe the same real-world entity over time and so be able to trace the history of that entity.

2   State of the Art

Quality Aspect in Linked Data

The assessment of data quality is considered as a continuous cycle involving four major steps: the definition of quality dimensions, measuring these dimensions through sound and measurable metrics, and analyzing the results [21]. Starting from some previous works which describe six most important classifications of quality dimensions, it is possible to define a basic set of data quality dimensions, including accuracy, completeness, consistency, and timeliness. Concerning the Web, a model that associates quality information with Web data is proposed in [18]. Several dimensions are considered, such as volatility, completability, and semantic and syntactic accuracy.

---

[1] http://dblp.l3s.de/d2r/

Linked Data, as all the data driven application need a thorough assessment. But the assessment of Linked Data poses a number of unique challenges: due to the structured nature of Linked data published in an open environment such as the web. A comprehensive study of various problems related to the quality in Linked Data have been conducted in [13]. In fact the use of incompatible levels of abstraction makes complex a true context-sensitive analysis in elaborating query answering and visualization scenarios. Broken links in Linked Data or the ambiguous use of owl:sameAs are some of the data quality errors that can reduce the usability of a Linked Data approach. Broken Linked Data appears when it is impossible to retrieve the content of structured data due to server errors or the general unavailability of a reference. The owl:sameAs property is used to connect different data element to support semantic data integration. That is, any two URI references connected by owl:sameAs should be the same thing. But in reality, the correctness cannot be ensured and some analysis in the literature underline the different semantic associated by different designers to the owl:sameAs properties. Therefore, the authors in [9] conduct an empirical study and proposed several components of a general strategy for integrating and fusing information from the URIs in an owl:sameAs network. An approach has been proposed for quality and trustworthiness assessment based on provenance information in [10]. A description of dimensions and related metrics for the assessment of Linked Data are partly a contribution of Web community [1]. With the goal to evaluate a quality-driven information filtering a framework is proposed in [3] which supports information consumers in their decision whether to accept or reject information. This framework requires the contribution of the consumer on writing the policies.

Temporal Linking Aspects in Linked Data

Record linkage considers a set of records as input and discovers which of them refers to the same real-world entity, even if the records are not identical. They typically relies on string comparison techniques which compare multiple properties of the entities that are to be interlinked. The first studies were introduced in the statistic community [7]. Record linkage, also called identity resolution or duplicate detection, is a well-known problem in database community [5] as well as in the ontology matching community [6].

Recently, this approach is finding application in a new community such as the Linked Data. The usual approach uses automatic or semi-automatic record linkage heuristics to generate links between data sources. Silk − a Link Discovery Framework is a toolkit used for discovering and maintaining data links between Web data sources [20].

Considering the time evolution of entities in the the record linkage, some approaches have been developed [16]. The value evolution over time has been addressed in [16] by introducing the concept of decay applied in a global fashion. This approach could be also employed in our solution. Within the Semantic Web community, the representation of temporal information encodes the semantics into a time ontology which describes the temporal content of Web pages and the temporal properties of Web services [12]. Some other existing approaches include *temporal RDF* for the representation of temporal information which introduce time in RDF by assigning a number t for the temporal validity of a triple [8]; *versioning* which suggests that the ontology has different versions, one per instance of time [15]; *named graph* to implement temporal graphs which were designed to handle statements temporal validity [19], etc. There is also another approach based on tracing knowledge evolution over time which extract temporal facts to build a large-scale temporal information system [22].

3   Proposed Approach and Methodology

The contribution of this PhD work is twofold: (i) enrich the actual quality assessment framework defined in [3] by adding a new component composed by a set of quality dimensions, new measures and validation tools for higher quality of the published data; (ii) define new algorithms for performing linking between two data sets considering the temporal aspect of entities in the Linked Data.

In the following, we give an overview of the methodology which we want to follow to come up with the aforementioned contributions.

New Component for Data Quality Assessment

While there are significant overlaps, our approach focuses on assessing the quality of structured data on the web by defining a "filtered" set of quality dimensions which fit better to the user or application requirement, rather than a continuously creation of new dimensions and separated methodologies. In this context we introduce the concept of Data Profiling (DP) defined as the application of data analysis techniques to existing data sources for the purpose of determining their quality. Therefore, the aim of this work is to define a data quality framework as a set of guidelines and techniques that, starting from input information describing a given application context, defines a rational process to assess and improve the quality of published data. Furthermore, our intent is to drive through an automatic or semi-automatic data quality approach. The achievement of an automatic data quality framework raises complex research issues and challenges, which we intend to tackle in this PhD. More precisely, we will focus on the creation of a repository of quality dimensions that are interesting for Linked Data purpose (both consumer and producer view point), for each dimension at least one metric (subjective or objective) and a framework able to analyse the data sources by means of probes that implements the above defined metrics. Results of this analysis will be shown in a dashboard so that it could be easier to understand the quality of exposed data source.

As a first step work we introduce a framework based on green engineering aspects where we have extracted only 9 of the original 12 principles with a short description, the dimensions in which they expand and measures for the assessment of linked data on the Web. [14]. The evaluation of the data will be done through validators which will consist of open source or off-the-shelf algorithms offered in the Web of Data community, as well as new validators (e.g. to check comprehensibility) that we are implementing.

Temporal Linking in Linked Data

The approach followed in this paragraph is inspired by legacy related work for structured data from the database community since the record linkage techniques are used to assess and improve the quality. In fact, if data quality issues are related to the accuracy and completeness dimensions which represent a quality aspect of the data then the improvement method is targeted to the record linkage technique.

However, the temporal record linkage gain an important role since it goes further from the traditional record linkage for the quality assessment. Therefore, considering the temporal aspects within the Linked Data is an essential step in order to provide connection of the same entities expressed in different time stamp.

A first step of defining a temporal record linkage will be targeted at transferring techniques from relational databases to the Linked Data environment [17]. For instance, it has been observed that the record linkage is a well-known problem in database community [5] and many of the techniques from these fields are directly applicable in the Linked Data context [11]. Thus, we will appropriately adapt the use of "time decay", which aims to capture the effect of time elapse on entity value evolution. As an example, let us consider RDF triples that describe paper authors. Let consider two real world persons: A1 and A1'.

In Figure 1 we have author A1 who was at "The Open University" in year Y1; then A1 moved to "University of Milan Bicocca" in year Y2; A1' describe another entity, this author moved from "University of Milan Bicocca" in year Y3 to "Karlsruhe Institute of Technology" in year Y4.

Despite the challenges, temporal information does present additional evidence for linkage. We can notice that the if we consider A1 and A1' as the same person he is moving back and forth from one university to another. Exploring such evidence would require a global view of the facts with the time factor in mind. In particular we want to apply the concept of *false positive* and *false negative* related to the time decay. In particular, we consider *false negative* when there are changes on a value then it is not necessarily considered that these values are referring to different entities; we consider *false positive* when a value remain the same with a long time gap then it is not referring to the same entity. Afterwards we plan to learn decay from labelled data and apply it when computing links between entities.

An overview of what will be followed during the research is briefly described below. First, when creating explicit data links between entities, the traditional link discovery technique reward high value similarity and penalize low value similarity. However, as time slip away, values of a particular entity may evolve. Meanwhile, different entities are more likely to share the same value(s) with a long time gap. Thus, decay is introduced to reduce the penalty for value disagreement and reward for value agreement over a long period. We expect to have better results by applying decay in similarity computation.

```
<http://dblp.l3s.de/.../publications/P1> dc:creator <http://dblp.l3s.de/.../authors/A1>
<http://dblp.l3s.de/.../authors/A1> opus:has_affiliation <http://open.ac.uk>
<http://dblp.l3s.de/.../authors/A1> foaf:homepage <http://www.open.ac.uk/A1>
<http://dblp.l3s.de/.../publications/P1> dcterms:issued "Y1"
<http://open.ac.uk> rdfs:label "The Open University"
```
1

```
<http://dblp.l3s.de/.../publications/P2> dc:creator <http://dblp.l3s.de/.../authors/A1>
<http://dblp.l3s.de/.../authors/A1> opus:has_affiliation <http://www.disco.unimib.it>
<http://dblp.l3s.de/.../authors/A1> foaf:homepage <http://www.disco.unimib.it/A1>
<http://dblp.l3s.de/.../publications/P2> dcterms:issued "Y2"
<http://www.disco.unimib.it> rdfs:label "University of Milano-Bicocca"
```
2

```
<http://dblp.l3s.de/.../publications/P3> dc:creator <http://dblp.l3s.de/.../authors/A1'>
<http://dblp.l3s.de/.../authors/A1'> opus:has_affiliation <http://www.disco.unimib.it>
<http://dblp.l3s.de/.../authors/A1'> foaf:homepage <http://www.disco.unimib.it/A1>
<http://dblp.l3s.de/.../publications/P3> dcterms:issued "Y3"
<http://www.disco.unimib.it> rdfs:label "University of Milano-Bicocca"
```
3

```
<http://dblp.l3s.de/.../publications/P4> dc:creator <http://dblp.l3s.de/.../authors/A1'>
<http://dblp.l3s.de/.../authors/A1'> opus:has_affiliation <http://www.kit.edu>
<http://dblp.l3s.de/.../authors/A1'> foaf:homepage <http://www.kit.edu/A1>
<http://dblp.l3s.de/.../publications/P4> dcterms:issued "Y4"
<http://www.kit.edu/> rdfs:label "Karlsruhe Institute of Technology"
```
4

Fig. 1. An example of RDF triples that describe the evolution of the entity author

4   Results and Conclusions

The PhD work is now in the first year. Current work involves analysing the data quality dimensions addressed by several research efforts. The results so far are: literature study on data quality methodologies by a comparative description, create a framework to be able to provide a set of dimensions and their respective measures for helping the consumer or the publisher on evaluating the quality of their Linked Data. Related to the second contribution we firstly considered a comparison of interlinking approaches. The idea was to verify if there exist a full cover of all steps presented in the traditional record linkage activity. The approach we considered was the following: (i) define the main steps present in the record linkage (ii) evaluate the current works based on those steps (iii) define the problems and future works.

In Figure 2 we can see the works[2] considered in this comparison which operate principally on the instance matching level. The basic idea is that only one or two of them cover all the steps used in the record linkage task. A key aspect has been underestimated so far in the research in the linking task, in particular the Search Space Reduction step. This step has been deeply investigated only in two works. Therefore, in the Search Space Reduction there are no many methods proposed to reduce efficiently the number of entity comparisons. Finally, we can conclude that a lot of works need to be done to improve the quality of linking techniques

---

[2] http://dl.dropbox.com/u/2500530/Linked%20Data%20vs%20Record%20Linkage.pdf

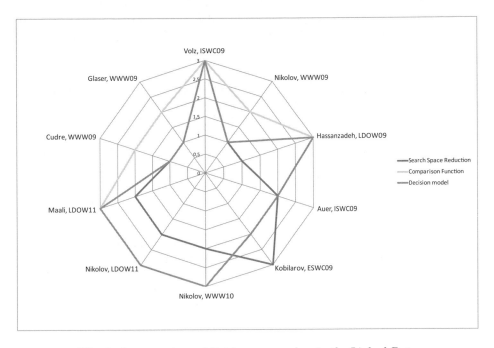

Fig. 2. A comparison of linking approaches in the Linked Data

in the Linked Data. With the perspective of improving the interlinking between data sets we concentrate on linking with temporal information such that to capture the effect of elapsed time on entity value evolution. We define a use case for applying the approach explained earlier. We consider that this work will be very beneficial for the above reasons.

In our future work, we will focus on the effectiveness of the assessment measures defined in the framework based on their nature (subjective or objective). We especially would be interested to investigate the role of the temporal linking technique in the overall framework, e.g., how some of the dimensions defined in the framework will be processed by the temporal linking technique to achieve better results.

Acknowledgements. I would like to thank my supervisor Andrea Maurino and my tutor Prof. Carlo Batini for their suggestions, and the anonymous reviewers for their helpful comments.

References

1. Quality criteria for linked data sources (2010),
   http://sourceforge.net/apps/mediawiki/trdf/
   index.php?title=quality-criteria-for-linked-data-sources

2. Batini, C., Cappiello, C., Francalanci, C., Maurino, A.: Methodologies for data quality assessment and improvement. Proceedings of the ACM Comput. Surv., 16:1–16:52 (2009)
3. Bizer, C., Cyganiak, R.: Quality-driven information filtering using the wiqa policy framework. Web Semantics (2009)
4. Bizer, C., Heath, T., Berners-Lee, T.: Linked Data - The Story So Far. IJSWIS 5(3), 1–22 (2009)
5. Elmagarmid, A., Ipeirotis, P., Verykios, V.: Duplicate record detection: A survey. IEEE Transactions on Knowledge and Data Engineering, 1–16 (2007)
6. Euzenat, J., Shvaiko, P.: Ontology Matching. Springer-Verlag New York, Inc. (2007)
7. Fellegi, I., Sunter, A.: A theory for record linkage. Journal of the American Statistical Association, 1183–1210 (1969)
8. Gutierrez, C., Hurtado, C.A., Vaisman, A.: Introducing time into rdf. IEEE Trans. on Knowl. and Data Eng., 207–218 (2007)
9. Halpin, H., Hayes, P.J., McCusker, J.P., McGuinness, D.L., Thompson, H.S.: When owl:sameAs isn't the same: An analysis of identity in linked data. In: Patel-Schneider, P.F., Pan, Y., Hitzler, P., Mika, P., Zhang, L., Pan, J.Z., Horrocks, I., Glimm, B. (eds.) ISWC 2010, Part I. LNCS, vol. 6496, pp. 305–320. Springer, Heidelberg (2010)
10. Hartig, O., Zhao, J.: Using web data provenance for quality assessment. In: Proceedings of the International Workshop on Semantic Web and Provenance Management, Washington DC, USA (2009)
11. Heath, T., Bizer, C.: Linked Data: Evolving the Web into a Global Data Space. Morgan & Claypool (2011)
12. Hobbs, J.R., Pan, F.: An ontology of time for the semantic web, pp. 66–85 (2004)
13. Hogan, A., Harth, A., Passant, A., Decker, S., Polleres, A.: Weaving the pedantic web. In: International Workshop on LDOW at WWW (2010)
14. Hoxha, J., Rula, A., Ell, B.: Towards green linked data. Submission on the Semantic Web-ISWC (2011)
15. Klein, M., Fensel, D.: Ontology versioning on the semantic web, pp. 75–91. Stanford University (2001)
16. Li, P., Dong, X., Maurino, A., Srivastava, D.: Linking temporal records. In: Proceedings of the VLDB Endowment, vol. 4 (2011)
17. Ozsoyoglu, G., Snodgrass, R.T.: Temporal and real-time databases: A survey. IEEE Trans. on Knowl. and Data Eng., 513–532 (1995)
18. Pernici, B., Scannapieco, M.: Data Quality in Web Information Systems. In: Spaccapietra, S., March, S., Aberer, K. (eds.) Journal on Data Semantics I. LNCS, vol. 2800, pp. 48–68. Springer, Heidelberg (2003)
19. Tappolet, J., Bernstein, A.: Applied Temporal RDF: Efficient Temporal Querying of RDF Data with SPARQL. In: Aroyo, L., Traverso, P., Ciravegna, F., Cimiano, P., Heath, T., Hyvönen, E., Mizoguchi, R., Oren, E., Sabou, M., Simperl, E. (eds.) ESWC 2009. LNCS, vol. 5554, pp. 308–322. Springer, Heidelberg (2009)
20. Volz, J., Bizer, C., Gaedke, M., Kobilarov, G.: Discovering and Maintaining Links on the Web of Data. In: Bernstein, A., Karger, D.R., Heath, T., Feigenbaum, L., Maynard, D., Motta, E., Thirunarayan, K. (eds.) ISWC 2009. LNCS, vol. 5823, pp. 650–665. Springer, Heidelberg (2009)
21. Wang, R.Y.: A product perspective on total data quality management. Commun. ACM, 58–65 (1998)
22. Wang, Y., Zhu, M., Qu, L., Spaniol, M., Weikum, G.: Timely yago: harvesting, querying, and visualizing temporal knowledge from wikipedia. In: Proceedings of the International Conference on Extending Database Technology. ACM (2010)

DC Proposal: Towards a Framework for Efficient Query Answering and Integration of Geospatial Data*

Patrik Schneider**

Institut für Informationssysteme, Technische Universität Wien
Favoritenstraße 9-11, A-1040 Vienna, Austria
patrik@kr.tuwien.ac.at

Abstract. Semantic Web technologies are becoming more interleaved with geo-spatial databases, which should lead to an easier integration and querying of spatial data. This is fostered by a growing amount of publicly available geospatial data like OpenStreetMap. However, the integration can lead to geographic inconsistencies when combining multiple knowledge bases. Having the integration in place, users might not just issue a points-of-interest search, but rather might be interested in regions with specific attributes assigned to them. Though, having large amounts of spatial data available, standard databases and reasoners do not provide the means for (quantitative) spatial queries, or struggle to answer them efficiently. We seek to combine spatial reasoning, (nonmonotonic) logic programming, and ontologies for integrating geospatial databases with Semantic Web technologies. The focus of our investigation will be on a modular design, on efficient processing of large amounts of spatial data, and on enabling default reasoning. We propose a two-tier design related to HEX-programs, which should lead to a plausible trade-off between modularity and efficiency. Furthermore, we consider suitable geo-ontologies to semantically annotate and link different sources. Finally, the findings should lead to a proof-of-concept implementation, which will be tested for efficiency and modularity in artificial and real-world use cases.

1   Background and Problem Statement

Fostered by a popular demand for location-aware search applications, linking and querying spatial data has become an active research field. At the same time, governments open up their official datasets for public use, and collaborative projects like OpenStreetMap (OSM) are becoming large sources of spatial data (http://www.openstreetmap.org/). In this context, geospatial databases are the backbone for storing and querying these data. Hence they have been extensively studied by the Geographic Information Systems (GIS) community (cf. [9,18]).

Geospatial databases often have the drawback that querying them is complicated, inference mechanisms are virtually non-existent, and extending them is difficult. In response, Semantic Web technologies are becoming more interleaved with geospatial databases [5,24], which should lead to an easier integration and querying of spatial

---

* Supported by the Austrian Research Promotion Agency (FFG) project P828897, the Marie Curie action IRSES under Grant No. 24761 (Net2), and the EC project OntoRule (IST-2009-231875).
** Advisors: Thomas Eiter and Thomas Krennwallner.

L. Aroyo et al. (Eds.): ISWC 2011, Part II, LNCS 7032, pp. 349–356, 2011.
© Springer-Verlag Berlin Heidelberg 2011

data. The integration should happen on several levels. First, the different data sources have to be linked to concepts and roles of different (often predefined) ontologies. After having several ontologies and their assertions in place, they have to be merged or linked in a certain manner. However, on the spatial level, linking can cause geographic inconsistencies, e.g., if the same place is located on different coordinates because of imprecise data. On the logical level, inconsistencies arise by introducing contradictions in the joint knowledge base (KB). Searching and retrieving location-based information is possible with a working integration. But users might be interested in *areas* with specific attributes assigned to them instead of searching plain or semantically annotated points-of-interest (POIs). For example, the walk-ability of a certain neighborhood in a city could be of interest (cf. http://www.walkscore.com/). However, having spatial data of larger cities or even countries, standard tableaux-based reasoners do not provide the means for (quantitative) spatial queries or struggle to answer them efficiently.

We seek to combine spatial reasoning, (nonmonotonic) logic programming rules (such as Answer Set Programming [6], Prolog [2], or Semantic Web Rule Language (SWRL) [17]), and ontologies for interleaving the different approaches. Besides having rules to formalize spatial integrity constraints [1], rules opens a way to qualitative spatial reasoning with the Region Connection Calculus (RCC) being applied on top of ontologies and spatial data [15]. Several authors have considered these combinations. We distinguish between heterogeneous and homogeneous combinations, where heterogeneous combinations can be separated into loose couplings and tight integrations. Grütter *et al.* focus on enhancing ontologies with spatial reasoning based on RCC and SWRL-rules to capture dependencies between different administrative regions [15]. In [1], a combined framework based on Description Logic Programs (DLP) was introduced. This approach considers the coupling with a spatial databases and the focus on the formulation of spatial constrains. With PelletSpatial and DLMAPS, there has been proof-of-concept implementations of qualitative spatial reasoning in a Description Logics (DL) reasoner, featuring consistency checking and spatial query answering [27,28].

Modularity as a design goal enables the integration of external computation sources, which could be a DL reasoner, spatial data sources, or computational geometry engines. With efficiency as a goal, we have to identify efficient external computation sources, optimize information flow between them, and prune intermediate results. We need to put a particular focus on the trade-off between expressivity and performance first, and between pre-computation and on-demand calculation of queries thereafter. We will try to capture nonmonotonic notions such as reasoning by default to express exceptions. For example, it is important to express statements such as "by default all restaurants in a city are non-smoking," but we like to state some exceptions with a smoking permission. Concluding from the points above, we will focus on the following research questions:

- Given the focus of our investigation on modular design, efficiency, and nonmonotonic reasoning, which will be the most suitable architecture for a framework of combining spatial data, rules, and ontologies?
- What methods are feasible to infer certain regions from a selected point set? And, how to deduce attributes from the created regions? For example, we would like to investigate which point sets (e.g., cafés, restaurants, or pubs) make up a suitable neighborhood for dining.

- Considering several spatial data sources like OSM or Open Government Data (OGD), how can we combine this sources using logic programs as an integration mechanism?
- How does the framework behave for query answering over large data sets (e.g., the OSM data set of Austria)? And how does it perform in an artificial test environment and further under real-life conditions in an e-government and public transit system.

2 Related Work

A well known top level ontology is GeoOWL, which keeps a strict distinction between the geographic object, called *Feature*, and its footprint, called *Geometry* (http://www.w3.org/2005/Incubator/geo/XGR-geo/). Furthermore, GeoNames is a feature-centric geographical database containing about 7.5 million unique features (http://www.geonames.org/). In particular the categorization in nine top-feature (such as area feature, road feature, building feature, etc.) and corresponding sub-features (e.g., street, railroad, trail) are of interest. The creators of GeoNames also created an OWL ontology, which is very instance-heavy, with just a few concepts defined. Coming from the field of pervasive computing, the Standard Ontology for Ubiquitous and Pervasive Applications (SOUPA) is a general ontology, which covers the domain of space, time, actions, and agents [8]. The top concept *SpatialThing* is divided into the sub-concepts *GeographicalSpace* and *RCCSpatialRegion*. Different from other frameworks, the RCC calculus is considered as a concept instead of a set of transformation rules.

RCC8 is a fragment of RCC, where eight binary predicates are defined for representing the relationships between two regions [3,23]. Drawing from the close connection between DL and Modal Logics, and between *RCC8* and Modal Logics, Katz and Cuenca Grau defined a translation from RCC8 into DL. This is achieved by defining a DL concept for every region and a set of translation rules for every RCC8 constructor [19].

Smart *et al.* and Abdelmoty *et al.* [26,1] address the need for rules to extend geo-ontologies and facilitate spatial reasoning. This is mainly due to the limited expressivity of OWL. They identify two possible extensions, namely the formulation of spatial integrity constraints and rules for spatial relationships between objects in space. They developed a geo-ontology framework which is split into three components: a geo-ontology management system, a spatial reasoning engine, and an error management system [26,1]. In [28] the authors describe a framework which focuses on four different ideas. The first approach deals with compiling the spatial relationships to the ABox and using nRQL as a query language. Another approach is a hybrid concept, where the ABox is associated with a *Space Box* (SBox) containing a set of spatial ground atoms which represent the whole spatial information. For querying the SBox, nRQL [16] is used, which is extended with *spatial query atoms*. In another approach, the ABox is extended again with an SBox, but spatial assertions are computed by means of inspection methods and materialized on the fly. The last method uses a standard ABox and exploits qualitative spatial reasoning, which is usable through spatial query atoms. All approaches were incorporated in the DL reasoner RacerPro for the DLMAPS system. PelletSpatial is a proof-of-concept implementation of RCC8 with a DL reasoner, featuring consistency checking and query answering. The authors extended the DL reasoner with a hybrid RCC8 reasoning engine, which is based on a path-consistency algorithm and a RCC8 composition table [27]. Finally, in the work of Grütter *et al.* a web search is enhanced with DL and spatial reasoning based on RCC8. RCC8 is encoded in SWRL-rules to capture spatial dependencies between different administrative regions [15].

SWRL was one of the first proposals for combining rules and ontologies. The rule layer in SWRL was set on top of an OWL KB by allowing material implication of OWL expressions [17]. Heterogeneous loose coupled approaches keep the rule base and DL KB as separate, independent components. The knowledge exchange is managed by an interface between the components. As a prominent example Description Logic Programs (dl-programs) can be taken, which were introduced by [11] and combine DL and normal logic programs under stable model semantics. Later they were extended in [10] to well-founded semantics. The concept of plug-ins in dl-programs was further generalized to HEX-programs [12] and lead to the successful development of the dlvhex reasoner.[1] In heterogeneous tight integrated approaches the combining of rules and DL is based on the integration of their models, where each model should satisfy its domain and agree with the other model. *CARIN* [20] and $\mathcal{DL}$+log [25] represent this approach. Full integrated approaches do not have any separation between the two vocabularies, this could either be achieved by a bidirectional translation of the different vocabularies or by rewriting both vocabularies to an overlapping formalism. Description Logic Programs (DLP) [14] and Hybrid MKNF knowledge bases [22] can be counted to this approach.

In the wider scope of our interest are Semantic Web search engines. For example, the authors of [13] developed two prototypes of a search engine. In the proposed systems an additional annotation step is used together with a domain specific ontology, with this step semantics is added to the elements of a web page.

3   Expected Contributions

Our contribution will be partly on the formal level and partly will include practical aspects. Derived from the research questions, we identify the following objectives:

- Creating a rule-based framework for combining heterogeneous spatial data, ontological reasoning, and spatial reasoning with focus on modular design and efficiency. Furthermore, we will consider non-monotonic features such as exception handling.
- Qualitative spatial reasoning will be considered, first by inferring regions out of points, and second by defining spatial relations among the regions. The spatial relations could be expressed in the well-know calculus RCC8. Furthermore, we will investigate the qualitative attributes of the inferred regions.
- The data integration should consider the semantical annotation and linking of heterogeneous data, and suitable ontologies for OSM, OGD, and other sources are needed. We will evaluate whether a modular or a centralized approach is more appropriate.
- Finally, we will provide a proof-of-concept implementation, which will be evaluated for query answering on large data sets and benchmarked against existing tools like PelletSpatial or DLMAPS.

We recognize, that objectives are quite challenging, particularly to find a good trade-off between a modular design and efficiency.

---

[1] http://www.kr.tuwien.ac.at/research/systems/dlvhex/

4   Proposed Methods

To fulfill the objectives, we have to put our focus on two issues. First, we need *a formal representation of the different abstraction levels* of spatial data. Based on the representation, *an inference mechanism for constraint checking and query answering* can be defined. Second, we need to outline an *architecture*, which is built around a multi-tiered reasoning engine. A central part of the architecture will be a top-level geo-ontology, which acts as an central repository for linking the different spatial data sources. The data sources could be defined in their own specific ontologies, which could be linked to the top-level ontology. The data sources will mostly be points in the metric space, but also other geometrical objects as lines and polygons appear.

4.1   Query Rewriting and Reasoning

We distinguish two different models for spatial data, namely point-based and region-based spatial models. The related spatial logics are considered either with reasoning about topological relations (interpretation over topological spaces) or about distances over metric spaces. We will start with point-based spaces and extend them by transformation to region-based spaces. These transformations could be calculated by a convex hull or by Voronoi tessellation [4]. The transformations require first metric spaces, but in addition also topological spaces by describing the relations between regions in RCC8. We identify a two-tier design to achieve a good trade-off between modularity and scalability.

Tier 1. The first tier is concerned with DL reasoning extended with point- and range-based spatial queries. At this point, there is no qualitative spatial reasoning, just a transformation (similar to the first step of [19]) of points from the spatial model to ABox assertions. Furthermore, we already consider the later described semantical annotation of points, which links the spatial data to DL concepts and roles.

Tier 2. The second tier is responsible for advanced transformation like Voronoi tessellation and qualitative spatial reasoning like RCC8. We propose to use HEX-programs, which facilitate declarative meta-reasoning through higher-order atoms. HEX-programs are an interesting candidate, because external atoms (e.g., description logic (dl) atoms) offer a query interface to other external computation sources (e.g., DL reasoning). Following from the idea of dl-atoms, we could extend HEX-programs with spatial atoms, which encapsulate access to the first tier and enables qualitative reasoning capabilities.Furthermore, by using HEX-programs under the answer-set semantics, we will be able to perform default and closed-world reasoning, translating and manipulate reified assertions, exception handling, and search in the space of assertions [12].

Dealing with Inconsistencies. Contrary to [26], we omit an error management system and perform inconsistency checks as a preprocessing step. ABox inconsistency is checked for linked spatial assertions by simply using the underlying DL reasoner. But as mentioned in [26], we have to deal with topological, directional, and duplicate inconsistencies. Stocker *et al.* [27] propose to check topological inconsistencies by calculating an $n \times n$ matrix M, where n are the different regions (represented as polygons), and using the *path-consistency algorithm* on M to approximate consistency. Duplicate inconsistencies occur with spatial objects of different sources, which are *intentionally* the same, but are not aligned by `owl:sameAs`. We use custom heuristics to determine the similarity between objects, where the objects names, directions, shapes, and locations are considered. We leave directional inconsistencies for further research activities.

4.2 Architecture

Regarding technical aspects of the framework, we propose an architecture consisting of the following four parts:

Geo-Ontology. The OWL2 profiles OWL2 QL [7] and OWL2 EL [21] are interesting candidates for representing our spatial ontology. There has been a considerable effort of developing efficient query answering algorithms over ontologies using an RDBMS for both DLs. We draw on already defined work with GeoOWL for modeling a suitable geo-ontology. For our needs, the feature concept of GeoOWL is too general, thus we adopt a more detailed categorization based on GeoNames and OSM. There has been a large community effort to categorize the geospatial information in OSM, so we can derive with some additions an ontology out of the categorization. It is open how to unify the various OGD sources under a common ontology. However, most of the considered OGD sources of the city of Vienna are covered by the existing OSM concepts.

Knowledge Base. Heterogeneous sources of spatial data like OSM, OGD, or even local food guides could be integrated and linked by a central DL KB, which is part of our first tier. The KB should be based on our geo-ontology and act as a repository for the different sources keeping the vocabulary needed for concept and role queries. In particular, the annotated information should be kept in the KB, so we fulfill the modularity criteria and do not alter the original data sources. For example, for querying restaurants, the type of cuisine and atmosphere will be kept in the KB, but the geospatial information will be kept in the spatial database. The different spatial data sources can be stored in database like PostGIS keeping their native projections. Hence we will use a projection function to convert points to a predefined reference coordinate system.

Reasoning Engine. The first tier of the engine will incorporate a DL reasoner and a proprietary implementation for spatial predicates like *Near* or *Along*. From the vocabulary of our geo-ontology in combination with spatial functions, a join of a spatial and DL query will be created. We will exploit the rewriting techniques of OWL2 QL by compiling TBox and query into a SQL statement, which can be evaluated by an RDBMS over the ABox [7]. The second tier of the engine will mainly be build around dlvhex, an implementation of HEX-programs. We need at least one plugin to a computational geometry engine, one plugin to the first tier, and one plugin, which evaluates RCC8 relations on existing or on-demand calculated regions. Depending on the external data sources, a further plugin for accessing directly an RDBMS might be desirable.

Annotating Engine. A priori, the spatial data are not linked to any DL KB. Hence this step is crucial for integration them and extending the vocabulary of the queries. The spatial data is *linked* to the DL KB concepts and roles by asserting spatial objects to the ABox. For the OSM data, a straightforward task is to find related concepts, because our geo-ontology is partly derived from the OSM categories. For other data sources, we have to develop domain specific heuristics to assign spatial objects to our categorization (e.g. restaurant which are named *pizzeria* belong to Italian restaurants).

5 Conclusion and Future Work

We have proposed a novel framework for combining heterogeneous spatial data, on-tological reasoning, and spatial reasoning. Our focus will be mostly on modularity, scalability, and non-monotonic features as default reasoning. Query Answering will not just cover standard POIs searches, but queries based on qualitative spatial reasoning (e.g., RCC8), which considers the inference of regions and the calculation of properties and spatial relations among the regions. For the data integration we consider the se-mantical annotation and linking of OSM, OGD, and further data sources by a top-level geo-ontology, which acts as a central repository. The above finding should lead to a proof-of-concept implementation with HEX-programs, which should be integrated into a larger research prototype containing a routing services and POI exploring facilities.

The ability to formulate constraints and defaults is a common goal for a knowl-edge representation formalism and increases expressivity quite dramatically. Using the loose-coupling approach to combining rules with external computation sources helps by facilitating modularity and allowing to integrate different DL reasoners and compu-tational geometry engines quite naturally. However, compared to a tight coupling ap-proach, using external computation sources usually has a negative effect on efficiency, as the external sources could be intractable, and certain optimizations and structural dependencies are easier to detect in homogeneous KBs.

Our future work will be along two parallel paths. Along the theoretical path, we have to investigate query rewriting and reasoning. Particularly the combination of modular HEX-programs, qualitative spatial reasoning, and DL needs to be addressed. We will also investigate how well qualitative spatial reasoning can be expressed in a rule-based languages. Furthermore, we will refine our centralized geo-ontology and investigate on a more modular design. Along the practical path, we will develop several dlvhex plu-gins, which enables access to spatial data and computational geometry functions. Fur-thermore, we will consider an implementation, which is more geared towards tractable evaluation. Finally, our implementation will be tested for efficiency and modularity in an artificial test environment and further under real-life conditions in an e-government and public transit system. The implementation is part of a larger project concerned with e-government and public transit systems.

References

1. Abdelmoty, A., Smart, P., El-Geresy, B., Jones, C.: Supporting Frameworks for the Geospa-tial Semantic Web. In: Mamoulis, N., Seidl, T., Pedersen, T.B., Torp, K., Assent, I. (eds.) SSTD 2009. LNCS, vol. 5644, pp. 355–372. Springer, Heidelberg (2009)
2. Apt, K.R., Warren, D.S., Truszczynski, M.: The Logic Programming Paradigm: A 25-Year Perspective. Springer, New York (1999)
3. Bennett, B.: Spatial reasoning with propositional logics. In: KR 1994, pp. 51–62 (1994)
4. de Berg, M., Cheong, O., van Kreveld, M., Overmars, M.: Computational geometry: Algo-rithms and Applications. Springer, Heidelberg (2008)
5. Bishr, Y.A.: Geospatial semantic web: Applications. In: Encyclopedia of GIS, pp. 391–398. Springer, Heidelberg (2008)

6. Brewka, G., Eiter, T., Truszczyński, M.: Answer set programming at a glance. Commun. ACM (to appear, 2011)
7. Calvanese, D., De Giacomo, G., Lembo, D., Lenzerini, M., Poggi, A., Rodriguez-Muro, M., Rosati, R.: Ontologies and databases: The *DL-lite* approach. In: Tessaris, S., Franconi, E., Eiter, T., Gutierrez, C., Handschuh, S., Rousset, M.-C., Schmidt, R.A. (eds.) Reasoning Web 2009. LNCS, vol. 5689, pp. 255–356. Springer, Heidelberg (2009)
8. Chen, H., Perich, F., Finin, T.W., Joshi, A.: Soupa: Standard ontology for ubiquitous and pervasive applications. In: MobiQuitous 2004, pp. 258–267 (2004)
9. DeMers, M.N.: Fundamentals of geographic information systems, 4th edn. Wiley (2008)
10. Eiter, T., Ianni, G., Lukasiewicz, T., Schindlauer, R.: Well-founded semantics for description logic programs in the semantic web. ACM Trans. Comput. Log. 12(2), 11 (2011)
11. Eiter, T., Ianni, G., Lukasiewicz, T., Schindlauer, R., Tompits, H.: Combining answer set programming with description logics for the semantic web. Artif. Intell. 172(12-13), 1495–1539 (2008)
12. Eiter, T., Ianni, G., Schindlauer, R., Tompits, H.: Effective Integration of Declarative Rules with External Evaluations for Semantic-Web Reasoning. In: Sure, Y., Domingue, J. (eds.) ESWC 2006. LNCS, vol. 4011, pp. 273–287. Springer, Heidelberg (2006)
13. Fazzinga, B., Gianforme, G., Gottlob, G., Lukasiewicz, T.: Semantic Web Search Based on Ontological Conjunctive Queries. In: Link, S., Prade, H. (eds.) FoIKS 2010. LNCS, vol. 5956, pp. 153–172. Springer, Heidelberg (2010)
14. Grosof, B.N., Horrocks, I., Volz, R., Decker, S.: Description logic programs: combining logic programs with description logic. In: WWW 2003, pp. 48–57. ACM (2003)
15. Grütter, R., Scharrenbach, T., Waldvogel, B.: Vague spatio-thematic query processing: A qualitative approach to spatial closeness. Trans. GIS 14(2), 97–109 (2010)
16. Haarslev, V., Möller, R., Wessel, M.: Querying the semantic web with racer + nrql. In: ADL 2004 (2004)
17. Horrocks, I., Patel-Schneider, P.F.: A proposal for an owl rules language. In: WWW 2004, pp. 723–731 (2004)
18. Jones, C.B.: Geographical information systems and computer cartography. Prentice Hall (1997)
19. Katz, Y., Cuenca Grau, B.: Representing qualitative spatial information in owl-dl. In: OWLED 2005 (2005)
20. Levy, A.Y., Rousset, M.C.: Combining horn rules and description logics in carin. Artif. Intell. 104(1-2), 165–209 (1998)
21. Lutz, C., Toman, D., Wolter, F.: Conjunctive query answering in the description logic el using a relational database system. In: IJCAI 2009, pp. 2070–2075 (2009)
22. Motik, B., Rosati, R.: A faithful integration of description logics with logic programming. In: IJCAI 2007, pp. 477–482 (2007)
23. Renz, J.: Qualitative Spatial Reasoning with Topological Information. LNCS (LNAI), vol. 2293. Springer, Heidelberg (2002)
24. Rodríguez, M.A., Cruz, I.F., Egenhofer, M.J., Levashkin, S.: GeoS 2005. LNCS, vol. 3799. Springer, Heidelberg (2005)
25. Rosati, R.: Dl+log: Tight integration of description logics and disjunctive datalog. In: KR 2006, pp. 68–78 (2006)
26. Smart, P.D., Abdelmoty, A.I., El-Geresy, B.A., Jones, C.B.: A Framework for Combining Rules and Geo-Ontologies. In: Marchiori, M., Pan, J.Z., Marie, C.d.S. (eds.) RR 2007. LNCS, vol. 4524, pp. 133–147. Springer, Heidelberg (2007)
27. Stocker, M., Sirin, E.: Pelletspatial: A hybrid rcc-8 and rdf/owl reasoning and query engine. In: OWLED 2009. Springer, Heidelberg (2009)
28. Wessel, M., Möller, R.: Flexible software architectures for ontology-based information systems. J. Applied Logic 7(1), 75–99 (2009)

DC Proposal: Automatically Transforming Keyword Queries to SPARQL on Large-Scale Knowledge Bases

Saeedeh Shekarpour*

Universität Leipzig, Institut für Informatik, AKSW,
Postfach 100920, D-04009 Leipzig, Germany
lastname@informatik.uni-leipzig.de
http://aksw.org

Abstract. Most Web of Data applications focus mainly on using SPARQL for issuing queries. This leads to the Web of Data being difficult to access for non-experts. Another problem that will intensify this challenge is when applying the algorithms on large-scale and decentralized knowledge bases. In the current thesis, firstly we focus on the methods for transforming keyword-based queries into SPARQL automatically. Secondly, we will work on improving those methods in order to apply them on (a large subset of) the Linked Data Web. In an early phase, a heuristic method was proposed for generating SPARQL queries out of arbitrary number of keywords. Its preliminary evaluation showed promising results. So, we are working on the possible improvements for applying that on the large-scale knowledge bases.

1 Introduction

Web of Data is growing at an astounding rate (currently amounting to 28 Billion triples[1]) and contains a wealth of information on a large number of domains which enables automated agents and other applications to access the information on the web more intelligently. The vendor-independent standard SPARQL has been specified and allows to query this knowledge easily. Yet, while SPARQL is very intuitive for SQL-affine users, it is very difficult to use for lay users who are not familiar with the concepts behind the Semantic Web. Because a naive user needs to acquire both knowledge about the underlying ontology and proficiency in formulating SPARQL queries to query the endpoint. Also, with the huge amount of background knowledge present in Linked Data, it is difficult for naive as well as proficient users to formulate SPARQL queries. Consequently, there is a blatant need for another mean to query the Semantic Web that is appealing for novice users. In other word, the naive user prefers to hold the traditional paradigm of interaction with search engines which is based on natural language or keyword-based queries.

---

* Supervisor: Sören Auer.

[1] http://www4.wiwiss.fu-berlin.de/lodcloud/state/ (June 19th, 2011)

L. Aroyo et al. (Eds.): ISWC 2011, Part II, LNCS 7032, pp. 357–364, 2011.
© Springer-Verlag Berlin Heidelberg 2011

Nowadays, keyword-based search is the most popular and convenient way for finding information on the Web. The successful experience of keyword-based search in document retrieval and the satisfactory research results about the usability of this paradigm [23] are convincing reasons for using the keyword search paradigm to the Semantic Web.

Document retrieval approaches use probabilistic foundations combining with ranking algorithms for selecting documents containing those keywords which practically works well. Since the nature of RDF data is different and potentially, information need is more complex, we require novel and efficient algorithms for querying Data Web. RDF data has a directed and multiple edges graph structure with typed vertices and labeled edges. The fundamental base of that is entity which has three main features as type, attributes and relations with other entities. Therefore, objective of search on this data is retrieval of entities or maybe along with some attributes and relations.

Although it is around one decade that semantic search has been targeted by researchers, to the best of our knowledge, still there is not any service which practically obviate this need. Services such as *Sindice* [25], *Sig.ma* [24], *Swoogle* [6] or *Watson* [5] offer simple search services[2], but are either restricted to the retrieval of single RDF documents or in the case of Sig.ma to the retrieval of information about a single entity from different sources.

In essence, the problem which we are addressing in the proposal can be stated as follows:

How can we interpret a user query in order to locate and exploit relevant information for answering the user's query using large-scale knowledge bases from the Linked Data Web?

Figure 1 shows a birds-eye-view of the envisioned research. Based on a set of user-supplied keywords, first, candidate IRIs (Internationalized Resource Identifier) for each of the keywords issued by the user is computed. Then, by using an inference mechanism, a subgraph based on the identified IRIs is extracted and represented to the user as the answer.

2   Research Challenges

To query Web of data by naive user, the NL (Natural Language) queries should be interpreted in accordance to user's intention. Because natural language inherently includes ambiguity, precise interpretation is difficult. For instance, relations of terms in an NL query are prevalently either unknown or implicit. In addition, the retrieval of knowledge from a large-scale domain such as Linked Data Web needs efficient and scalable algorithms. Some algorithms and applications can be robust in a small or medium scale, while they may not be suitable when applied to large scale knowledge bases present in the entire Linked Data Web. Therefore, a suitable approach for querying the Linked Data Web is essential for retrieving information both efficiently and precisely in a way that can be easily utilized

---

[2] These systems are available at: http://sindice.com, http://sig.ma, http://swoogle.umbc.edu, http://kmi-web05.open.ac.uk/WatsonWUI

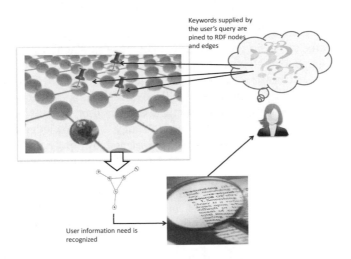

Fig. 1. Birds-eye-view of the envisioned research

by a naive user. Based on this observation, we will investigate 3 fundamental questions:

- Can we (semi-)automatically interpret user queries using background knowledge from the Linked Data Web possibly involving the user in a feedback loop?
- How can we efficiently retrieve meaningful answers for the user query from large scale knowledge bases on the Linked Data Web?
- How can we cluster and represent the retrieved answers to the user in a meaningful way?

With regard to the first problem, we will use the keyword-based query that a user enters in a traditional interface, as a model. This model involves some difficulties due to the ambiguity, because in natural language, a query may stand for different meanings and be interpreted in different ways. This ambiguity in the case of keyword-based query is even intensified. Therefore, the main task is to interpret the user query using background knowledge. In the case of the second question, the user query needs to be transformed to the formal SPARQL query syntax and sent to different knowledge bases for retrieving relevant instances.

For tackling the third question, we will investigate clustering methods and extend an approach which can be applied on the graph nature of the answers. Therefore, the instances will be delivered in different categories.

Another problem that will intensify the aforementioned challenges is when applying the algorithms on large-scale and decentralized datasets, which will pose several difficulties. For example, mapping of user query terms to various ontologies causes severe ambiguity due to redundancy in entities and terminologies.

3   Methodology

Figure 2 indicates the main components of this thesis. The first component consists of processing and analyzing the keyword-based query employing a linguistic component. We will use the GATE infrastructure in order to pre-process the query. The output of this component will be a series of terms associated with the input query. Afterwards, extracted terms of the user query will be mapped to entities which exist in the underlying ontology and knowledge base such as DBpedia. The mapping process will be performed by applying some similarity metrics with regarding the context and using synonyms provided by lexical resources such as WordNet.

In the second component, the relation of the mapped entities will be recognized by using a heuristics approach over a graph traversal method. We aim to make a comparison between the existing approaches and our proposed method. We will design an interface that enables a user to interact with the system without any special knowledge of vocabularies or structure of the knowledge base. We need to mention that various interpretations of the user query are created since different options for mapping user terms to entities and connecting these mapped entities exist. With regard to tools, we use Jena and Virtuoso. The results of both components can be evaluated based on a comparison with a gold standard (i.e. a hand-crafted collection of natural language queries and their equivalent SPARQL queries with automatically constructed SPARQL queries obtained from the inference engine). Some accuracy and performance metrics such as precision, recall, fscore, MMR and runtime will be chosen. Another component is related to clustering of the instances. We will investigate and develop a clustering method based on context and background knowledge for organizing instances.

At the last phase, we will work on improving the previous components and develop them in order to apply them on (a large subset of) the Linked Data Web. For this, we will use a benchmark method for development. For this, a set of datasets which are large in terms of both the number of triples and variety of the background knowledge will be chosen. In an iterative phase, each of the earlier components will be individually applied on a selected set of datasets (the size of the selected set of datasets will increase in each phase). By comparing the performance metrics in each phase, appropriate revisions will be done on the algorithms and methods.

4   A Synopsis of the Current Situation

In an early phase, a novel method for generating SPARQL queries based on two user-supplied keywords was proposed [21]. Since this method is based on simple operations, it can generate SPARQL queries very efficiently. Also it is completely agnostic of the underlying knowledge base as well as its ontology schema. We currently use DBpedia as the underlying knowledge base, but this method is easily transferable to the whole Data Web. The implementation of this method

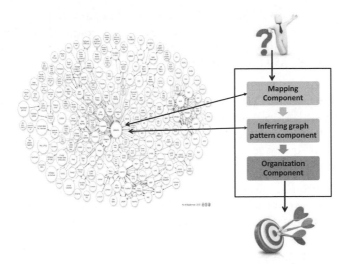

Fig. 2. Architecture of the envisioned semantic search algorithm

is publicly available at: http://lod-query.aksw.org. Table 1 shows some samples of keywords for which the application is capable to retrieve suitable results. These keywords were categorized based on the type of queries which they can answer in three categories, i.e. similar instances, characteristics of an instance and associations between instances.

Table 1. Samples of keywords and results

Keywords	Answers
	Instance characteristics.
Kidman spouse	`d:Kidman  dp:spouse  Keith Urban .`
Iran language	`d:Iran  dp:Language  d:Persian_language .`
	Associations between instances.
Obama Clinton	`d:Obama  dp:predecessor  d:Bush .  d:Bush  dp:predecessor  d:Clinton .`
Volkswagen Porsche	`d:Volkswagen_Group  dp:subsidiary  d:Volkswagen .`
	Similar instances.

Germany Island	`1. d:Germany  dp:Islands  d:Rgen. d:Rgen a do:Island.`
	`2. d:Germany  dp:Islands  d:Fhr. d:Fhr  a do:Island.`
	`3. d:Germany  dp:Islands  d:Sylt.  d:Sylt   a do:Island.`

Lost Episode	`1. d:Raised_by_Another  dp:series  dbp:Lost.`
	`   d:Raised_by_Another  a       do:TVEpisode.`
	`2. d:Homecoming        dp:series  dbp:Lost.`
	`   d:Homecoming        a       do:TVEpisode.`
	`3. d:Outlaws           dp:series  dbp:Lost.`
	`   d:Outlaws           a       do:TVEpisode.`

In the next phase, we proposed a method for generating SPARQL queries for arbitrary number of keywords [Submitting Article to WSDM 2012]. Its base is an inference mechanism embedded in a graph traversal approach which both accurately and efficiently builds SPARQL queries for the straightforward inter-pretations of the user's queries. Implementation of this method is also available at: http://sina.aksw.org. This application generats SPARQL query for those keyword-based queries which can be converted to a conjunctive query. A conjunctive query is a conjunction of triple patterns (based on the standard notions of the RDF and SPARQL specifications). For instance, consider the example shown below:

Example 1. Let suppose we have the keywords *people, birthplace, deathplace* and *Southampton.* A straightforward interpretation of these keywords is *people whose birthplace and deathplace are in southampton.* It can be expressed by a conjunctive query as $(?p\ a\ Person) \wedge (?p\ birthplace\ Southampton) \wedge (?p\ deathplace\ Southampton)$

Because of the promising results of the preliminary evaluation, we are currently working on this method for improving the overall functionality with regard to accuracy and efficiency so as to run that over more number of knowledge bases.

5   State of the Art

With the advent of the Semantic Web, information retrieval and question answering approaches were adapted for making use of ontologies. We can roughly divide related work into ontology-based information retrieval, ontology-based question answering and keyword search on structured data.

Ontology-based information retrieval. Approaches falling into this category annotate and index documents using a background ontology. The retrieval process is subsequently carried out by mapping user query terms onto these semantic document annotations. The approaches described in [26,24,6,5,25,3,9,19] are examples of this paradigm. All these approaches use background knowledge to enhance the retrieval accuracy, however, they do not utilize the background knowledge for semantically answering user queries.

Ontology-based question answering. Approaches falling into this category take a natural language question or a keyword-based query and return matching knowledge fragments drawn from the knowledge base as the answer. There are two different methods: (1) Using linguistic approaches for extracting complete triple-based patterns (including relations) from the user query and matching these triples to the underlying ontology (e.g. PowerAqua [16], OntoNL [12] and FREyA [4]). (2) Detecting just entities in the user query and discovering relations between these entities by analysing the knowledge base. Examples for this second group are KIM [18] and OntoLook [13] and [20,7,27,17]. In these two approaches the RDF data is considered to be a directed graph and relations among entities are found through sequences of links (e.g. using graph traversal). Sheth [22] introduced the term *semantic association* for describing meaningful and complex relations between entities. Our work differs from these approaches, since it is completely independent of the underlying schema. Furthermore, schema information is in our approach just *implicitly* taken into account, so a complex induction procedure is not required.

Keyword search on relational and XML data. With the beginning of the millennium, research on keyword search on relational and XML data attracted research interest. Meanwhile there exist many approaches such as [1], [11], [10], [14] for the relational domain and [8], [15], [2] for the XML domain. Especially the relational domain is relevant to our work due to the similarities to the RDF datamodel. All these approaches are based on *schema graphs* (i.e. a graph where tables and their primary-foreign key relations are represented as nodes and edges, respectively). In our work, we do not rely on an explicitly given schema, which is often missing for datasets on the Web of Data. However, achieving sufficient performance for instant query answering is more an issue in the RDF case, which is why our approach is currently limited to two keywords.

References

1. Agrawal, S., Chaudhuri, S., Das, G.: Dbxplorer: A system for keyword-based search over relational databases. In: ICDE, pp. 5–16. IEEE Computer Society (2002)
2. Chen, L.J., Papakonstantinou, Y.: Supporting top-k keyword search in xml databases. In: ICDE, pp. 689–700. IEEE (2010)
3. Crestani, F.: Application of spreading activation techniques in information retrieval. Artif. Intell. Rev. 11(6), 453–482 (1997)
4. Damljanovic, D., Agatonovic, M., Cunningham, H.: FREyA: an Interactive Way of Querying Linked Data Natural Language. In: Proceedings of 1st Workshop on Question Answering over Linked Data (QALD-1), Collocated with the 8th Extended Semantic Web Conference (ESWC 2011), Heraklion, Greece (2011)
5. D'aquin, M., Motta, E., Sabou, M., Angeletou, S., Gridinoc, L., Lopez, V., Guidi, D.: Toward a new generation of semantic web applications. IEEE Intelligent Systems 23(3), 20–28 (2008)
6. Ding, L., Finin, T.W., Joshi, A., Pan, R., Scott Cost, R., Peng, Y., Reddivari, P., Doshi, V., Sachs, J.: Swoogle: a search and metadata engine for the semantic web. In: Grossman, D.A., Gravano, L., Zhai, C., Herzog, O., Evans, D.A. (eds.) CIKM, pp. 652–659. ACM (2004)
7. Guha, R.V., McCool, R., Miller, E.: Semantic search. In: WWW, pp. 700–709 (2003)
8. Guo, L., Shao, F., Botev, C., Shanmugasundaram, J.: Xrank: Ranked keyword search over xml documents. In: Halevy, A.Y., Ives, Z.G., Doan, A. (eds.) SIGMOD Conference, pp. 16–27. ACM (2003)
9. Holi, M., Hyvönen, E.: Fuzzy View-Based Semantic Search. In: Mizoguchi, R., Shi, Z.-Z., Giunchiglia, F. (eds.) ASWC 2006. LNCS, vol. 4185, pp. 351–365. Springer, Heidelberg (2006)
10. Hristidis, V., Gravano, L., Papakonstantinou, Y.: Efficient ir-style keyword search over relational databases. In: VLDB, pp. 850–861 (2003)
11. Hristidis, V., Papakonstantinou, Y.: Discover: Keyword search in relational databases. In: VLDB, pp. 670–681. Morgan Kaufmann (2002)
12. Karanastasi, A., Zotos, A., Christodoulakis, S.: The OntoNL framework for natural language interface generation and a domain-specific application. In: Digital Libraries: Research and Development, First International DELOS Conference, Pisa, Italy, pp. 228–237 (2007)
13. Li, Y., Wang, Y., Huang, X.: A relation-based search engine in semantic web. IEEE Trans. Knowl. Data Eng. 19(2), 273–282 (2007)

14. Liu, F., Yu, C.T., Meng, W., Chowdhury, A.: Effective keyword search in relational databases. In: Chaudhuri, S., Hristidis, V., Polyzotis, N. (eds.) SIGMOD Conference, pp. 563–574. ACM (2006)
15. Liu, Z., Chen, Y.: Reasoning and identifying relevant matches for xml keyword search. PVLDB 1(1), 921–932 (2008)
16. Lopez, V., Uren, V.S., Motta, E., Pasin, M.: Aqualog: An ontology-driven question answering system for organizational semantic intranets. J. Web Sem. 5(2), 72–105 (2007)
17. Ning, X., Jin, H., Jia, W., Yuan, P.: Practical and effective ir-style keyword search over semantic web. Inf. Process. Manage. 45(2), 263–271 (2009)
18. Popov, B., Kiryakov, A., Kirilov, A., Manov, D., Ognyanoff, D., Goranov, M.: Kim - semantic annotation platform. Journal of Natural Language Engineering 10(3-4), 375–392 (2004)
19. Rocha, C., Schwabe, D., de Aragão, M.P.: A hybrid approach for searching in the semantic web. In: WWW, pp. 374–383. ACM (2004)
20. Schreiber, G., Amin, A., Aroyo, L., van Assem, M., de Boer, V., Hardman, L., Hildebrand, M., Omelayenko, B., van Osenbruggen, J., Tordai, A., Wielemaker, J., Wielinga, B.: Semantic annotation and search of cultural-heritage collections: The MultimediaN E-Culture demonstrator. Journal of Web Semantics 6(4), 243–249 (2008)
21. Shekarpour, S., Auer, S., Ngomo, A.-C.N., Gerber, D., Hellmann, S., Stadler, C.: Keyword-driven sparql query generation leveraging background knowledge. In: International Conference on Web Intelligence (2011)
22. Sheth, A., Aleman-Meza, B., Budak Arpinar, I., Halaschek-Wiener, C., Ramakrishnan, C., Bertram, Y.W.C., Avant, D., Sena Arpinar, F., Anyanwu, K., Kochut, K.: Semantic association identification and knowledge discovery for national security applications. Journal of Database Management 16(1), 33–53 (2005)
23. Tran, T., Mathäß, T., Haase, P.: Usability of Keyword-Driven Schema-Agnostic Search. In: Aroyo, L., Antoniou, G., Hyvönen, E., ten Teije, A., Stuckenschmidt, H., Cabral, L., Tudorache, T. (eds.) ESWC 2010. LNCS, vol. 6089, pp. 349–364. Springer, Heidelberg (2010)
24. Tummarello, G., Cyganiak, R., Catasta, M., Danielczyk, S., Delbru, R., Decker, S.: Sig.ma: Live views on the web of data. J. Web Sem. 8(4), 355–364 (2010)
25. Tummarello, G., Delbru, R., Oren, E.: Sindice.com: weaving the open linked data, pp. 552–565 (2007)
26. Wang, H., Liu, Q., Penin, T., Fu, L., Zhang, L., Tran, T., Yu, Y., Pan, Y.: Semplore: A scalable ir approach to search the web of data. J. Web Sem. 7(3), 177–188 (2009)
27. Zenz, G., Zhou, X., Minack, E., Siberski, W., Nejdl, W.: From keywords to semantic queries – incremental query construction on the semantic web. Web Semantics 7(3), 166–176 (2009)

DC Proposal: Enriching Unstructured Media Content about Events to Enable Semi-automated Summaries, Compilations, and Improved Search by Leveraging Social Networks

Thomas Steiner*,**

Universitat Politècnica de Catalunya
Department LSI
08034 Barcelona, Spain
tsteiner@lsi.upc.edu

Abstract. Mobile devices like smartphones together with social networks enable people to generate, share, and consume enormous amounts of media content. Common search operations, for example searching for a music clip based on artist name and song title on video platforms such as YouTube, can be achieved both based on potentially shallow human-generated metadata, or based on more profound content analysis, driven by Optical Character Recognition (OCR) or Automatic Speech Recognition (ASR). However, more advanced use cases, such as summaries or compilations of several pieces of media content covering a certain event, are hard, if not impossible to fulfill at large scale. One example of such event can be a keynote speech held at a conference, where, given a stable network connection, media content is published on social networks while the event is still going on.

In our thesis, we develop a framework for media content processing, leveraging social networks, utilizing the Web of Data and fine-grained media content addressing schemes like Media Fragments URIs to provide a scalable and sophisticated solution to realize the above use cases: media content summaries and compilations. We evaluate our approach on the entity level against social media platform APIs in conjunction with Linked (Open) Data sources, comparing the current manual approaches against our semi-automated approach. Our proposed framework can be used as an extension for existing video platforms.

Keywords: Semantic Web, Linked Data, Multimedia Semantics, Social Networks, Social Semantic Web.

1 Introduction

Official statistics [15] from YouTube, one of the biggest online video platforms, state that more than 13 million hours of video were uploaded during 2010, and

---

* Thomas Steiner is partly funded by the European Commission under Grant No. 248296 for the FP7 I-SEARCH project.
** Advisors: Joaquim Gabarró Vallés (UPC) and Michael Hausenblas (DERI).

L. Aroyo et al. (Eds.): ISWC 2011, Part II, LNCS 7032, pp. 365–372, 2011.
© Springer-Verlag Berlin Heidelberg 2011

35 hours of video are uploaded every minute. The mostly text-based video search engine behind YouTube works mainly based on textual descriptions, video titles, or user tags, but does not take semantics into account: it does not get the meaning of a video, for example whether a video tagged with "obama" is about *the* Obama, or about a person that just happens to have the same name. We speak of the semantic gap in this context. In [11], Smeulders et al. define the semantic gap as *"The semantic gap is the lack of coincidence between the information that one can extract from the visual data and the interpretation that the same data have for a user in a given situation"*. Our thesis presents an approach for bridging the semantic gap for media content published through social networks by adding proper semantics to it, allowing for summaries, compilations, and improved search.

The remainder of this paper is structured as follows: Section 2 lists related work, Section 3 presents the proposed approach, Section 4 provides an evaluation of our work so far and gives an outlook on the evaluation plan for future work, which is detailed in Section 5. We conclude this paper with Section 6.

2   Related Work

We refer to previous work in a number of related areas, including enriching unstructured media content, event illustration, and video summarization.

The W3C Ontology for Media Resources [7] defines a set of mappings for many existing metadata formats. It aims to foster the interoperability among various kinds of metadata formats currently used to describe media resources on the Web. Time-based semantic annotations are possible using the `relation` property by linking to an RDF file or named graph containing annotations for a media resource (or fragment [14]), but currently there is no solution for embedding a set of RDF triples directly into one of the properties of the ontology. Similarly, the MPEG-7 [5] standard deals with the storage of metadata in XML format in order to describe multimedia content. Several works like for example [1] by Celma et al. have already pointed out the lack of formal semantics of the standard that could extend the traditional text descriptions into machine understandable ones. The authors explain that semantically identical metadata can be represented in multiple ways. Efforts [3] have been made to translate MPEG-7 into an ontology to enhance interoperability. These efforts, however, did not gain the traction their authors had hoped for.

In [8], Liu et al. present a method combining semantic inferencing and visual analysis for automatically finding photos and videos illustrating events with the overall objective being the creation of a Web-based environment that allows users to explore and select events and associated media, and to discover connections between events, media, and people participating in events. The authors scrape different event directories and align the therein contained event descriptions using a common RDF event ontology. The authors use Flickr and YouTube as media sharing platforms and query these sources using title, geographic coordinates, and upload or recording time. In order to increase the precision, the authors either use (title and time), or (geographic coordinates and time) as combined search queries. Liu et al. prune irrelevant media via visual analysis. While the authors of [8] start with curated event directory descriptions and the two media sources Flickr and YouTube, we focus on leveraging social networks and a broad range of media content sharing platforms specialized in live-streaming. Limiting the event scope to concerts, Kennedy et al. describe a system for synchronization and organization of user-contributed content from live music events in [6]. Using audio fingerprints, they synchronize clips from multiple contributors such that overlapping clips can be displayed simultaneously. Furthermore, they use the timing and link structure generated by the synchronization algorithm to improve the representation of the event's media content, including identifying key moments of interest. In contrast to us, Kennedy et al. focus mainly on content analysis, without revealing the origin of the considered media content, where we focus on semantically enriching media content coming from social networks.

Shaw et al. present a platform for community-supported media annotation and remix in [9], and describe how community remix statistics can be leveraged for media summarization, browsing, and editing support. While our approach is semi-automatic, their approach is manual.

3   Proposed Approach

Our objective for this thesis is to create a framework that allows for the semi-automatic generation of summaries or compilations of several pieces of media content covering a certain event and leveraging social networks. The term "event" is defined[1] by WordNet as *"something that happens at a given place and time"*. Our proposed process of generating a summary or compilation for an event includes the following steps:

Event-Selection: Decide on an event that shall be summarized.

Micropost-Annotation: Find relevant microposts on social networks containing links to media content about the event, and annotate these microposts one by one using RDF, leveraging data from the LOD cloud[2].

Media-Content-Annotation: Retrieve the pieces of media content and the accompanying metadata one by one, and annotate them using RDF, leveraging data from the LOD cloud.

Media-Content-Ranking: Rank and order the pieces of media content by relevance, creation time, duration, and other criteria.

Media-Content-Compilation: Based on the ranking, suggest a summary or compilation taking into account user constraints such as desired duration, composition (videos only, images only, combination of both), etc.

4   Evaluation

In the following we introduce our evaluation plan and present already existing evaluation for each of these steps.

---

[1] http://wordnetweb.princeton.edu/perl/webwn?s=event
[2] http://lod-cloud.net/

Event-Selection. For this task we currently have preliminary results only. We have selected two recent events, the Semantic Technology Conference 2011 (SemTech) and the Apple Worldwide Developers Conference 2011 (WWDC), both located in San Francisco. For WWDC there is a DBpedia page (db:Apple_Worldwide_Developers_Conference) besides the official event website (http://apple.com/wwdc), for SemTech there is just the event website (http://sem-tech2011.semanticweb.com/). We have disambiguated both event names with their locations using an API from previous work [12], [13]. The plaintext labels of the extracted named entities for WWDC are "Apple", "San Francisco", "Apple Inc.", "iPhone OS", and "Apple Worldwide Developers Conference". Each named entity is uniquely identified by a URI in the LOD cloud, which allows for ambiguity-free exploration of related entities, and is also indispensable in cases where no unique event hashtag is known. Having links to the LOD cloud is very important for the discovery of microposts on social networks that might have links to media content for the task *Micropost-Annotation*.

Micropost-Annotation. We have implemented a generic framework based on several Natural Language Processing (NLP) Web services in parallel for the on-the-fly enrichment of social network microposts [13]. This framework has been successfully tested on overall 92 seven-day active users. The context of the tests was the detection of news trends in microposts, however, the general contribution of this framework is the extraction and disambiguation of named entities from microposts. Figure 1a shows an example. By using Google Analytics, named entity occurrences can be easily tracked over time. As an example, Figure 1b shows the occurrences graph generated by Analytics of the named entity "tsunami" (db:tsunami). Japan was hit by a tsunami on March 11, exactly where the peak is on the graph. In general the occurrences graphs also for other examples indeed correspond to what we would expect from the news headlines of the considered days, which implies the correct functioning of our micropost annotation framework.

Media-Content-Annotation. For this task we currently have preliminary results only. At present we have implemented an interactive Ajax application called SemWebVid [12] that allows for the automatic annotation of videos on YouTube with RDF. Based on the same API that powers SemWebVid, we have implemented a command line version of the annotation mechanism that in the future can be used to batch-process videos. For now, we have annotated videos with the interactive Ajax application and reviewed the annotations manually, however, at this point, have not yet compared the annotations to a gold standard. We use the Common Tag [2] vocabulary to annotate entities in a temporal video fragment [14]. An example can be seen in Figure 2. This simple and consistent annotation scheme will make the comparison with a gold standard easier. Using Common Tag, both the video per se, as well as video fragments of the whole video can be annotated in the same way.

Whole Video Annotation. From our experiences so far, video annotation of the whole video works accurately. We have tested our approach with keynote and

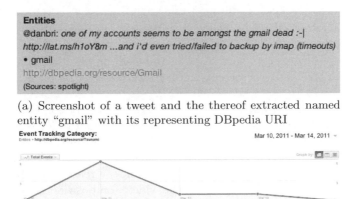

(a) Screenshot of a tweet and the thereof extracted named
entity "gmail" with its representing DBpedia URI

(b) Popularity of the named entity "tsunami" from March
10 - 14 in tweets. Japan was hit by a tsunami on March 11,
at the peak

Fig. 1. Tweet annotation and popularity of a named entity over time

```
<http://www. youtube.com/watch?v=hzFp3rovfY0#t=171,177>
  a ma:MediaFragment ;
  ctag:tagged
    [ a ctag:Tag ;
      ctag:label "Commodore 64" ;
      ctag:means <http://dbpedia.org/resource/Commodore_64>
    ] .
```

Fig. 2. Annotated named entity in a video fragment

conference session videos (for examples from the Google I/O events in 2010 and
2011), political speeches (for example Obama's inauguration address), but also
more underground video productions such as a video[3] about the music artist
Timbaland being accused of stealing a tune from the Commodore 64 scene.

In-Video Annotation. Results for the subtask of annotating in-video fragments
are currently still sparse in most test cases. We found that sending smaller input
texts increases the recall of the NLP Web services that we use without lowering
the precision. In consequence we are now considering splitting up the to-be-
analyzed data into smaller pieces, at the cost of processing time. We found
the sweet spot between recall, precision, and processing time to be around 300
characters. This figure was also publicly confirmed by Andraž Tori, CTO of
Zemanta, at a keynote speech.

Media-Content-Ranking. We have no results yet for this task. We plan to
let the user tweak the ranking criteria interactively and see the effect on the

---

[3] http://www.youtube.com/watch?v=hzFp3rovfY0

ranking immediately. This could happen via sliders, where a user could change the weights of criteria like view count, duration, recency, etc. The evaluation will happen based on user feedback from a test group. We will have to test whether video genre-specific weights have to be introduced, or whether common weights across all genres already reveal satisfactory results.

Media-Content-Compilation. We have no results yet for this task. Similar to the previous task *Media-Content-Ranking*, we plan to let the user adjust media composition criteria on-the-fly. For the desired duration, a slider seems adequate UI-wise, however, we have to do some experiments whether the video compilation can work fast enough to make the slider's reaction seem interactive. The same speed constraint applies to the video composition selection (videos only, images only, combination of both). Obviously the quality of the final video summaries needs to be evaluated by a test group, ideally against a gold standard of human-generated *event x in n seconds* videos. A typical example is the YouTube video `http://www.youtube.com/watch?v=skrz1JsxnKc`, which has a 60 seconds summary of Steve Jobs' keynote at WWDC. The problem, however, with these user-generated summaries is that they usually use official professionally produced video footage and not user-generated content. This allows for high quality audio and video quality, whereas user-generated content from mobile devices typically suffers from problems like noisy environments when recorded from the middle of an audience, overexposure when recorded against stage lighting, or a lack of detail when recorded from too far away. We will see in how far this is an issue once we have a working prototype of the whole framework.

5   Future Work

We present future work for each of the previously introduced steps.

Event-Selection. We will keep an eye on a wide variety of events, such as concerts, conferences, political demonstrations, elections, speeches, natural disasters, festivities, but also non-publicly announced events such as private parties, always given there is enough social media coverage and media content available. We will archive social network communication produced around these events for further analysis.

Micropost-Annotation. So far we have implemented a solid framework capable of annotating microposts. Future work in this task will be to further increase recall and precision by incorporating more Natural Language Processing engines both for English and non-English languages. While English is covered quite well by the existing engines, other key languages such as the so-called FIGS languages (French, Italian, German, Spanish) are still not optimally covered. Our work here will focus on the integration and the alignment of the output formats of the various NLP services, both commercial and non-commercial. The main constraint here will be the processing time, and depending on the event the sheer amount of potentially available microposts within a short period of time (compare nation-wide elections with a private party). We will also work on improving entity consolidation and ranking algorithms when different NLP services have agreeing or contradicting results for the same input text.

Media-Content-Annotation. At present we have implemented both a command line and an interactive Ajax version of the media content annotation mechanism tailored to the YouTube video platform. Future work in this task will be to improve precision and recall by the same improvement steps as in the *Micropost-Annotation* task. In addition to these steps, a major improvement will come from piece-wise rather than all-at-once analysis of the available unstructured metadata, taking into account the 300 characters sweet spot mentioned before. The constraint with this task is processing time, especially the more NLP services are involved in processing the data. Our current approach will have to be broadened to support other popular social media video and photo sharing platforms, some of them covered in [10]. In addition to that we will work to support Facebook's and Twitter's native photo and video sharing features.

Media-Content-Ranking. This task has not started yet. It will consist of development efforts in order to create a testing framework for the interactive ranking and re-ranking of user-generated media content.

Media-Content-Compilation. This task has not started yet. In a first step, the task consists of application development using JavaScript, HTML5, and CSS in order to generate media content compilations. We will make heavy use of the HTML5 media elements interface [4] for the `video` and `audio` elements as defined in the HTML5 specification. In a second step, an evaluation framework has to be developed in order to objectively judge the generated results. We will also investigate in how far existing third party manually generated summaries can be used as a gold standard.

6   Conclusion

We have had a look at related work from the fields of enriching unstructured media content, event illustration, and video summarization. In continuation we have introduced the required steps for our proposed approach and have evaluated our work so far, considering the tasks *Event-Selection, Micropost-Annotation, Media-Content-Annotation, Media-Content-Ranking*, and *Media-Content-Compilation*. Finally, we have provided an outlook on future work for each task.

Keeping in mind our objective for this thesis, we have the basic bricks in place, both for media content annotation, and for social network communication enrichment. Now we need to put the two pieces together in order to get a working product. The main research question is *"how can semantically annotated media content linked to from semantically annotated microposts be ranked, key moments of interest be detected, and media content fragments be compiled in order to get a compelling summary?"*. We are envisioning both a stand-alone Web application where a user can select an event and get a custom-made video summary, but also Web browser extensions for existing media content sharing platforms where users can start off with one piece of media content and see its broader context.

References

1. Celma, Ò., Dasiopoulou, S., Hausenblas, M., Little, S., Tsinaraki, C., Troncy, R.: MPEG-7 and the Semantic Web (2007),
   http://www.w3.org/2005/Incubator/mmsem/XGR-mpeg7-20070814/
2. Common Tag. Common Tag Specification (January 11, 2009)
   http://commontag.org/Specification
3. García, R., Celma, O.: Semantic integration and retrieval of multimedia metadata. In: Proc. of the ISWC 2005 Workshop on Knowledge Markup and Semantic Annotation, vol. 185, pp. 69–80 (2005)
4. Hickson, I.: HTML5 W3C Editor's Draft, Media elements (October 25, 2007),
   http://www.w3.org/TR/html5/video.html#media-elements
5. IEEE MultiMedia. MPEG-7: The Generic Multimedia Content Description Standard, Part 1. IEEE MultiMedia 9, 78–87 (2002)
6. Kennedy, L., Naaman, M.: Less talk, more rock: automated organization of community-contributed collections of concert videos. In: Proc. of the 18th Int. Conference on World Wide Web, pp. 311–320. ACM, New York (2009)
7. Lee, W., Bürger, T., Sasaki, F., Malaisé, V., Stegmaier, F., Söderberg, J.: Ontology for Media Resource 1.0 (June 2009)
8. Liu, X., Troncy, R., Huet, B.: Finding media illustrating events. In: Proc. of the 1st ACM Int. Conference on Multimedia Retrieval, pp. 58:1–58:8. ACM, New York (2011)
9. Shaw, R., Schmitz, P.: Community annotation and remix: a research platform and pilot deployment. In: HCM 2006: Proc. of the 1st ACM Int. Workshop on Human-centered Multimedia, pp. 89–98. ACM Press (2006)
10. Levine, S.: How People Currently Share Pictures On Twitter (June 2, 2011),
    http://blog.sysomos.com/2011/06/02/
11. Smeulders, A.W.M., Worring, M., Santini, S., Gupta, A., Jain, R.: Content-Based Image Retrieval at the End of the Early Years. IEEE Trans. Pattern Anal. Mach. Intell. 22, 1349–1380 (2000)
12. Steiner, T.: SemWebVid - Making Video a First Class Semantic Web Citizen and a First Class Web Bourgeois. In: 9th International Semantic Web Conference, ISWC 2010 (November 2010)
13. Steiner, T., Brousseau, A., Troncy, R.: A Tweet Consumers' Look At Twitter Trends. In: Workshop Making Sense of Microposts at ESWC 2011, Heraklion, Crete (May 30, 2011), http://research.hypios.com/msm2011/posters/steiner.pdf
14. Troncy, R., Mannens, E., Pfeiffer, S., Deursen, D.V.: Media Fragments URIs. W3C Working Draft (December 8, 2010),
    http://www.w3.org/2008/WebVideo/Fragments/WD-media-fragments-spec/
15. YouTube.com. Official Press Traffic Statistics (June 14, 2011),
    http://www.youtube.com/t/press_statistics

DC Proposal: Ontology Learning from Noisy Linked Data

Man Zhu[*,**]

School of Computer Science & Engineering, Southeast University, Nanjing, China
mzhu@seu.edu.cn

Abstract. Ontology learning - loosely, the process of knowledge extraction from diverse data sources - provides (semi-) automatic support for ontology construction. As the 'Web of Linked Data' vision of the Semantic Web is coming true, the 'explosion' of Linked Data provides more than sufficient data for ontology learning algorithms in terms of quantity. However, with respect to quality, notable issue of *noises* (e.g., partial or erroneous data) arises from Linked Data construction. Our doctoral researches will make theoretical and engineering contribution to ontology learning approaches for noisy Linked Data. More exactly, we will use the approach of Statistical Relational Learning (SRL) to develop learning algorithms for the underlying tasks. In particular, we will learn OWL axioms inductively from Linked Data under probabilistic setting, and analyze the noises in the Linked Data on the basis of the learned axioms. Finally, we will make the evaluation on proposed approaches with various experiments.

1 Motivation

Ontology learning refers to the task of providing (semi-) automatic support for ontology construction [3], and can overcome the knowledge acquisition bottleneck brought by the tedious and cumbersome task of manual ontology construction [17]. Recent ontology learning approaches have attempted to learn ontology from various types of data sets, such as text, xml, and database, but they seldom explore learning from Linked Data. Based on URIs, HTTP and RDF, the Linked Data project [2] aims to expose, share and connect related data from diverse sources on the Semantic Web. Linked Open Data (LOD) is a community effort to apply the Linked Data principles to data published under open licenses. With this effort, a large number of LOD data sets have been gathered in the LOD cloud, such as DBpedia, Freebase and FOAF profiles. LOD has gained rapid progress and is still growing constantly. Until May 2009, there are 4.7 billion RDF triples and around 142 million RDF links [2]. After that, the total has been increased to 16 billion triples in March 2010 and another 14 billion triples have been published by the AIFB according to [21].

---

[*] Advisor: Zhiqiang Gao, School of Computer Science & Engineering, Southeast University, Nanjing, China, zqgao@seu.edu.cn

[**] Advisor: Zhisheng Huang, Department of Mathematics & Computer Science, Vrije University, Amsterdam, The Netherlands, huang@cs.vu.nl

L. Aroyo et al. (Eds.): ISWC 2011, Part II, LNCS 7032, pp. 373–380, 2011.
© Springer-Verlag Berlin Heidelberg 2011

The advantages of learning from Linked Data, and what distinguishes it from learning from other resources, are depicted in Table 1. The most common used learning resource is HTML documents, which emerged and developed ever since the invention of the Web. The distinguishing feature thereof is that they constitute a large-scale data set and are generally publicly accessible. However, the structures inside are formed through simple HTML tags, and the HTML documents are linked to each other on document level. Compared to HTML documents, XML documents (made to be the origin of comparison in Table 1) are far more easily accessible to machines. XML can overcome the shortcomings of HTML (highly human interpretable contents, not for machines) to some extent, because XML documents contain certain structural information. The characters of glossaries are similar to that of XML documents. Besides the links among words (phrases), the structures inside are simple. The biggest problem of learning from database is that it is limited in both contents and accessibility. Learners can only learn from databases of specific domain. According to the description and statistics described in the last paragraph, we conclude that compared with other resources Linked Data is superior in that it is publicly available, highly structured, relational, and large with respect to learning.

The other side of the Linked Data coin poses the challenges we are going to cope with during the doctoral research: First, due to the publishing mechanism of the Linked Data, it contains noises inherently [4,1]. Hogan et al. analyzed the types of noises which exist in the Linked Data [9]. We are particularly interested in handling two types of noises: *partiality* and *error*. Partiality means that concept assertions or the relationships between named individuals are actually true but missed, and error means that the RDF triples are not correct (with respect to some constraints). Take a family ontology for example. The declarations of 'Heinz is a father' and 'Heinz is a male' exist in the RDF triples, then Heinz should have a child, however it is not declared in the ontology. This is an example of partiality. Besides, if we know Anna has a child, and she is a female, then Anna should not be a father, but in the ontology Anna is incorrectly declared to be a father. This illustrates the error case. Second, the ontologies in Linked Data are generally inexpressive. For example, one of the most popular ontologies, DBpedia ontology[1], is claimed as a shallow ontology. The TBox of this ontology mainly includes a class hierarchy [10].

In our doctoral researches, we endeavor to inductively learn ontologies using statistical relational learning (SRL) models. The development of SRL has been driven by real-world needs of handling noises, relations (Figure 1). In the early days, ML community have been focused on learning deterministic logical concepts. However, those methods failed to fit perfectly for noises and large-scale circumstances, which leads to statistical methods that ignored relational aspects of the data, such as neural networks, generalized linear models. On the other hand, inductive logic programming (ILP) is designed to learn first-order rules directly which are much more expressive [18]. It is argued in [4] that inductive learning methods, could be fruitfully exploited for learning from Linked Data.

---

[1] http://wiki.dbpedia.org/Ontology

Table 1. Comparison of learning from various resources. XML Document is made to be the origin of comparison, indicated by '0'. '+' and '−' denote degree of the corresponding character (above/below the origin), '++' and '−−' denote stronger degree than '+' and '−'.

	Publicly Available	Structured	Linked	Large
HTML Document	+	−−	0	++
Glossary	0	−	0	0
XML Document	0	0	0	0
Database	−	+	+	−
LOD	+	+	+	+

For the last few years, the ILP community and the statistical machine learning community have been incorporating aspects of the complementary technology (machine learning, probability theory, and logic), which leads to the emerging area of SRL. It attempts to represent, reason, and learn in domains with complex relational and rich probabilistic structure [8]. Using SRL, two characters of Linked Data, which distinguish Linked Data from other data sets, can be easily handled: 1) Linked Data are highly structured due to the relations between entities and the underlying ontology. 2) Linked Data contains noises, here, as described above, we refer particularly to partiality and error.

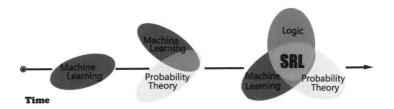

Fig. 1. The evolution of Statistical Relational Learning (SRL). SRL integrates technologies from machine learning, probability theory, and logic.

2   Related Work

There is an important body of previous work that our work builds on. We note a subset of them here. Lehmann J. et al. have done a series of work on learning Description Logics, and the algorithms proposed are implemented in DL-Learner. DL-Learner is a framework for learning Description Logics and OWL from positive (and negative) examples (in ILP, ground literals of target concept are called examples, if the ground literal is true, it is positive, negative on the contrary), and supports several learning algorithms (CELOE, random guesser learning algorithm, ISLE, brute force learning algorithm) based on ILP and machine learning [11]. [13] learned Description Logics $\mathcal{ALC}$, and [14] learned Description Logic $\mathcal{ALCQ}$ using a learning algorithm based on refinement operators, and the algorithm is implemented and evaluated in the DL-Learner. AutoSPARQL is a most recent work, which makes use of the individual assertions in the ABox, and can be used to learn descriptions for individuals [12]. [19] proposes a log-linear Descriptions Logics based on $\mathcal{EL}++$. It integrates log-linear model with Description Logic $\mathcal{EL}++$, and can be used to learn coherent ontologies. In [20], Völker J. and Niepert M. propose a statistical approach, to be specific, association rule mining, for learning OWL 2 $\mathcal{EL}$ from Linked Data. Fanizzi N. et al. proposed a specific concept learning algorithm by extending FOIL algorithm, called DL-FOIL [7]. [6] works on the induction and revision of terminologies from metadata. Lisi F.A. et al. have done a series of work on learning rules. In [15] hypotheses are represented as $\mathcal{AL}$-log rules, and the coverage relations are defined on the basis of query answering in $\mathcal{AL}$-log. Correspondingly, [16] learns $\mathcal{DL}$-log rules, besides the differences in the expressive power of the target language, it also reformulate the coverage relation and the generality relation as satisfiability problems.

3   Proposed Approaches

In the doctoral researches, the following issues will be studied:

How to Learn from Noisy Linked Data? We propose to learn ontologies from Linked Data by SRL methods. Generally speaking, SRL models combine relational representations and probabilistic learning mechanisms such as graphical models. The majority of proposed SRL models can be categorized according to several dimensions: 1) the representation (logic or frame-based) formalisms. 2) probabilistic semantics (Bayesian networks, Markov networks, stochastic grammar etc.). We will learn ontologies under probabilistic setting, where the learning problem is transformed into finding an optimum axiom A satisfying certain function, such as $A = \arg\max_A P(A, E)$, E denotes all assertions (facts) in the original ontology. In SRL models, different probabilistic semantics are used for modeling the probability distributions, and random variables corresponding to assertions (containing partiality and errors) can encode probabilistic information, thus is suitable for handling noises. For example, in Markov logic, the facts and the terminology axioms correspond to a Markov network. The joint probability distribution is defined according to this Markov network, where nodes represent assertions with probabilities. In the current work (c.f. Sect. 4.2), the axioms are attached with weights. By maximizing the joint probability, the weights can be learned. Each step the candidate axiom with the largest weight is selected to be further expanded in the next step if the joint probability still increases. The process of finding the optimum axiom can be viewed as searching in a predetermined hypotheses space. This process can go in a top-down or bottom-up manner. Top-down algorithms start from the most general hypothesis $target \sqsubseteq \top$, and iterate to select one from candidates according to a performance criteria and add to the hypothesis until the stop criteria is reached. In bottom-up approaches the iteration begins at the most specific hypothesis whose right-hand is the intersection of all possible literals.

We will propose a SRL model suitable for learning from noisy data. Currently, a number of SRL models have been proposed from various research fields with different application background. For example, Markov logic [5] is one of the most recent SRL model, which combines first-order logic with undirected graphical models (Markov networks). According to one of our recent works (c.f. Section 4.2), Markov logic can be applied to ontology learning from noisy data. Still the results can be better. We argue that the performance of applying the currently proposed SRL models to learn from Linked Data can be further improved by proposing a SRL model particularly for this task concerning the following aspects: 1) As we all know, OWL builds on Description Logic basis. Today's SRL models use either frame-based, which contains simple relations, or logic representation (e.g. FOL, which is more complex). In terms of expressing power, they are not the best fit for OWL. 2) Currently SRL models are still weak in analyzing the independencies inside the probabilistic model. However, independencies play an important role in saving parameter space as well as improving the computing efficiency, which should be studied carefully. Thus we will propose a more suitable model for learning from noisy Linked Data with the goal of improving both the learning accuracy and the learning efficiency.

How to Guide the Search? We will propose methods to structure the hypotheses space. In ILP algorithms such as FOIL [18], which learns first-order rules, the rules are generated through adding literals to the current rule. New literal can be of the form $Q(v_1, \ldots, v_r)$ (at least one v_i already exists in the rule), $equal(v_j, v_k)$, or the negation of either of the first two forms. Using this kind of approaches, it is still unknown that whether the following can be guaranteed: 1) will adding a new literal lead to a more specific concept? 2) can all concepts be traversed? Another approach, named refinement operator, defines a mapping $S \mapsto 2^S$ on a quasi-ordered space S, thus it structures the hypotheses space according to quasi-ordering relations, such as subsumption. A number of refinement operators have been proposed, such as $\mathcal{ALC}$ refinement operator [13] and $\mathcal{ALCQ}$ refinement operator [14]. The properties of refinement operators, such as (weakly) complete, ensure that if an axiom should be correct according to the Linked Data, it can be reached by the refinement operator. Current refinement operators will be improved by, firstly, at each step, the hypotheses generated by the refinement operator should be finite, and secondly, the refinement operator should be designed for OWL and its profiles according to specific models. For example, if Markov logic is chosen as the model, then in each step in the iteration, the dependencies between the candidate hypotheses should be minimized, so as to guarantee that the weights learned truly reflect the confidence of the hypothesis (c.f. Section 4.2).

How Many Partiality and Errors Are There? In [9], Hogan discusses common errors that can be systematically detected in RDF publishing. The results provide a significant basis for our motivations. We are still interested in analyzing the Linked Data more semantically. The axioms we learned contain two parts. One part of them already exist in the ontology, which can be evaluated automatically by comparing with the original ontology. The other part of them are not in the ontology. Given the observation that the ontologies in Linked Data are generally inexpressive (c.f. Sect. 1), this part of axioms are not necessarily incorrect. We will evaluate them manually. Finally, we will have a set of correct axioms. We want to answer the question of "How many partiality and errors are there in Linked Data?". By querying the original ontology and comparing the results with the learned axioms, we will propose algorithms to analyze the data in ABox to know whether some of them are missed or some of them are wrongly stored in the Linked Data.

4   Results and Evaluation

4.1   DLP Learning from Uncertain Data

The origin of this work can be found in [22], where we focused on learning description logic programs (DLP) from explicitly represented uncertain data. DLP is an intermediate knowledge representation that layers rules on top of ontologies. DLP is an expressive and tractable subset of OWL, and plays an important role in the development of the Semantic Web. We modified the performance evaluation criteria based on pseudo-log-likelihood in the designed ILP like algorithm PIDLP. With the new performance evaluation criteria, uncertainties are handled and meanwhile DLPs can be learned. We also tested the algorithm in two datasets, and the results demonstrated that the approach is able to automatically learn a rule-set from uncertain data with reasonable accuracy. However, in many cases, uncertainties exist implicitly, such as in Linked Data. In what follows, we transfer our attention to learning from noisy Linked Data without handling explicitly specified uncertainties.

4.2   Learning $\mathcal{ALCI}$ from Noisy Data by Markov Logic

$\mathcal{ALCI}$ contains inverse role constructor in addition to the basic Description Logic $\mathcal{ALC}$. In our most recent work, we examine learning $\mathcal{ALCI}$ from noisy Linked Data, attempting to take the first step towards our first proposed approach. The procedure of learning can be viewed as searching for the optimize hypotheses (axiom) in the hypotheses space composed of all possible axioms according to certain criterion. In this work, Markov logic is used for handling noises. More specifically, hypotheses are accompanied with a weight which indicate the degree of consistency between the hypotheses and the RDF triples in the data set. In each iteration, the weights are learned with the target of joint probability maximization, and we choose the hypothesis with the largest weight. The iteration runs until performance stops to improve. We evaluate the approach on 4 data sets in addition to a small data set illustrating the functionality of learning definitions. The results demonstrate that the method performs well under noises, and is capable of learning $\mathcal{ALCI}$ with an average precision of 0.68 and recall 0.59.

4.3   Evaluation

The evaluation goes in two-fold (semi-)automatically. Firstly, using ontologies as gold standard, such as EKAW ontology, we separately treat the TBox and ABox in the Linked Data as testing and training set. We learn ontologies from the ABox, and evaluate the results learnt according to the TBox. Widely used IR measures *precision, recall,* and *F1-score* are adopted here. This way, the performance of the proposed approaches can be observed. Nevertheless the axioms in TBox may still not be complete [10], and the learned axioms not in the TBox are not bound to be wrong. Thus secondly, the part of learned axioms not in the TBox will be manually evaluated. We assign 3 numbers to indicate the correctness of the axioms: 1-low, 2-medium, and 3-high. For each axiom, a group of people will determine whether or not they think it is correct, and offer a judgement represented as a number. The final result will be an average.

5   Future Works

In the future, we plan to do the following works: firstly, we will further improve our approaches of learning OWL axioms from Linked Data (such as DBPedia) by adopting new SRL models. By proposing mechanisms for representing OWL axioms with probabilistic graphical models, and analyzing the independencies inside, more accurate results and more efficient algorithms can be found. Secondly, we will work towards answering the question "How many partiality and errors are there?". According to the preliminary results we got from the work (c.f. Sect. 4.2), a number of axioms we learned that are not in the Linked Data should be correct. In addition to the analyzes carried out by Hogan et al. in [9], we will propose algorithms to analyze the noises in the Linked Data.

Acknowledgements. We would like to thank Jiahong Shi, Yunxia Sun, and Yuan Si who contribute to this work. Additionally, we gratefully acknowledge funding from the National Science Foundation of China under grants 60873153, 60803061, and 61170165.

References

1. Auer, S., Lehmann, J.: Creating knowledge out of interlinked data. Semantic Web 1(1), 97–104 (2010)
2. Bizer, C., Heath, T., Berners-Lee, T.: Linked data-the story so far. Int. J. Semantic Web Inf. Syst. 5(3), 1–22 (2009)
3. Cimiano, P.: Ontology learning and population from text: algorithms, evaluation and applications. Springer, Heidelberg (2006)
4. d'Amato, C., Fanizzi, N., Esposito, F.: Inductive learning for the semantic web: What does it buy? Semantic Web 1(1-2), 53–59 (2010)
5. Domingos, P., Lowd, D.: Markov Logic: An interface layer for artificial intelligence. Synthesis Lectures on Artificial Intelligence and Machine Learning 3(1), 1–155 (2009)

6. Esposito, F., Fanizzi, N., Iannone, L., Palmisano, I., Semeraro, G.: Knowledge-intensive induction of terminologies from metadata. In: McIlraith, S.A., Plexousakis, D., van Harmelen, F. (eds.) ISWC 2004. LNCS, vol. 3298, pp. 441–455. Springer, Heidelberg (2004)
7. Fanizzi, N., d'Amato, C., Esposito, F.: DL-FOIL concept learning in description logics. In: Železný, F., Lavrač, N. (eds.) ILP 2008. LNCS (LNAI), vol. 5194, pp. 107–121. Springer, Heidelberg (2008)
8. Getoor, L., Taskar, B.: Introduction to statistical relational learning. The MIT Press (2007)
9. Hogan, A., Harth, A., Passant, A., Decker, S., Polleres, A.: Weaving the pedantic web. In: 3rd International Workshop on Linked Data on the Web (LDOW 2010), in conjunction with 19th International World Wide Web Conference, CEUR (2010)
10. Ji, Q., Gao, Z., Huang, Z.: Reasoning with noisy semantic data. In: Antoniou, G., Grobelnik, M., Simperl, E., Parsia, B., Plexousakis, D., De Leenheer, P., Pan, J. (eds.) ESWC 2011. LNCS, vol. 6644, pp. 497–502. Springer, Heidelberg (2011)
11. Lehmann, J.: Dl-learner: Learning concepts in description logics. The Journal of Machine Learning Research 10, 2639–2642 (2009)
12. Lehmann, J., Bühmann, L.: AutoSPARQL: Let users query your knowledge base. In: Antoniou, G., Grobelnik, M., Simperl, E., Parsia, B., Plexousakis, D., De Leenheer, P., Pan, J. (eds.) ESWC 2011, Part I. LNCS, vol. 6643, pp. 63–79. Springer, Heidelberg (2011)
13. Lehmann, J., Hitzler, P.: A refinement operator based learning algorithm for the ALC description logic. In: Blockeel, H., Ramon, J., Shavlik, J., Tadepalli, P. (eds.) ILP 2007. LNCS (LNAI), vol. 4894, pp. 147–160. Springer, Heidelberg (2008)
14. Lehmann, J., Hitzler, P.: Concept learning in description logics using refinement operators. Machine Learning 78, 203–250 (2010)
15. Lisi, F.A.: Building rules on top of ontologies for the semantic web with inductive logic programming. Theory and Practice of Logic Programming 8(03), 271–300 (2008)
16. Lisi, F.A.: Inductive logic programming in databases: From Datalog to DL+ log. Theory and Practice of Logic Programming 10(03), 331–359 (2010)
17. Maedche, A., Staab, S.: Ontology learning for the semantic web. IEEE Intelligent Systems 16(2), 72–79 (2001)
18. Mitchell, T.: Machine learning. McGraw-Hill, New York (1997)
19. Niepert, M., Noessner, J., Stuckenschmidt, H.: Log-linear description logics. In: IJCAI, pp. 2153–2158 (2011)
20. Völker, J., Niepert, M.: Statistical schema induction. In: Antoniou, G., Grobelnik, M., Simperl, E., Parsia, B., Plexousakis, D., De Leenheer, P., Pan, J. (eds.) ESWC 2011, Part I. LNCS, vol. 6643, pp. 124–138. Springer, Heidelberg (2011)
21. Vrandecic, D., Krötzsch, M., Rudolph, S., Lösch, U.: Leveraging non-lexical knowledge for the linked open data web. The Fifth RAFT 2010 The yearly bilingual publication on nonchalant research 5(1), 18–27 (2010)
22. Zhu, M., Gao, Z., Qi, G., Ji, Q.: DLP learning from uncertain data. Tsinghua Science & Technology 15(6), 650–656 (2010)

DC Proposal: Capturing Knowledge Evolution and Expertise in Community-Driven Knowledge Curation Platforms

Hasti Ziaimatin

eResearch Lab, School of ITEE,
The University of Queensland, Australia
h.ziaimatin@uq.edu.au

Abstract. Expertise modeling has been the subject of extensive research in two main disciplines - Information Retrieval (IR) and Social Network Analysis (SNA). Both IR and SNA techniques build the expertise model through a document-centric approach providing a macro-perspective on the knowledge emerging from large corpus of static documents. With the emergence of the Web of Data, there has been a significant shift from static to evolving documents, characterized by micro-contributions. Thus, the existing macro-perspective is no longer sufficient to track the evolution of both knowledge and expertise. The aim of this research is to provide an all-encompassing, domain-agnostic model for expertise profiling in the context of dynamic, living documents and evolving knowledge bases. Our approach combines: (i) fine-grained provenance, (ii) weighted mappings of Linked Data concepts to expertise profiles, via the application of IR-inspired techniques on micro-contributions, and (iii) collaboration networks - to create and enrich expertise profiles in community-centered environments.

Keywords: Expertise profiling, Linked Data, Semantic Web, fine-grained provenance, micro-contributions.

1 Introduction

Acquiring and managing expertise profiles represents a major challenge in any organization, as often, the successful completion of a task depends on finding the most appropriate individual to perform it. The task of expertise modeling has been the subject of extensive research in two main disciplines: information retrieval (IR) and social network analysis (SNA). From the IR perspective, static documents authored by individuals (e.g., publications, reports) can be represented as bags-of-words (BOW) or as bags-of-concepts (BOC). The actual expertise location is done by associating individual profiles to weighted BOWs or BOCs either by ranking candidates based on their similarities to a given topic or by searching for co-occurrences of both the individual and the given topic, in the set of supporting documents. Such associations can then be used to compute semantic similarities between expertise profiles. From the SNA perspective, expertise profiling is done by considering the graphs connecting individuals in different contexts, and inferring their expertise from the shared domain-specific topics.

L. Aroyo et al. (Eds.): ISWC 2011, Part II, LNCS 7032, pp. 381–388, 2011.
© Springer-Verlag Berlin Heidelberg 2011

With the emergence of the Web of Data (which has evolved from the increasing use of ontologies, via the Semantic Web [1] and Web 2.0 [2]) there has been a significant shift from static documents to evolving documents. Wikis or diverse knowledge bases in the biomedical domain (e.g., Alzforum[1]) are examples of environments that support this shift by enabling authors to incrementally refine the content of the embedded documents to reflect the latest advances in the field. This gives the knowledge captured within them a dynamic character, and generating expertise profiles from this knowledge raises a new and different set of challenges. Both the IR and the SNA techniques build the expertise model through a document-centric approach that provides only a macro-perspective on the knowledge emerging from the documents (due to their static, final nature, i.e., once written, the documents remain forever in the same form). However, the content of living documents changes via micro-contributions made by individuals, thus making this macro perspective no longer sufficient when tracking the evolution of both the knowledge and the expertise. As a result, such dynamic content requires a novel method for representing underlying concepts and linking them to an expertise profile. This method involves capturing and analyzing micro-contributions by experts through a fine-grained provenance model.

1.1 Aims and Objectives

A comprehensive, fine-grained provenance model, able to capture micro-contributions in the macro-context of the host living documents, will facilitate expertise profiling in evolving knowledge bases or environments where the content is subject to ongoing changes.

The focus of the research will be on addressing the following questions:

- How can we model the provenance and evolution of micro-contributions in a comprehensive and fine-grained manner?
- How can we bridge the gap between Linked Data [3] domain concepts and their lexical representations, by taking into account *acronyms*, *synonyms* and/or *ambiguity*?
- What is the best IR-inspired model for consolidating the domain concepts present in micro-contributions and for computing ranked maps of weighted concepts describing the expertise profiles?
- How can we enrich expertise profiles using existing collaboration networks?
- What is the appropriate methodology for evaluating emerging expertise profiles from evolving micro-contributions?

These research questions can be further specified into the following objectives:

O1: development of a comprehensive model for capturing micro-contributions by combining coarse and fine-grained provenance, change management and ad-hoc domain knowledge;

O2: development of a profile building mechanism by computing ranked maps of weighted (Linked Data) concepts and consolidating (Linked Data) concepts via IR-inspired techniques;

O3: development of a profile refinement mechanism by incrementally integrating the knowledge and expertise captured within given social professional networks.

---

[1] http://www.alzforum.org/

The outcomes will be applied and evaluated in the context of the SKELETOME project[2] (a knowledge base for the skeletal dysplasia domain), the iCAT project[3] (a collaboratively engineered ontology for ICD-11) and the Alzforum (Alzheimer disease) knowledge base.

1.2 Significance and Innovation

The main innovation of our research lies in the acquisition and management of the temporal and dynamic characteristics of expertise. Tracking the evolution of micro-contributions enables us to monitor the activity performed by individuals, which in turn, provides a way to show not only the change in personal interests over time, but also the maturation process (similar to some extent to the maturation process of scientific hypotheses, from simple ideas to scientifically proven facts) of an expert's knowledge. Using well-grounded concepts from widely adopted vocabularies or ontologies, such as the ontologies published in the Linked Data Cloud, enables a straightforward consolidation of the expertise profiles. As a result, the overhead imposed by performing co-reference entity resolution (to consolidate the expertise concepts to a shared understanding) will be reduced to a minimum. From an academic perspective, a shift in the scientific publishing process seems to gain momentum, from the current document-centric approach to a contribution-oriented approach in which the hypotheses or domain-related innovations (in form of short statements) will replace the current publications. Examples of this new trend can be seen via nano-publications [4] or liquid publications [5]. In this new setting, mapping such micro-contributions to expertise will become essential in order to support the development of novel trust and performance metrics.

2 Related Work

Expertise profiling is an active research topic in a wide variety of applications and domains, including bio-medical, scientific, education. In this section we present a brief overview of the related efforts, with particular accent on the Information Retrieval and the Semantic Web domains.

The two most popular and well performing types of approaches in TREC (Text Retrieval Conference) expert search task are profile-centric and document-centric approaches. These studies use the co-occurrence model and techniques such as Bag-of-Words or Bag-of-Concepts on documents that are typically large and rich in content. Often a weighted, multiple-sized, window-based approach in an information retrieval model is used for association discovery [6] or the effectiveness of exploiting the dependencies between query terms for expert finding is proved [7]. Other studies present solutions through effective use of ontologies and techniques such as *spreading* to include additional related terms to a user profile by referring to an ontology (Wordnet or Wikipedia) [8]. Such traditional techniques work well with large corpuses as word occurrence is high and frequency is sufficient to capture the semantics of the document. However, when dealing with shorter texts such as

---

[2] http://itee.uq.edu.au/~eresearch/projects/skeletome/
[3] http://icat.stanford.edu/

micro-contributions within evolving knowledge bases, these traditional techniques are no longer reliable. Their heavy dependency on statistical techniques (e.g., TF/IDF) cannot be applied on micro-contributions, because these don't offer sufficient context to capture the encapsulated knowledge.

A different approach is adopted by the ExpertFinder framework, which uses and extends existing vocabularies on the Semantic Web (e.g., FOAF, SIOC) as a means for capturing expertise [9]. Algorithms are also proposed for building expertise profiles using Wikipedia by searching for experts via the content of Wikipedia and its users, as well as techniques that use semantics for disambiguation and search extension [10]. We intend to leverage these studies in order to enable the straightforward consolidation of expertise profiles to generate a shared understanding by using widely adopted vocabularies and ontologies. This will also lead to a seamless aggregation of communities of experts.

In the context of micro-blogging, a proposed framework includes the capacity to link entities within each blog post to their disambiguated concept on the Semantic Web [11]. However, this approach relies on the author to manually annotate the entities in the post. Our proposed approach could alleviate this manual linking of the entities. Early results in discovering Twitter users' topics of interest are proposed by examining, disambiguating and categorizing entities mentioned in their tweets using a knowledge base. A topic profile is then developed, by discerning the categories that appear frequently and cover the entities [12]. Although this study analyzes short texts, there are fundamental differences between micro-contributions in the context of online knowledge bases and Twitter messages. These differences include: shortening of words, usage of slang, noisy postings and the static nature of twitter messages.

Finally, other studies focus on methods for finding experts in social networks and online communities via a conceptual framework that uses ontologies such as FOAF, SIOC or SKOS [13]. A propagation-based approach to finding experts in a social network is proposed that makes use of personal local information and relationships between persons in a unified approach [14]. Most studies focus on connecting experts based on their profiles and social networks. However, our focus is on leveraging existing collaboration networks and enriching expertise profiles based on the relatedness of an expert's profile and the profiles of his/her collaborators.

3   Approach and Methodology

The objectives listed in Sect. 1.1, and depicted in Fig.1 represent the building blocks of our research methodology. In the following sections we detail the provenance and change management model for micro-contributions, the profile building and the profile refinement phases.

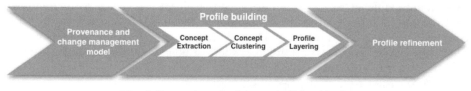

Fig. 1. Research methodology - building blocks

3.1 Provenance and Change Management Model

The first step of our research involves the development of a comprehensive provenance and change management model for micro-contributions. The resulting ontology will combine coarse and fine-grained provenance modeling (using the SIOC ontology [15]) with change management aspects captured by the generic W7 model [16] and SIOC actions [17]. The Annotation Ontology [18] will then be used to bridge the lexical grounding and the ad-hoc domain knowledge, represented by concepts present in the Linked Data Cloud, via ontologies such as SNOMED. As a result, the final model will have a layered structure where *micro-contributions* will *annotate* the contributed *text* and will be linked via the same *annotations* to domain knowledge. Instances of the model will not only be useful for expertise profiling, but will also act as a personal repository of micro-contributions, to be published, reused or integrated within multiple evolving knowledge bases.

3.2 Profile Building

The Profile building phase comprises three steps: (i) concept extraction, i.e., extracting the domain concepts from micro-contributions, (ii) concept clustering, i.e., consolidating the domain concepts around particular centroids, and (iii) profile layering, i.e., building the actual expertise profile by using the temporal dimension intrinsically associated with micro-contributions.

Concept Extraction. This step bridges the gap between the domain knowledge and evolving documents. While in general our methodology is domain-agnostic, in the context of our target use cases, we intend to utilize the NCBO Annotator[4] to identify and annotate concepts within the micro-contributions. For every identified concept, we will define its corresponding lexical chain of terms from the results of the annotation. Assuming that terms having more relations with other terms are semantically more important, we will attach a weight to each term based on its relations with other terms such as identity, synonymy and their weights in the lexical chain. We will use WordNet [19] to determine the relatedness of terms. The concept weight is then obtained by summing the scores of all terms in the lexical chain. The output of this step is therefore, a weighted vector of terms in the lexical chain representing each concept identified by the annotator in a micro-contribution.

Concept Clustering. In this step, the aim is to maintain a collection of concept clusters such that as each concept is presented, either it is assigned to one of the current clusters, or it starts off a new cluster, while two existing clusters are merged into one [20]. We propose a real-time, incremental and unsupervised algorithm for clustering concepts resulting from the previous step. The concept-cluster similarity is measured using the classic cosine similarity between the reference point Rp (representative concept) of each cluster and a concept identified in the previous step. If the similarity value is higher than a predefined similarity threshold, the concept is assigned to the nearest cluster; otherwise incoming concepts are examined to create a new cluster [22]. If the concept already exists in the identified cluster, its weighted vector is adjusted according to the weighted vector of the assigned concept; otherwise, the concept and its associated weighted vector will be assigned to the cluster. The weight associated

---

[4] http://www.bioontology.org/annotator-service

with a concept in a cluster will therefore represent the overall weight of the concept across all micro-contributions for an expert. As concept weights in clusters are subject to change, we will recalculate a cluster's reference point (using a method similar to the Evolving Clustering Method (ECM) [21]), each time the weight of a concept is changed or a concept is added to a cluster. As this is a completely separate process, it will not affect the performance of our proposed clustering algorithm [22].

Profile Layering. We introduce a temporal dimension to user profiles by splitting and combining concepts on a timeline. Thus, the user profiles will be multi-layered; static, session, short-term and long-term, with each layer reflecting a user's interests within a certain period. This approach will not only reflect the changeability of user interests but also maintain the steadiness of persistent preferences [22]. Once concepts are extracted, weighted and clustered, we will detect any change of context and assign the latest as the current context (currency criterion) to the session layer. The short-term layer consists of the most frequently updated and used concepts (frequency criterion), which are in turn chosen from the most recent concepts in the concept currency list. The long-term layer is derived from the concepts of the short-term layer (currency and frequency criteria), whose *persistency factor, PF,* is high. PF is a measure to infer an expert's continuous interests by combining a concept's frequency count with its evidence of being a constituent of the user's short-term layer [22]. The emergence of a new model for a user is not determined by predefined parameters, such as the fixed time period after which a new model should be created. It is rather driven by natural dynamics of changing user interests, signified by change of the concepts in terms of their ranking or new concepts in the short-term profile layer [22].

3.3 Profile Refinement

As a final step, we will look at existing collaboration networks and refine profiles based on the collaboration structure and collaborators' expertise. With regard to collaboration structure, we will take into consideration the type of collaboration *(e.g. co-authorship)* and its strength. For a given expert, we will retrieve a dense sub-graph of his/her collaborators, through measuring the connectedness of direct neighbours to the expert node *(clustering coefficient)*. We will then establish the collaboration strength *(strong, medium, weak, extremely weak, unknown)* by measuring the minimum path length that connects two nodes in the network; i.e. *Geodesic*. The expert's profile will then be refined based on the profiles of experts with whom there is a strong collaboration. The refinement is performed by considering the similarity in expertise through measuring the cosine similarity between the weighted vector of expertise concepts and the type and strength of collaboration between experts.

4 Evaluation

We will specifically evaluate the concept clustering, overall profile building and profile refinement phases using the use cases outlined in Section 1.1. *Precision* and *recall* will be used as metrics in all of the evaluation steps.

The proposed concept clustering method will be compared against the ECM algorithm [21] since (as with our method) it does not limit the number of clusters and the threshold value determines the ranges within which concepts assigned to a particular cluster must lie. In ECM the threshold value is a distance, whereas in our proposed method, it's a similarity. Therefore in our experiments, the distance threshold for the algorithm will be converted into similarity [22]. We will process the same use cases with ECM and use cosine similarity as a similarity measure. The evaluation will compare our method against ECM, using a number of different similarity thresholds. We will also record the following quantities as additional evaluation measures for this step: the number of concepts correctly assigned to the relevant cluster (*true positives*), the number of concepts incorrectly assigned to a cluster (*false positives*), the number of concepts correctly identified as irrelevant to all clusters, hence their assignment to the pool (*true negatives*) and the number of concepts relevant to a cluster, incorrectly identified as irrelevant (*false negatives*) [22].

The evaluation of the profile building phase will be performed using a range of domain experts from each of our use cases. The chosen participants will have significantly different interests – in order to capture a variety of expertise with variable change rates. The profile resulting from the building phase will be compared with manually generated and maintained expert profiles over the same period. This comparison will specifically target new profile generation/update as a result of new concepts and concept rank changes, as well as the precision and recall of our method.

For profile refinement, we will target existing collaboration networks specific to our domain, such as BiomedExperts[5] – an online community connecting biomedical researchers through the display and analysis of the networks of co-authors with whom each investigator publishes scientific papers. As in the previous step, we will select a number of experts from our use cases with a variety of interests and co-authorship activities. The refined profile will be compared against the expertise profiles generated by the collaboration network.

5   Conclusion

In this paper, we have proposed a methodology for building expertise profiles from micro-contributions in the context of *living* documents and evolving knowledge bases. This methodology consists of three building blocks: (i) a comprehensive provenance and change management model for micro-contributions, (ii) a profile building phase that includes expertise concept extraction and clustering, and (iii) a profile refinement phase that takes into account existing social professional networks. This research will lead to two significant outcomes: time-dependent expertise profiling, as well as adaptive and novel trust and performance metrics in incrementally changing knowledge environments.

Acknowledgements. The work presented in this paper is supported by the Australian Research Council (ARC) under the Linkage grant SKELETOME - LP100100156.

References

1. Lee, T.B., Hendler, J., Lassila, O.: The semantic web. Scientific American 284(5), 34–43 (2001)
2. O'Reilly, T., Musser, J.: Web 2.0 principles and best practices. 20, 2008 (retrieved March, 2006)

---

[5] http://www.biomedexperts.com/

3. Bizer, C., Heath, T., Berners-Lee, T.: Linked data-the story so far. Int. J. Semantic Web Inf. Syst. 5(3), 1–22 (2009)
4. Mons, B., Velterop, J.: Nano-Publication in the e-science era. In: Proceedings of the Workshop on Semantic Web Applications in Scientific Discourse, co-located with ISWC 2009, Chantilly, Virginia, US (October 2009)
5. Casati, F., Giunchiglia, F., Marchese, M.: Liquid publications, Scientific Publications Meet the Web. Technical Rep. DIT-07-073, Informatica e Telecomunicazioni, University of Trento (2007)
6. Zhu, J., Song, D., Rüger, S.: Integrating multiple windows and document features for expert finding. Journal of the American Society for Information Science and Technology 60(4), 694–715 (2009)
7. Yang, L., Zhang, W.: A study of the dependencies in expert finding. In: Proceedings of the 2010 Third International Conference on Knowledge Discovery and Data Mining. IEEE Computer Society, Washington, DC, USA (2010)
8. Thiagarajan, R., Manjunath, G., Stumptner, M.: Finding experts by semantic matching of user profiles. In Technical Report HPL-2008-172, HP Laboratories (October 2008)
9. Aleman-Meza, B., Bojārs, U., Boley, H., Breslin, J.G., Mochol, M., Nixon, L.J., Polleres, A., Zhdanova, A.V.: Combining RDF Vocabularies for Expert Finding. In: Franconi, E., Kifer, M., May, W. (eds.) ESWC 2007. LNCS, vol. 4519, pp. 235–250. Springer, Heidelberg (2007)
10. Demartini, G.: Finding experts using wikipedia. In: Proceedings of the Workshop on Finding Experts on the Web with Semantics (FEWS 2007) at ISWC/ASWC 2007, Busan, South Korea (November 2007)
11. Passant, A., et al.: Microblogging: A semantic and distributed approach. In: Proceedings of Workshop on Scripting for the Semantic Web (2008)
12. Michelson, M., Macskassy, S.A.: Discovering users' topics of interest on twitter: a first look. In: Proceedings of the 4th Workshop on Analytics for Noisy Unstructured Data in conjunction with the 19th ACM CIKM Conference, pp. 73–80 (2010)
13. Breslin, J.G., et al.: Finding experts using Internet-based discussions in online communities and associated social networks. In: First International ExpertFinder Workshop, Berlin, Germany (2007)
14. Zhang, J., Tang, J., Li, J.: Expert Finding in a Social Network. In: Kotagiri, R., Radha Krishna, P., Mohania, M., Nantajeewarawat, E. (eds.) DASFAA 2007. LNCS, vol. 4443, pp. 1066–1069. Springer, Heidelberg (2007)
15. Breslin, J.G., et al.: SIOC: An approach to connect web-based communities. The International Journal of Web-based Communities 2(2), 133–142 (2006)
16. Ram, S., Liu, J.: Understanding the Semantics of Data Provenance To Support Active Conceptual Modeling. In: Chen, P.P., Wong, L.Y. (eds.) ACM-L 2006. LNCS, vol. 4512, pp. 17–29. Springer, Heidelberg (2007)
17. Orlandi, F., Champin, P.A., Passant, A.: Semantic Representation of Provenance in Wikipedia. In: Proceedings of the SWPM, Workshop at the 9th International Semantic Web Conference, ISWC 2010 (2010)
18. Ciccarese, P., et al.: Ao: An open annotation ontology for science on the web. In: Proceedings of Bio Ontologies 2010, Boston, MA (2010)
19. Fellbaum, C.: WordNet: An electronic lexical database. The MIT press, Cambridge (1998)
20. Charikar, M., et al.: Incremental clustering and dynamic information retrieval. In: Proceedings of the 29th Annual ACM Symposium on Theory of Computing, pp. 626–635 (1997)
21. Kasabov, N.K.: Evolving connectionist systems: the knowledge engineering approach. Springer, London (2007)
22. Shtykh, R.Y., Jin, Q.: Dynamically constructing user profiles with similarity-based online incremental clustering. International Journal of Advanced Intelligence Paradigms 1(4), 377–397 (2009)

Keynote:
10 Years of Semantic Web Research: Searching for Universal Patterns

Frank van Harmelen

Vrije Universiteit Amsterdam, NL

Abstract. At 10 years of age, there is little doubt that the Semantic Web is an engineering success, with substantial (and growing) take-up in business, government and media. However, as a scientific field, have we discovered any general principles? Have we uncovered any universal patterns that give us insights into the structure of data, information and knowledge, patterns that are valid beyond the engineering of the Semantic Web in its current form?

If we would build the Semantic Web again, surely some things would end up looking different, but are there things that would end up looking the same, simply because they have to be that way?

L. Aroyo et al. (Eds.): ISWC 2011, Part II, LNCS 7032, p. 389, 2011.
© Springer-Verlag Berlin Heidelberg 2011

Keynote:
Building a Nervous System for Society: The 'New Deal on Data' and How to Make Health, Financial, Logistics, and Transportation Systems Work

Alex Pentland

Massachusetts Institute of Technology (MIT), US

Abstract. Most of the functions of our society are based on networks designed during the late 1800s, and are modelled after centralized water systems. The rapid spread of ubiquitous networks, and connected sensors such as those contained in smartphones and cars, allow these networks to be reinvented as much more active and reactive control networks - at the scale of the individual, the family, the enterprise, the city and the nation. This will fundamentally transform the economics of health, finance, logistics, and transportation. One key challenge is access to the personal data at scale to enable these systems to function more efficiently. In discussions with key CEOs, regulators, and NGOs at the World Economic Forum we have constructed a 'new deal on data' that can allow personal data to emerge as accessible asset class that provides strong protection for individuals. The talk will also cover a range of prototype systems and experiments developed at MIT, outline some of the challenges and growth opportunities, focusing on how this new data ecosystem may end up strongly promoting but also shaping the semantic web.

L. Aroyo et al. (Eds.): ISWC 2011, Part II, LNCS 7032, p. 390, 2011.
© Springer-Verlag Berlin Heidelberg 2011

Keynote:
For a Few Triples More

Gerhard Weikum

Max-Planck-Institute for Informatics (MPI), DE

Abstract. The Web of Linked Data contains about 25 billion RDF triples and almost half a billion links across data sources; it is becoming a great asset for semantic applications. Linked Data comprises large general-purpose knowledge bases like DBpedia, Yago, and Freebase, as well as many reference collections in a wide variety of areas, spanning sciences, culture, entertainment, and more. Notwithstanding the great potential of Linked Data, this talk argues that there are significant limitations that need to be overcome for further progress. These limitations regard data scope and, especially, data quality. The talk discuss these issues and approaches to extending and enriching Linked Data, in order to improve its scope, quality, interpretability, cross-linking, and usefulness.

L. Aroyo et al. (Eds.): ISWC 2011, Part II, LNCS 7032, p. 391, 2011.
© Springer-Verlag Berlin Heidelberg 2011

Author Index